DON'T TREAD ON ME

2012: The Elite Deadline and the Coming World Economy Crash

My Government Education with a Spook Known as "The Clown"

DANTE KALI DAS

Don't Tread On Me 2012: The Elite Deadline and the Coming World Economy Crash

Published by: Kali Das Publishing

ISBN: 978-3-00-034078-9

Printed in The United Kingdom

Cover and Interior Design: 1106 Design

DEDICATION AND THANK YOU

Dedicated to Mother Meera: Who has kept me going for all these long hard years.

To Babaji: And the Path of Truth, Simplicity, and Love. The Path of "Humanitarianism".

And to Muniraji, and Andreas for their influence as well.

And of course for my child, who has "Come Back" a "Second Time" for my "Aid and Support". Who I thank for the Love, the Courage, the Strength, and the Support that my kid is, and gives: there is no better, I Love You.

And of course for my girlfriend who has screwed up royally many a time for Corrupt-Liars, but who has "Stayed through it All".

AND WITH LOVE FOR MEERA AND DIPSEY: Our two pets who were killed by Whorbat "The Clown" and Her Corrupt-Agents, "The Buffoons". Meera deceased December 6th 2009. And Dipsy deceased April 8th 2011.

Also: For Those Within Government, Agencies, and Law Enforcement who are Not Corrupt. Including "Kylie".

Special Thanks To: Green Day for the song "Holiday", and Charles H. Ferguson and Matt Damon for the movie "Inside Job". Thanks to Jeff Rense, and Project Camelot, for the "Job" they do as well. Last but not least, Special Thanks to "Sasha" and the song "Hide and Seek", this is about "The Three Investigators", this is my song.

Also dedicated to David Icke and Linda (Thanks for listening, Linda).

A special thank you to Dinder and her wolf print art.
Thank you for your improvement and contribution,
without it the book cover is just not the same, nor as good.

For all those in MOBBING being Attacked, and for all those being Attacked by COINTELPRO. And for all those being Attacked with this Electromagnetic-Technology. YOU ARE NOT ALONE, THERE ARE MANY OF US, AND WE ARE ONE PEOPLE. YOU ARE "MY PEOPLE". DO NOT BE "VICTIMS": BE STRONG, BE LOUD, BE SELF-EMPOWERED, AND FIGHT BACK.

From You Tube: BinBinDinDinGsta: 2pac—Holla if Ya Hear Me. The "Version" of this song that catches the "Feeling and Sentiment".

"I guess cause I'm black born
I'm supposed to say peace, sing
songs, and get capped on
But it's time for a new plan, BAM!
I'll be swingin like a one man, clan
Here we go, turn it up, don't stop
To my homies on the block gettin
dropped by cops
I'm still around for ya
Keeping my sound on the ground for ya"

"And now I'm like a major threat
Cause I remind you of the things
you were made to forget
Bring the noise, to all my boyz
Know the real from the bustas and the decoys
And if ya hustle like a real G
Pump your fists if ya feel me,
holler if ya hear me
Learn to survive in the nine-tre'
I make rhyme pay, others make crime pay
Whatever it takes to live and stand
Cause nobody else'll give a damn
So we live like caged beasts
Waitin for the day to let the rage free
Still me, till they kill me
I love it when they fear me—
Holler if ya hear me!

From the song "Holler if ya hear Me", by Tupac Shakur.
Tupac Shakur deceased September 13th, 1996.

Tupac Shakur killed by COINTELPRO, which is another name for GOVERNMENT GANGS. There is no "East Coast-West Coast Rivalry", only "Divide and Conquer" and Murder, by GOVERNMENT GANGS created by a Government Program known as COINTELPRO and MOBBING.

TABLE OF CONTENTS

INTRODUCTION 1

CHAPTER 1: 11
Lobbying in America

CHAPTER 2: 12
Guantanamo Bay

CHAPTER 3: 13
Water-Boarding—Is It Torture

CHAPTER 4: 16
Are You a Republican or a Democrat? If You Are, You Are Divided: People Need to Focus on the Root Problem of America Which is Corruption in Government and the "Apathy" of the Masses: How About "A War Against Corruption"?

CHAPTER 5: 18
Democracy and Pseudo-Democracy

CHAPTER 6: 20
The American Government and the American Constitution

CHAPTER 7: 22
The Difference Between a Patriot and Nationalist: "We The People" Vs. "We the Government"

CHAPTER 8: 23
Corruption: A New Definition to be Considered, a Mental Disease, a Mental-Disorder

CHAPTER 9: 24
Guantanamo Bay Part 2: A Test-Case and Public-Profiling to Check Public Apathy in Regards to Torture and Concentration Camps

CHAPTER 10: 27
The "Patriot Act": Illegal-Laws/Grey Laws—Laws With Double-Meaning: My "Situation"/Mobbing: Illegal-Laws Passed by Corrupt-"Illegal-People": When "We the Government" Attacks "We The People"

CHAPTER 11: 32
The "Terrorism-Scare" Equals the "Red-Scare" our Constitution Trampled by Secret and Illegal-Laws and Programs

CHAPTER 12: 36
The Path of Experience vs. The Path of Consciousness

CHAPTER 13: 41
The National-Debt Part 1: 2011–2012, Not "If", but When and Why the American Economy will Collapse, and the World-Economy with it: A Reality Check for Everyone who Likes their Money

CHAPTER 14: 59
Why Such a Good Country and Powerful Country is Going Down-Hill: America the "Standard-Bearer of the World" is Being Deliberately Reverse-Engineered into the Ground

CHAPTER 15: 62
The Elite = Powerful People who Feel they Can Do What they Want: The Mentality of the Elite: A Down-To-Earth Look and Explanation of these "People": A Look at these Elite from a "Laymans Perspective"

CHAPTER 16: 74
Flying the Friendly Skies—The Restricting and Shutting Down of the Airways: The "No-Fly-List"—People Have Not Changed, the Government Has

CHAPTER 17: 79
The Internet Shutdown in 2012 : The Net Has Brought Unity of The People; It's Got to Go: and the Weeding out of Alternative Thinking and Ideas; So Much for the Free-Flow of Ideas

CHAPTER 18: 81
Religion: A Divided People is a Conquered People

CHAPTER 19: 88
The World-Economy Part Two

CHAPTER 20: **90**
Hidden Technology for Hidden Government Programs: Electromagnetic-Technology / Exotic-Technology: My "Introduction" to Exotic-Technology: A Laymans Explanation of it: Its Exotic-Misuses for Government and the Elite

CHAPTER 21: **144**
Normal Climate Change and "Elite Climate Change": Exotic-Technologies for Manipulating Weather, Prophecies, and The Masses: "Guaranteed Bad Weather" Down the Road Time-Wise

CHAPTER 22: **158**
Theory of Corrupt and Powerful People—Why Doesn't the Average Person Believe?

CHAPTER 23: **161**
The Rich and Powerful: The "Corrupt Rich and Powerful": The "Corrupt Rich and Powerful United Together"—The Elite: An Easier Way of Looking at and "Seeing" these People

CHAPTER 24: **172**
The Elite—What Separates the Elite from "Other Crime and Criminals" Is their Knowledge of Science and "Meta-Physics": The "Team" of the Elite—the "Henchmen" that Get the Dirty Work Done: the "Team" Working for it is a Fascist-Cult, a World-Cult: the Elite's "Sales-Pitch" to this Cult—a "Pyramid-Scheme"

CHAPTER 25: **183**
The Elite Deadline: 2012 and Why: a Cosmic-Cycle, Yugas and a "Central-Sun": an "Electromagnetic Space Cloud", called "Local-Fluff" by Astronomers—which Contains "The Space Ribbon"—a Cloud of Light: an Opening to "Units of Consciousness", the Path of Consciousness or the Path of Experience?

CHAPTER 26: **198**
Prophecy: Negative and Positive Energy: Consciousness: the "Path of Experience" vs. the "Path Of Consciousness"
How to Change the Future in a Positive Manner

CHAPTER 27: **207**
The Elite's Hijacking of a "Cosmic-Event"—"Sold as a Sign" for the Appearance of a "World-Leader": an Appearance of Light Which Astronomers Refer to as the "Space Ribbon", What Many Call the

"Rapture or Photon Belt": The Elite's Plans to Install a "Puppet", One Who Many Call the "Anti-Christ or Dajjal"

CHAPTER 28: **232**
"My History": How I Got in this "Mobbing/Situation": a Brief-Simplified Version

CHAPTER 29: **283**
A List of Government-Agent-Tactics Which Have Been Used on Me Which Can Help You See the World Around You More Accurately: A Guide to Spotting Corruption in the Media and World Around You

CHAPTER 30: **331**
Mobbing: This is Not an Investigation, but Let Me Tell You What this is, the Bigger Picture: COINTELPRO—A "Dirt Creating Mission" against the Target for Manipulating the Public: Mobbing—The "Next Level of COINTELPRO": The Creation of a "Fascist Movement" and "MK-ULTRA" for Creating Citizens Compliant to Government: also "Reversed-Legality" for Illegally Creating "Dirt" in Regards to a Target

CHAPTER 31: **382**
Corruption: A Summary "Of What to Do": Citizens, The People, Agents and Politicians: "Uniting in Goodness" and Fighting Back: a Matter of Energy: "Human Physics"

CHAPTER 32: **413**
A List: Tips and Pointers for the Two Agencies Involved, Being Manipulated and Used, by "The Clown"

CHAPTER 33: **420**
"Don't Tread on Me": A Letter of Chastisement to Three Different Government Agencies

NOTE TO READERS

As a result of my "Circumstances" this Book has been "Rushed a Bit". Quality of my Writing and Editing Mistakes are a result of this. Future Endeavors will no doubt meet my standards, "Unique Circumstances" are a major factor with this Book. Read on to understand why.

INTRODUCTION

"We're the Best in the Business"

—The Clown and Her Buffoons

I originally began this book due to my circumstances involving Government Agents, and a "Situation" often referred to as "Mobbing or Street Theater": I call it ugly and it is what I also call "Localized Fascism". This "Mobbing" involves the Nasty Tactics from COINTELPRO, and is in my opinion "The Next Level" of COINTELPRO among other things. COINTELPRO stands for Counter Intelligence Program, was a Government Program exposed in the seventies, declared "Non-American", and was supposedly ended. It not only has NOT ended, but it looks like it has "Escalated and Evolved" into a "New Form": "Mobbing". Having said that, regardless of "Laws", it is Illegal.

Mobbing is a "Misuse of Citizens", a small "Movement, Wave, Fascist-Movement" that centers around the "Target", in this case me—and here in Europe this has extended to My Family as well. It involves getting Citizens to engage in Criminal Behavior, based simply on what they have been Told and have Heard, by, and due to, Government Agents. Their "Perceptions" are Altered so that the Public "Sees, Hears, and Experiences" what they are TOLD to "See, Hear, and Experience". This is called "Programming" and it is Modern Day Brain-Washing, which is what Cults use.

This involves "Blind-Compliance to Authority" without having all of the Facts. It involves the Psychology used in Creating Fascist-Movements, based on Government Studies, and involves the Psychology developed and acquired by another Government Study, for "Traumatizing the Target" to achieve "Compliance"—this last Government Program was known as MK-ULTRA. In both cases, the goal is "Compliance to Authority" through Psychology, Programming, and other means: Fascism IS when Government becomes Cult. This "Mobbing" is ugly, Illegal, and Hidden, yet goes on and exists out in the open.

"Mobbing" was Created by Corrupt-Politicians for "Greater Control" of The People. Based on what I have learned as a result of this "Situation", this Book will help those who are interested in seeing the World around them More Clearly, in regards to Corruption and Tactics of Corrupt-Government. This Book is for anyone interested in knowing about the coming World Economy-Crash, and those Corrupt-Politicians and Heads of Industry, who I refer to as the Elite. This Book is also for anyone in this "Mobbing", and anyone who is interested in, or believes in an "Idea of 2012"—Prophecies. If you are in this "Mobbing", this book can only help. If you believe in Corrupt-Politicians who do what they want for their own Self-Gain, an Elite, this Book can only help. And if you believe in an "Idea of 2012" this Book can only help. These "Elite" have a "Plan" for more Control and Power in place, for 2012, which will affect us all. Thanks to The Clown, and this Mobbing, I have sorted many things out.

This Mobbing in my life has gone on for eighteen years, has followed me across Three Continents and Three Countries, and no doubt at this point has cost MILLIONS of dollars—and there is no justifiable reason for this. I've learned quite a bit about how Government works, Agents and their Tactics, and the Manipulation of Individuals and the Masses, as a result of this. Mostly because the "person" in charge of my life for the last nine years, the Government Agent who I call The Clown throughout this book, has pushed me into understanding quite a few things I would not have contemplated otherwise. As a result of this Demented-Jackass and my "Spook Education" I see the World very differently now.

In no small part I understand quite a bit the "Average Person" does not, because this Agent has perpetuated my Mobbing, refusing to let it end, and based on Personal-Reasons. There are "Rules, Protocols, and Standards" for this "Mobbing" and after a host of "Different People in Charge", and different Agencies, I have come to realize what this Mobbing is about and "The Rules" of it. So, even after it was "Supposed to be Over", and based on "The Rules", I have been held in by this "Dementoid". Say what you like but this "Behavior" is anything but Professional, and the reasons, the Excuses, this Clown has "Held Me In" for have been Ridiculous to say the least.

I have "Passed" and should be Free, and should have been Free seven years ago—at least. As it is I have "Passed", and by any Agency's Standards, and this "Mobbing" has merely gone on due to "Personal Reasons" on The Clown's end. Her not getting Her own "Personal Way" and "Self-Fulfillment", initially. As I began this Book, which Exposes The Clown among other things, this became further Motivation for her "Keeping Me In" this Mobbing, to try to "Cover-Up" her Mistakes and Corruption. The words "Non-Professional, Snob, Princess, and Corrupt" come to mind, as do others. As a result I have been pushed into writing this book, and I have included my "Understanding of the World around Me" in this book, to make the most of a Negative-Experience, to Educate and Inform others, and to help others—there is no point in letting my "Education" go to waste, never mind Tax-Dollars. At least the Public can "Benefit" from all of this.

So I eventually came to realize The Clown was not going to stop interfering with my life, and my Families, or stop Torturing me and my Family. Or Free Me and my Family, seeing as how She "Held me In this Mobbing" repeatedly by "Selling Cheap Excuses" and Lies to those around me—those Citizens aware of my "Mobbing" and some close to me (like the Feds back in America did), including other Field-Agents who are Not Corrupt. I "Kept Passing" by her "Agencies Standards" which remained unknown to others around me, people taking part in this Mobbing, this Clown kept Lying and Manipulating, to "Convince Them I Had Failed". I have watched this go on for almost eight years now, it is all one Big Lie and Deceit.

So I came to realize another point, how Corrupt, Non-Professional and Personal this Clown was, and is. But like a lot of people and Citizens, I believed in a "Stereotype of a Job-Title" for a long time, including in regards to "Government Agents" and Her: I expected them to be "Professional and Clean". In fact many of the themes throughout this book are based on my experiences with The Clown, Corrupt-Agents in America who started this "Mobbing", and this "Mobbing" itself. These things are the "Basis of my Government Education", and like an actor or director in a movie who understands how the "Movie is made behind the Scenes", I now see the World differently, and on a "Deeper-Level".

So knowing of The Clown's Corruption, I eventually started looking at the Facts and "Type and Quality" of person in charge. I began "Looking" and catching this Clown as She made more mistakes and blunders, and based on Her Personality. I began "Profiling" The People involved, and came to realize that the person I was dealing with was Corrupt, and also a Psychopath—quite literally. She is what I call a "Power-Monger", Corrupt and Abusive of Job-Title and others for the "Self-Fulfillment" she gets out of this. Her "Issues" are "Power and Abuse of Power" and the "Feeling" she gets from this. As I learned at the University, people who "Rape" have "Power Issues" not "Sex Issues" and this woman is a "Rapist Hiding behind a Job-Title"—"The Act" is different, the "Issues" NOT. This is why Whorbat (Changed Name), The Clown, chose this Job I realized. You would expect better from Government Agencies, and as hard as it was to believe, this was obvious once I realized this.

I did a "Rewind on my Past Dealings" with this Clown, and everything fit. Including Her tendency for "Showing Off", her Inability to "Perceive Consequences", Her lack of Empathy and Misuse of Force, Her Self-Aggrandizement, Lack of Self-Control, and mistakes so stupid, they were beyond belief: and a lot of them due to "Swollen-Ego". I will give myself some points of credit as well for being Observant and Intelligent. Having said that some of the mistakes were so stupid, I thought they were deliberate, which they weren't. Thus I named Her, The Clown. A "Show-Off" who's "Constant Need" to do so, has led to the dumbest of Mistakes and Blunders. A Bungling-Idiot.

Now there were other things "Noted by Me" as well, along the way, involving mistakes of The Clown and Her Buffoons—the Agents working for Her. A big one was the Excessive Misuse of Electromagnetic-Technology, "Exotic-Technology", which Government Agencies have and are Hiding from the Public. Quite obviously they are Hiding this Technology because they do not want Citizens knowing how it is Misused. Having said that it is only Technology, and less difficult to understand or believe, than most think. I feel that my chapter on this does a good job of explaining it, the principles involved, in "Laymans Terms", and will dispel a lot of Disbelief. As it is I live with the Reality of it, as does my Family. Thanks to The Clown, I now know about this.

In fact currently as I write this, and near to finishing this book, I am interfered with by this Exotic-Technology, and Constantly: which is why my writing in some places "Repeats", and is why this Book is released in a "Form not normally up to my Standards". As I try to "Edit" I am constantly attacked, so that mistakes remain. People, my readers, may not want to believe in this Technology, however I assure you these Agencies do believe in it, as they use it. In the end it is only Technology and the Public-Sector is very close to developing this anyhow—we are not talking about "Growing a Second Head". So this book has been delayed for Two Years, mostly with this Technology, and NOT because of an Agency but because of Agents not wanting to be Held-Accountable in any way, shape, or form. This is a matter of Sloppy-Criminals hiding behind their own Job-Title, and Misusing their Power and Authority to delay the inevitable. In fact these days I get to deal with Death-Threats and Innuendos, as well: I am being threatened to be Killed with this Technology at this point (Less than a week ago they killed our pet Dipsy with this Technology). Go figure, The Clown's idea of "Responsibility" apparently. But back to the initial point.

This Technology, an "External-Influence", became "Obvious in my Life" and eventually these Spooks "Came out of the closet and Told Me", and a few others in my life, that they had it—in 2006. Their Revealing and Showing-Off of this Technology in turn led to more Mistakes, and getting caught by me—eventually I realized this Technology had been used on me in America as well by another Agency. Actually Corrupt-Agents working for this Agency. I realized these Corrupt-Agents in America, who had Framed-Me to put me in this Mobbing, were worried I had "Figured Things Out Totally", were worried that I realized "To What Extent I had Been Framed", and so through whatever "Channels and Means" acquired this Exotic-Technology for finding out how much I knew. And for the sake of their own Butts I might point out. Just like The Clown does now, they Misused this Technology, to Avoid Being Caught for things they should not have been doing, and should not have done, in the first place.

How this "Mobbing" started, how long it has gone on, how much money has been spent on it, what these Corrupt-Agents in America got away with and for what, and what this Clown and Her Buffoons have gotten away with, all speaks for itself. This is not a matter of "Agencies" at this point, merely a matter of a "Situation" that was started by Corrupt-Lowlifes, and currently goes on due to totally Corrupt-Lowlifes. This "Mobbing" obviously has a "Life of it's Own", and quite literally at this point, continues on for no other reason than a Corrupt, Non-Professional-Employee, who is "Holding Me In" as a way of doing a Cover-Up of Her Mistakes. As I have come to realize, based on Observation, Her Agency is Clueless as to what is really going on out here.

Now once I realized this Clown to be a Corrupt-Psychopath, and realized some things in regards to the Agents in America, I decided to write a book to get to the bottom of it all, and to clear my name. As I did so, I began to find and discover, just how much Deceit, Manipulating, and Lying, had gone on in my life. The more I "Looked" and "Rewound on my Life", the more I found. For all intents and purposes I was a Detective (Like Jupiter Jones—for those "In the Know") putting a Large-Puzzle together, and a lot of the "Discoveries" I made were extremely unpleasant to say the least. On more than one occasion I was "Enraged" with my "Discoveries"—it is no wonder. And in the meantime as I wrote this book (a different one—more autobiographical), my life and my "Mobbing" with The Clown, continued on. It was like going through this "Mobbing" twice. Whatever it has been, it has not been easy.

Everything from how this "Mobbing" started, to The Clown, to Differences in Types of Agents, to Tactics Used and How, to Corruption and Laws, everything began to unravel and became More and More Obvious. The Clown and Her Buffoons watched with this Exotic-Technology, as everything began to "Fall Apart" on their end. Not only had I caught these Idiots on many a fine point, but now I was catching other Corrupt-Jerks from my Past as well, and even beginning to "Look at the Bigger Picture". In the meantime The Clown, no doubt "Slipping Mentally", continued to "Traumatize me and my Family" and in fact "Increased Things" as I began to put things together—and because of Her Mistakes. As I have put it many times, "She Stunk of Desperation".

So eventually, once I solved "Riddles from my Past" I began looking at the Bigger-Picture, in regards to all of this "Mobbing", this Illegal-Government-Program, and began "Applying my Government Education" to other things as well. Things which I could see happening in the News and World, and I decided to "Combine the two of These" for a book to let the Public and People know of what Government is getting away with—specifically Corrupt-People within Government that is, who hold the Power and Control Government. Power-Mongers, like The Clown, who Misuse their Job-Title for their own Self-Gratification, and regardless of cost to others. Everything in this World, I realized, that is WRONG, is a direct result of Corruption and of Corrupt-People. Corruption is a "Mental-Disease" that is destroying the World, and has certainly destroyed my life.

This Book helps to explain the "Mentality, Viewpoint, and Psychology" of Corrupt People, Power Mongers, the Elite, and helps point out what these Power-Mongers are "Up To". We are talking about Psychopaths with Job-Titles of Power, because the Power is what attracted them to the Job in the first place. They can do "What They Want", at Cost to Others and without getting caught, because of the Power they have. I have had no choice but to "Figure Out", and "Understand" this Type of Person, this Type of Personality: I have been Profiling one, and dealing with one, for years, and it is "Uncomfortable to wear the Viewpoint", to "Squeeze One's Mind Into This Warped-Viewpoint", to say the least. Thus we have my book, and it all began with The Clown, because She is a Power-Monger. I am able to explain to people, my readers, a Mentality, Psychology, and Viewpoint that most do not get because it is hard to do: that of the Power-Monger, the Elite.

The "Elite" being those in Government and Industry who feel "Entitled" to do as they please regardless of Laws and the Public.

As I put it within this Book, these Elite are the one group of people who should be held the MOST ACCOUNTABLE in regards to the Laws, because of the Power they are trusted with and wield—therefore posing the greatest threat to ones Country, Society, Culture, and People: yet are also held the LEAST ACCOUNTABLE. This is of course because of the Power they have, just like with The Clown.

Thanks to The Clown I am also able to "See, Understand, and Explain" a System that has been Created by these Power-Mongers, for Themselves, and their own Self-Gratification, as well. I began to realize that the System is Totally Screwed Up and Hogwash, and I came to realize "To What Extent". When people like this are put in charge, given positions of Authority, it is not "Isolated or a Mistake". I realized it comes from the top of the "Food-Chain" on down. It "Rubs Off" on The People underneath as they see "Corruption is Acceptable" and as they are exposed to it, many times of course quite deliberately as the "Corruption is Shared"—along with the money and "Benefits". Corrupt-People put Corrupt-People in charge, it's not hard to figure out. Corrupt-People within Government come up with Corrupt-Government-Programs like "Mobbing" and COINTELPRO for "Greater Control" and to Protect a Corrupt-System. For people at the "Top of the Food-Chain" it is about people wanting to maintain their lifestyle, and feeling safe from getting caught. When you are a criminal and put criminals in charge you need not worry. So I started following this chain upwards to Corrupt-Politicians, the Elite. But back to The Clown, and the end of this introduction, on a more personal note.

There were some mistakes made along the way here, by these Jackasses, in regards to me. For starters when you have what you say you want don't go on Torturing and "Ball-Kicking" (And not just me but my Kid as well). It makes what you want a Moot Point, and Totally Irrelevant. The next point is don't go on Torturing for the sake of Torture and you're own Personal-Dislikes, never mind your "Desire for Personal-Fulfillment". This makes you look Non-Professional which you are, Personal which you are, and Obvious which you are. Pulling the wool over the eyes of a bunch of Idiot-Citizens, and a bunch of Field-Agents who are "Less Qualified" than I am, hardly makes you a genius. Another point is that when you make an extreme point of Violating People, don't expect Respect. This points you out for the extreme Whacko-Ass-F-Nut that you are, never mind "Targeting someone's Child" and Pets as a way of "Getting to Them". There is something wrong with you and even without my help people will notice, including your own people and Agency. This book has come about because you have, and will, go on making Extreme Nonsense. Your Agency can "Deal with You", so much for "Dumping Blame on Me", and so much for "Avoiding Consequences and Responsibility".

Due to YOU Clown, you have created a whole series of problems for a lot of people, and gotten more people involved in YOUR MESS, as you have Misused Everyone, including Agencies, for Your Own Personal Cover-Up. This is for the World to see, including your own Agency who can contact me, in regards to all of your Deceit—and they can more than mark my words that you have no doubt "Edited your Files", to "Avoid Responsibility". And no doubt delaying my book has been a part of this so that you have had time to do so. You wanted it Personal, now you've got it. You wanted to be a "Show-Off", now you have it. You wanted Attention, now you have it. You wanted to Show-Off what you could "Get away With", Flaunting Yourself, now you are caught and exposed like the Brainless-Jackass you are. But at least you are getting "The Attention" you love so much.

You can "Talk all you Want" about "Being Professional", but it does not make you so. You should try actually "Being It", as opposed to "Speaking It". As a general rule, most "Leaders" do not put on "Personal Lesbian Sex-Shows" for the Target, or do you want to claim "Agency Protocol" and being "Professional" as well on this point? You're what, a Professional-Lesbian, following Agency Protocol in regards to the Target? You are The Clown, and for my readers I did not stick around for the "Performance Art" and "Apartment Theater"—although She can write it up any way She wants. No doubt she already has.

In regards to your Agency, THEY can deal with You and Your Nonsense. That they are oblivious to what you do and get away with is beyond me, and I am seriously irritated to say the least. That this Agency has allowed such an Idiot, and Psychopath, to run around free with a Job-Title of Rank and with this Electromagnetic-Technology, is absurd. Have you lost your damn minds? In general, as a general rule, I expect better from any damn Agency, never mind an American one, regardless of your "Reputation"—which would be better in the first place if you did not put Idiots like this in charge. Perhaps you as an Agency should contemplate the wisdom of hiring people with "Mental-Disorders". Either that or contemplate putting Charles Manson "In charge of my Life", as no doubt he would probably do a better job, and certainly make less mistakes. Where is the sense in Hiring Someone with an Extreme Mental Disorder? It ought to be clear.

You can give Charles Manson a Badge and Authority, and a real one at that, and he is still a Criminal—it says absolutely nothing about "His Character" and the "Type and Quality" of person that he is. It will also not "Explain his Actions", although he may be able to "Hide Them" under his Job-Title and Badge, if you get my point.

So start looking up this Delusional Jackass of yours and don't go blaming me. I expect Your Agency to "Remedy this Situation", and to do so quickly. And whatever you do, Do Not even think of trying to Dump Responsibility on me. I have had enough of Corrupt-Agents, Lying, Framing, Manipulating, Deceit, and Government-B.S., and the last thing I am going to do is wear "Anybody Else's Blame". I have had twenty years of it. Your Idiot The Clown is threatening my life at this point, in an assortment of ways, and I will be Filing an Official Complaint against this Demented-Jackass based on this alone, unless your Agency steps in and "Takes care of it's Corrupt Employees". Otherwise I will have "No Choice", never mind other "Options" which I will take as well. And I will put the "Results" in my Second Book, so that The Public is "Officially Aware" of any "Continuing Problems". Think I am going to trust anyone but myself on this point, forget it. You have Your Clown to thank for this, as there is no "Other Choice" based on the shoes I wear. Thanks for being a Responsible Agency, and "Monitoring Your Employees", which quite obviously you Do Not. And how do I know this?

Take care of Your Clown, take a good god damn look at how this Nonsense started, take a good damn look at how Absurd and Stupid this "Mobbing" has become (never mind the amount of money), and take care of your Imbeciles. With even the slightest bit of "spot checking" you'll realize this book came out as a result of Corrupt-Idiots, and my "Situation" has become, "Expansive" not quiet, as I have not been given a choice. Never mind the Death Threats that have been made as well, as this Book has neared completion—which I would have to be stupid to not take seriously. At some point I expect someone to take Responsibility here, someone Intelligent who does not see "Murdering Me" as a "Justifiable Solution" for making ones life easier. This woman is Your Employee, and Out of Control. As The Clown is incapable of Responsibility, and as Her Buffoons are merely Dependent-Ass-Kissers incapable of any Leadership Role (anybody can take a paycheck), or Good Decision Making, it will have to be someone within your Agency with a Brain. As one last further point, be glad I do not have any desire to "Sue your Agency" and if I did at this point, it would be for "Training your Agents". As it is you should know that these people are "Impersonating Agents of your Agency".

With that done Readers, lets begin.

Initially I began with isolated subjects and conspiracies, eventually I realized it's all interconnected. As you will note, a Major-Theme of my Book is Corruption, as this is what has Destroyed and Shaped my Life: that along with people being Brainless, Stupid, and Non-Questioning. This is where my Book Title comes from, as I have been "Walked Upon" and "Tread Upon" by many people, and by a Government and Corrupt-System that is walking upon the "Rights of The People". This Book is for All Real Americans and All Others within Other Lands, who are AGAINST CORRUPTION. So lets start with the obvious and work our way upwards, shall we?

CHAPTER 1

Lobbying in America

Lobbying is INSTITUTIONALIZED-BRIBERY, Institutionalized-Corruption. The American people have an "Issue" with this, as well they should, yet never seem to speak of this problem for what it is. "Lobbying" is a Deliberate-Mislabeling, a Tactic, to conceal the Truth—"Lobbying" is Institutionalized-Bribery, Corruption. Put it clearly and it should not even be an issue—BRIBERY IS BRIBERY. Once people start dealing with this Issue more accurately, start speaking of it for what it is, the problem itself will not be as "Hidden", and can be dealt with in a more Truthful Manner. IT WILL BE SEEN FOR THE DECEIT AND LIE IT IS. Who came up with the term "Lobbying"? Obviously POLITICIANS AND COMPANIES, The People engaging in this, who wanted the name changed to something less obvious. The People need to start labeling this problem for what it is, and confronting the POLITICIANS with this fact: BRIBERY IS BRIBERY AND CORRUPTION IS CORRUPTION. Which Politician is going to claim that "Lobbying" is a "Good-Thing", or a "Necessary Thing", or a "Part of the Process", once it is identified for what it is? No Politician is going to claim that Bribery is a "Good Thing" or a "Necessary Thing", or a "Part of the Process". If the American People really want to fix this problem then they need to address it for what it is... Plain and simple, see how quickly the reading goes with me?

CHAPTER 2

Guantanamo Bay

An experiment, "PUBLIC-PROFILING", a Test-Case and "Field-Testing", to check Public-Apathy in regards to Concentration-Camps and Torture. If your Government can get away with Torturing "Others" they can get away with Torturing YOU. THEIR rights are your rights. In a Democracy we should hold our Standards Higher, and if not, are we really acting as a Democracy?...

CHAPTER 3

Water-Boarding—Is it Torture?

This next piece came about as a result of being sick of hearing such an obviously stupid question being paraded by various "Media Outlets" as a subject of "Serious Philosophical Debate". This "Question" is sheer nonsense, "Junk-for-the-Mind", a TOTAL-DISTRACTION FROM THE TRUTH OF THE MATTER. An insult to the Intelligence of the Masses. If this subject is questionable for you, then I hope I clear things up for you...

Is Water-Boarding Torture?

You only ask the question if you're a damned-idiot, an Idiot-Savant without the "Savant". It's Torture when you EXPERIENCE it. Do you think it is an issue for The People who experience it? Until that time, if it remains a question in your head, you are an Idiot-Philosopher trying to "Look Intelligent" because that's the question you've heard over and over in the Media, and it is "Acceptable" to question such an obvious thing. In fact it "LOOKS" intelligent to do so. Having said that, if that is the case for you, for actual intelligent people, you look like an Idiot. If you are going to be Intelligent, you need to think intelligently, as opposed to "Looking It". Water-Boarding has only become an "Issue of Torture" because of the Media, until that time no one would even think to consider it anything but Torture. In fact until that time, before that time, if you had brought this up as an

"Issue of Debate and Consideration", people would have thought you were an Idiot for doing so. People are allowing themselves to be Manipulated by the Media on this point.

Your questioning Water-Boarding is a great "Philosophical Debate" in your own mind? This is "Objectivity"? This is Fake-Objectivity, this is "Media-Objectivity", this is FAKE-MEDIA-OBJECTIVITY. It is trying to find "Truth" inside of a false and incorrect answer—IT IS PHILOSOPHIZING for the sake of philosophizing. It is an insult to Intelligent, Conscious-People and a JOKE perpetrated on the Masses.

You can forget it as even being "An Issue" if you have a brain and can think for yourself. People who see the Truth of this "Issue" practice Independent-Thinking, people who see such an "Issue" as a matter of "Consideration and Debate", think what they are "Told to Think": which is what television is all about.

How hard is it to imagine on your end?

The only part of you "Questioning Water-Boarding" is your EGO and how Society might react, what SOCIETY MIGHT THINK, if you DO NOT contemplate the "Issue". Because Society has also had the same "mantra" repeated down it's throat. You are worried about what other "Like-Minded-Idiots" might think. You're worried about what OTHERS THINK, not what the Truth is. Your worries are EXTERNALIZED, not Internalized. Externalized is what others think, Internalized is ,"What if it was me?". This last question IS the answer. "Fake-Objectivity" comes from the "Outside-In", and is a matter of considering a "False-Viewpoint" because it has been "Crammed down your Throat" so many times.

This is "Divide and Conquer" with the Masses, plus an insult to the intelligence of all.

Of course for the seriously thick-headed, the best way to find out is to choose the "Path of Experience", and find out by being "Objective". In this case YOU, with some of your closest friends, can re-create the experience—after all "You will know it is True, when it happens to You". Consider yourself a "Scientist, Philosopher, Realist" in pursuit of the "Truth". How many "Dunkings", how many minutes, do you think you will need to figure out the Truth of the matter? How thick is your head? If Water-Boarding has been an issue of "Debate" then quite frankly I recommend you read something else other than this book, I prefer an Intelligent-Conscious reader. However if you are willing to find out the Truth of the matter, and need to "Find Out", consider me your "Friend"...

Do you share your "Media-Objectivity" in your personal life? How about with my writing?

To quote something I heard on the B.B.C., from someone in a foreign land who used to Torture People: "Torture is the art of killing a man without him dieing." And this is what I've been experiencing for the last 17 years, 18 in April, 2011. And especially the last eight years.

CHAPTER 4

Are You a Republican or a Democrat? If You Are, You Are Divided: People Need to Focus on the Root Problem of America Which is Corruption in Government and the "Apathy" of the Masses: How About "A War Against Corruption"?

> "Any American who turns a blind eye to Injustice and Corruption around him should not be allowed to call himself one—it is by definition fighting these things that makes us an American in the first place. In being an American we carry this Responsibility."
>
> —MR. LEE

Folks if we are going to live in a Democracy then we are going to have to take Responsibility for it. This means making sure that our Representatives are doing their job as they are meant to. This means making sure our Representatives are actually "Servants of The People". Turning a blind-eye, being apathetic, accepting what is wrong with our Land and with our People, accepting what is wrong with our Government and

Representatives, will not work. Acceptance is a great thing, but when it is used to hide Laziness of the Mind, and Apathy, it is not.

If we are going to live in a Democracy then we must be vigilant, fight for it, and protect it. Corruption is a sign of Decay, of a downward spiral for any country or empire, history speaks of this. Corruption is a sign, of any countries downfall. When we have Representatives that actually do their job properly, who actually represent "The People" and America, and not themselves (Corruption), then we will have a country THAT FUNCTIONS PROPERLY AND MORE SMOOTHLY.

Politicians are the one group of people where the letter of the law, justice, ought to be served: yet strangely enough, these people are held the LEAST ACCOUNTABLE within Our Land. Think this is a coincidence? You have to be blind AND stupid to think that it is. And these people are responsible for writing laws and "Representing our People". How intelligent are we as a "People" for allowing this to happen? How intelligent do we look to the rest of the World? Until people get over Political-Boundaries, and start focusing on the real issue, CORRUPTION, things will never change. People need to start working together, need to UNITE, and start dealing with this problem, taking our Country back from the Corrupt-Politicians. Corrupt-Politicians being a "Party Unto Themselves" who REPRESENT THEMSELVES, and who have "Hijacked America" for their own personal gain. As long as we remain divided on this issue, blaming it on POLITICAL-PARTIES, not on Corrupt-Individuals, they will continue to get away and misuse our Country. Corruption, A WAR AGAINST CORRUPTION, needs to be the issue. CORRUPTION exists on both sides of the fence. Political-Parties and their ideals are not the issue, weak-minded GREEDY-PEOPLE are. When you focus on the Real-Issue, there is no "Divided-People".

I RECOMMEND A WAR AGAINST CORRUPTION, within America, which will clean Our Country up. As opposed to a "War on Drugs" which will further divide the American People and allow the REAL CRIMINALS to get away. Generally speaking Corrupt-People are involved in the dealing of large amounts of drugs and transporting them, for money and power as an example. You can not clean up drug problems or anything else until you clean up Corruption within Government. How about a war AGAINST CORRUPTION? This will Unify The People around what is REALLY DESTROYING AMERICA, and Clean America up, making Her a Respected Nation again.

CHAPTER 5

Democracy and Pseudo-Democracy

Democracy is when power and the sharing and entitlement of power is spread out amongst the individuals, the Masses. Not when it is horded at the top of the Societal-Food-Chain: the rich, the Corporations, the Corrupt, the Elite. And not when its an "Exclusive Club", whose main requirement is being Corrupt. This is why more and more Government is being CENTRALIZED, it's to put the power in the hands of the few and to take away what little is left for the Masses, from the Masses. Last time I checked Police-States had this kind of power-structure for this very reason. Funny how people remain blind though, huh? The power-structure within American Government is changing. The Constitution, The PEOPLE'S PROTECTION, is being over-run, by Corrupt People who want MORE POWER BY CONSOLIDATING POWER, yet since the Country still wears the "Label of Democracy", well then it must be so. Is everyone blind? The label of "Democracy" for America is inaccurate these days, false, as in a Lie. Until "The People" actually retake an interest in the QUALITY OF THEIR REPRESENTATIVES, and the quality of their Country, reclaiming THEIR RESPONSIBILITY for Democracy, it will remain so. The inside is being "Eaten Away" by Corruption, while the "Surface remains the Same": like a tree being eaten away by termites. Until American's do so, reclaim their Responsibility for Democracy, it is just the "Shell of Democracy" that is left—"Appearances and Looks" until the "Tree Finally Breaks".

CHAPTER 6

The American Government and the American Constitution

The difference between the American Government and the American Constitution is that Government-Servants, Government is replaceable, THE CONSTITUTION IS NOT. There is something wrong with the POLITICIANS, NOT THE CONSTITUTION. The Politicians for the most part are hypocrites, they don't want to listen. When the System fails in America, and it will BY DESIGN (Folks America is reaching a point where all of the money it makes in a year can not pay the INTEREST owed on money borrowed—the National Debt. Do you honestly believe that Politicians are not aware of this IMPENDING REALITY?), there should not be an "Emotional-Reaction" from the Public as this will "Justify Martial Law", but a call for the ENTIRE REPLACEMENT of Government Employees, ie. POLITICIANS. When the entire System in America "Crashes", this will be a result of Corrupt-Politicians, not a result of the Constitution. Amongst all of them only a few are honest, and/or, not looking the other way due to fear of the Corrupt and Powerful amongst them. How about more Honest-Politicians with more back bone? How's that for a solution? In the end it is the only thing that will restore faith in the American System and America itself World-Wide. Again, any failings in America have to do with the QUALITY OF THE POLITICIANS, NOT THE CONSTITUTION. Any failings of America have to do with The People being Apathetic, not taking Action, in regards to the Up-Keep of their Country.

CHAPTER 7

The Difference between a Patriot and Nationalist: "We The People" vs. "We the Government"

Being a Patriot means putting the Constitution first, being a Nationalist means putting Government first. The Constitution, People, and Country, should ALWAYS come first. Without these basics the whole thing can fall apart due to Corruption... Being a Patriot means believing in Ideals which are infallible (like Freedom), being a Nationalist means believing in PEOPLE, Politicians, who ARE quite fallible. Gee, imagine that... To a very large extent, it is a fair case to say that it is a matter of those who are Self-Empowered vs those who ARE NOT Self-Empowered. The difference between those who "Think on their Own", vs. those who want to be "Told what to Think", and what to do. Keep this statement in mind, I will get to it later.

CHAPTER 8

Corruption: A New Definition to be Considered, a Mental Disease, a Mental-Disorder

Corruption: a Mental-Disease that eats Individuals, Cultures, Societies, and Countries from within. Do I mean literally a Mental-Disease like schizophrenia? Yes as a matter of fact I do, and it's about time that this was, and is, considered. However, like the common cold or influenza, it can spread from Person-to-Person, whichever person that it makes contact with. How strong is your own Immune-System? THE FINAL STAGE OF ANY SOCIETY, CULTURE, AND COUNTRY IS CORRUPTION. Corruption ALWAYS eats away at the very System, Culture, Society, and Country it exists within until it Implodes. Like a tree being eaten from within by insects, it looks fine until everything breaks and implodes at once due to the DECAY, Corruption Within. Take a good look at America and the World, ask yourself how much time is left.

CHAPTER 9

Guantanamo Bay Part 2: A Test-Case and Public-Profiling to Check Public Apathy in Regards to Torture and Concentration Camps

Guantanamo Bay is a test case, a profiling of the American Peoples Psyche, in regards to Torture and Prison-Camps. It's that plain and simple. It's also to Acclimate and Desensitize the American People to the idea of Torture-Camps, because that is what it is. No doubt there are more in store for America...

How many people in Guantanamo Bay are actually guilty? I doubt very many. Financial Incentives were and are offered as an incentive for turning people in. This creates a Greed-Witch-Hunt-Atmosphere. Under such circumstances people will turn in people they don't like, even their neighbors or total strangers for a change in life style and a few perks (desired objects) along the way: it's called GREED. The People initiating this Witch-Hunt know this, they have their own Goals and Objectives, which are also Greedy. Thus they spread the wealth based on Human-Weakness, based on Greed, which is a Weakness of Mind. It is a Corruption of the Mind.

It's a dumb way to go about catching "Bad Guys", especially when in this case, no real proof was needed. It's not like these people were going to court. But it is a good way of obtaining a lot of people, whether they are guilty or not. It is a good way of obtaining "Instant Results", a good way of obtaining a lot of people whether they are guilty or not, a good way of obtaining NUMBERS, which "Looks Good" to the American Public-Eye. This is about "Appearances". NEVER LETTING THESE PEOPLE GO TO COURT MAINTAINS THE LIE. None of this is coincidence so catch on and observe.

It is in fact a Farce and a Lie, never mind the Injustice.

This is why a lot of "Terrorist Cases" keep popping up in the Public-Eye in America. To maintain a Farce. Greedy-Informants and morally misaligned Federal-Agents, Corrupt-Agents (obviously not all—but applicable to these particular circumstances), are acting out of GREED AND EGO, targeting a certain group of people, or individuals, and encouraging them with ideas AND statements to Break the Law. Time and again it is the Informants leading the charge with these individuals. Never mind outright lying by these Informants. These Informants get thousands of dollars as a result.

THESE INFORMANTS ARE ACTUALLY ENCOURAGING AND PROMOTING TERRORISM for their bank account, because of Greed. The Feds, morally misaligned, in turn get to look good. This is how Ruby-Ridge happened as an example. This is why a lot of these cases make a "Media Sensation" but then later FALL FLAT on their face in court. But you don't hear the "Media Sensation" on those cases when they fall flat on their face, do you? WHEN THE TRUTH OF THE MATTER COMES OUT, YOU DO NOT HEAR ABOUT THAT. Take a look, it's the Initial-Farce that receives the attention, NOT THE REVELATION OF TRUTH LATER, of what it is, that gets the attention. Again it is about what is on the surface—what LOOKS GOOD and what SOUNDS GOOD in the "Moment". This is a matter of "Programming the Masses". Reality and the Truth get ignored when it does not fit this "Government and Media Agenda". When you see the Mainstream-Media practicing such a "Lack of Objectivity" we have to begin to question it, never mind the "Issue of Water-Boarding". Clearly the Mainstream-Media is being Used and Misused to affect and "Steer" the "Thinking and Thoughts" of the Masses. The Media is being Used to Affect the Perceptions of the Masses, this is called "Programming" and Programming is a form of "Modern Day Brainwashing". This is what "Propaganda" is, on many a fine point.

The Mistakes made, and the "Incorrectness" of parading around people who are "Guilty", when we later find out they are NOT, is never as "Popular with the Media", the Mistake of it all NEVER RECEIVES AS MUCH ATTENTION. Quite obviously the Mainstream-Media is Corrupt and Controlled and incapable of practicing "Objectivity" with others, with NON-GUILTY PEOPLE, and ITSELF. When they do acknowledge that "Terrorist Suspects" were Not Actually Guilty, it is brief and fleeting to "Protect the System". It is mentioned a few moments no doubt so they can "Claim Objectivity" without actually practicing it. And afterwords you do not see the Media going after the System, People, and Processes, which happened which led to Innocent People being charged. They make a "Big Sensation" when "Everything Looks Good", without having all of the Facts, but when the Truth comes out, and we find out it was all a Farce, they barely mention it. The whole thing is a Scam, Farce, and Lie. Both Government and the Media are guilty here, and the Masses are misled. A lot of what is going on resembles the Red-Scare and it is no "Coincidence".

I know what a Farce this is from my own life and the Government Agents in it. These Government Agents have THREE BASIC RULES as they have communicated to me. RULE ONE: First us then the Target, this means how they LOOK in any situation comes first REGARDLESS OF TRUTH. RULE TWO: How do we make it (any given situation) work for us? RULE THREE: If it works it is good. It's pretty cut and dry isn't it ? No Morals or Ethics there, but it does give you a very good basic idea of these people and their "Rules of Thumb". Being good and doing the right thing doesn't even begin to enter the picture with these people. These people lack Morals and Ethics and everyone else is an inanimate object to them. Of course I speak of Corrupt-Agents here, I have met more than a few, during nearly two decades of my "Mobbing" who were good, and about doing the right thing.

As I was saying I know what a sick joke, a Farce, it is from my own life. As usual the Agents get to "Look Good" and the Public is Deceived. This has been used repeatedly in my life. As an added note here when I get to the Exotic-Technology section, there will be a whole other series of questions to be asked in regards to Guantanamo Bay and the prisoners (and other situations as well). Like I've said before it is "Torture for Tortures Sake", and regardless of "Technique", Torture remains Torture regardless of the name.

CHAPTER 10

The "Patriot Act": "Grey-Laws"—Laws Deliberately Written with "Double Meaning" for Misuse: Illegal-Laws and Programs: My "Mobbing": Illegal-Laws Passed by Corrupt-People: when "We the Government" Attacks "We The People"

The Patriot Act is named after the very people it Targets. These are what I call "Illegal-Laws", "Grey Laws", and Violations. The Government Spooks (and Sell-Outs) I deal with in my life told me, "The Constitution is only a piece of paper with ink on it.", I countered back with "Well so are the"laws" you claim to be following.". They do what they do because they WANT TO do it (violate people), and like doing it, they are Corrupt as people, not because of Laws.

There is something wrong with the Government when it's Employees claim the Constitution is worthless paper, and see the Constitution as an OBSTACLE only. Patriots are those who believe in the Constitution and America as it is meant to be, and are therefore AN OBSTACLE to very Corrupt and Powerful people in Government: the "Elite", those who think they are above the Laws and People of America. If you want Total-Power and Total-Control, then things, Ideals, like Freedom and Democracy, and PEOPLE WHO BELIEVE IN THESE IDEALS, are an obstacle and in the way. Therefore so is our Constitution.

You can not have Corrupt-People in Government who want More Power, and All of the Power, but then expect them to leave "Power in the hands of The People", which you may note is what any Democracy is about. Democracy is the "Distribution of Power" amongst the Masses and People, yet Corrupt-People who want All of the Power, want All of the Power Centralized at the Top of Government and out of the "Hands of The People". These Systems are commonly called Dictatorships, and many times Dictatorships arise out of "Freedom Oriented Systems" because of Corruption. Of course Dictatorships "Target People" who are Outspoken, Independent, and who think for themselves—they Target people who see what is wrong with the System and "Critical of It".

As an added point these Government Agents, Spooks I deal with, really do what they do because of the money, power, and perks they get along the way, NOT because of the Laws. They are Sell-Outs, and Corrupt, they are not about "Ideals". If this were not so they would not be Breaking the Laws, and they would be defending the Constitution. They do what they WANT to do, FOR THEMSELVES, not out of any sense of Morals, Ethics, Honesty, "Good Intentions", or "Legal Reasons". These people by anyone's sense of the word are "PAID CRIMINALS", just like The People "Over Them": CORRUPT-POLITICIANS, who allow this to happen and who write "Illegal Laws", for the sake of THEIR OWN BENEFIT. For the sake of More Power.

Ironically there is one group of people who have been ABOVE THE LAW and will continue to be, they are called CORRUPT-POLITICIANS, the Elite. I find it ironic that the one group of people who should be held the MOST ACCOUNTABLE in regards to the Laws, because of the Power they are trusted with and wield—therefore posing the greatest threat to ones Country, Society, Culture, and People, are also held the LEAST ACCOUNTABLE. This is of course because of the Power they have...if you get my point. And this gets passed on down the Food-Chain in "miniature form" to Government Agents who are Corrupt, and then other branches of Law Enforcement. The whole damned System is backwards and running in REVERSE, is being deliberately REVERSED-ENGINEERED, this is the FORMULA for taking down Humanity and the World through Corruption. Again Corruption spreads like a Disease from "Weak-Mind to Weak-Mind", and eats up Countries from within as Corruption becomes more common place and "Acceptable".

The "Patriot Act" is for MISUSE, it is Corrupt, and it is for putting people in my "Mobbing" unless I am mistaken—and I do not think that I am. In fact it points to this.

My "Situation" which I get to later in this book, is a result of Corrupt-Politicians writing "Illegal Laws", and it is used to create "Compliance of The People" and subservience towards Government (If you want an idea of what my "Situation" is like see "The Game" with Michael Douglas). This "Mobbing" also exists IN QUIET, IN SECRET, because IT IS ILLEGAL regardless of what exists on paper—it exists outside of a "National Spotlight". Hidden-Laws and Programs, are Hidden for a reason: BECAUSE IF PEOPLE KNEW, PEOPLE IN LARGE NUMBERS WOULD OBJECT, which would result in these "Laws" being thrown out. It would also result in these "Secret Programs" against The People becoming ILLEGAL. Which is why they are HIDDEN, to keep them from becoming ILLEGAL. Think about this.

This is the purpose of "GREY LAWS" AND "ILLEGAL-LAWS". Many of these "Laws" are Deliberately Hidden, while others are written Deliberately-Vaguely, and with "Double-Meaning", so they can be Deliberately-Misused, like the Patriot-Act. And the Agents "Capable of Enforcing these Illegal-Laws and Programs" are no doubt Corrupt like The People who "Wrote Them". You have to be a certain "Type of Person" to do this in the first place. These Sick-Corrupt-Programs attract a certain "Type of Person" as you have to be Corrupt and Sick-Minded to take part in a lot of these Programs. So what you have is a Corrupt-System being Deliberately-Created by Corrupt-Politicians, for attacking it's own people. There may be Agents involved in these Programs who are not Corrupt, merely "Doing their Job" or "Simply Assigned" to what they do, but again no doubt these Illegal-Laws and Programs are "Attractive" to a "Certain Type of Person". This is mostly what I have "Experienced".

This makes these Agents enforcing these Illegal-Programs, criminal in Act and Intention. This means what these people are doing IS ILLEGAL. Having said that these "Illegal-Laws and Programs" are being HIDDEN because the Government wants these Illegal-Programs in place (to the point of Hiding Them) so they CAN NOT BE THROWN OUT by The People, who are in fact Targeted by these Laws. They want these "Illegal Laws and Programs" no matter what, which is why they pass these "Grey-Laws", Laws written with Double-Meaning and Intention. Technically it is all on paper, but no one has a clue as to what is really going on, or how these "Laws and Programs" are really being Used and Misused. This is why I refer to these "Laws" as GREY LAWS because they are written deliberately to be vague and have "Double Meaning", FOR INTENTIONAL MISUSE. They are NOT written clearly, with clear boundaries, in black and white. They are written for being Deliberately Misused.

When you look at the PATRIOT ACT, substitute the word PATRIOT for TERRORIST and you will see what I mean.

Just like my "Mobbing", the Patriot Act was designed for Criminal-Misuse and for Targeting the American-People, specifically PATRIOTS, not "Terrorists". IT GOES AGAINST THE CONSTITUTION. Check out the wording of this thing, it is obvious that "Patriots" are the Targets. This is why after 9/11, the "NO FLY LIST", suddenly became a list of thousands and thousands of people. In October 2006, the show "60 Minutes" did a news piece on the "No-Fly-List", and having a copy of this "No-Fly-List" reported that it had 44,000 NAMES ON IT. In 2008 the ACLU based on information they have seen, estimated the list of names is now OVER ONE MILLION. Folks there are not even 44,000 potential Terrorists running around America, never mind one million. Obviously this list is not about Terrorists...It is about PATRIOTS. It is about people with a "Particular Mentality", and one which has nothing to do with "Being Subversive", obviously.

Americans have not changed overnight, suddenly becoming a threat to the American-People, the GOVERNMENT HAS. The Government is Targeting Patriots not Terrorists. The "Situation" I am in, and refer to throughout this book, is the same thing. It is about Targeting Patriots and an assortment of others who Government Agents decide are to be "Attacked and Targeted" by Government, by way of using Citizens and Society. It is about "Dirtying the Hands of The People", and predisposing them to "Blind-Obedience towards Government", it is about "Compliance to Government". It is about when Government becomes Cult, "Laws, People's Rights, Justice, and Facts" go right out the window for "Serving Government Agents", for serving Government. It is disgusting and is about laying down the Groundwork for FASCISM. Fascism is when Government becomes Cult. "Programming" is Used by Cults, and it is being Used by Government on the Masses.

IT IS NO LONGER "WE THE PEOPLE", BUT "WE THE GOVERNMENT".

The Patriot Act, under "Emotional-Circumstances", and therefore people were not thinking clearly and easily Manipulated, was passed for the wrong reasons. It should have been well thought out, well scrutinized by The People, not HURRIED AND HIDDEN. WHICH IS WHY IT WAS PASSED WHEN IT WAS. When you look at some of the Laws involved with the PATRIOT ACT you realize that anyone can be a suspect, anyone with an "Opposing Viewpoint", anyone CRITICAL OF CORRUPT-GOVERNMENT. Strangely enough it can be "Misused" to Target people who are Out-Spoken and Critical of Government; it gives authority to Government to spy and take away the freedoms of ANYONE WHO HAS A VIEWPOINT WHICH "OPPOSES" THE GOVERNMENT. This is Democracy? This is a Farce. And as Corrupt as Government is these days, people are going to TRUST Government with these "Laws"? That is rather the point, you can be investigated for simply "Criticizing Government"—and that does not even begin to be about Democracy. This all resembles the "Red Scare". In the meantime as Government goes about violating peoples rights, it denies that it is doing so, which is ridiculous: it has written these "Laws" for that very purpose.

People had better wake up to the REALITY of these "Grey Laws", these "Illegal Laws", and had better REALIZE that these "Laws" ARE ACTUALLY A TACTIC. A Tactic for Creating "Laws" which are to be DELIBERATELY MISUSED. If you believe the "Patriot Act" is not being Misused then you have your head up your anatomy.

Anyone with a Viewpoint contrary to Government can be labeled "A Terrorist", it is set up that way deliberately—the PATRIOT ACT is an example of "GREY LAWS". It in fact TARGETS PATRIOTS. To top it off this piece of "Grey Legislation", these Grey Laws known as the PATRIOT ACT actually NAMES THE PEOPLE THAT IT TARGETS....How blind can Americans be? This is the World you live in, not the one you WANT to live in. Until we as Americans DECIDE TO BE RESPONSIBLE AND SELF-EMPOWERED, become proud again of our American-Heritage, our Country will continue to go downhill, and Corrupt-Politicians will continue to encroach on our rights as a People. PEOPLE HAVE GOT TO PUT THE CONSTITUTION FIRST. As things are now these Corrupt-Politicians have become so Fat and Powerful they are now Targeting an entire portion of the population who are a threat to THEIR CORRUPT-LIFESTYLE. Patriots are the FIRST PEOPLE to speak out and institute change in REGARDS TO A CORRUPT-GOVERNMENT, in regards to CORRUPT-POLITICIANS, because of the Love for their Country. Patriots are the FIRST ONES to protect their Country. How God-Damned Stupid, or Blind can anyone be? This is what this "Patriot Act" is about, other people's SELF-INTEREST. If Americans do not wake up soon, it will only be "We the Government".

CHAPTER 11

The "Terrorism-Scare" Equals the "Red-Scare" Our Constitution Trampled by Secret and Illegal, Laws and Programs

This "Mobbing" I am in, I believe, is about a Future Planned for New World Order. It is an Illegal-Program that is a "Product of Illegal Laws". If you do not want to believe that last statement, knock yourself out, but it sure as hell is not about Democracy; and if you look on the Internet this "Situation", this "Mobbing", is prevalent across the World, and seems to be "Growing in Volume". More people are being "Targeted", and these days "Terrorist-Attacks" are the "False-Pretense" for "Investigating" Harmless Citizens: good people who care about the "State of Things" with an "Opinion or Viewpoint" contrary to Corrupt-Government. More Illegal-Laws and Programs have been created for getting people Targeted by COINTELPRO and in this "Mobbing": because of, "Terrorism". The "System" and Government, are becoming more Fanatical.

In Reality this is not about Terrorism, this "Terrorism Scare", like the Red Scare, is about Manipulating People Emotionally, through Fear, for Greater Control. Corrupt-Politicians are the ones who have written and "Passed" these Illegal Laws, Implementing Illegal Programs, to "Protect their Power-Base and Self-Interests" as they "Prepare to Implement Plans in

2012". These Laws and Programs give Government the Power to "Investigate and Interfere" with anyone, and you have No Rights or Options: they want that they can "Monitor, Interfere, and Investigate" anyone. This is a Sign of a Government preparing for a Police-State. It is Fanatical. Like this "Situation/Mobbing" I am in. This "Terrorism Scare" has been used as an excuse for Creating Corrupt-Programs and Laws which "Target Individuals", and gets the Public's hands Dirty on behalf of Corrupt-Government. It is all very much like the Red Scare, and I see very little difference. The ONLY difference I see, is that this time Government is being much more careful, subtle, and invisible, to "Make Sure" that it's "Encroachment On Our Rights" is not "Detected and Exposed" for being as Bad and Severe as it is. The Government has "Learned and Perfected" since the Red Scare.

These Corrupt-Politicians, and their Elite Friends Heading Corporations and Companies, are going after "Citizens who are in the Way". They are "Preparing the Way" for their "Next-Step", a World-Economic-Crash, and New World Order. They are "Designing and Creating" a Population which attacks others within it, who are against what Corrupt-Government "Says or Does". In this way they are "Training the Population" to "Conform to Government" and to "Attack those who are Critical of It" and to Attack those who are for "Positive-Change". In essence those Without a Brain, Idiot-Citizens, are being "Trained to Attack" those who are "Capable of Thinking for Themselves". There are "Those who Do as They are Told" attacking "Those who Do and Think on Their Own". You have those "Who do not Think for Themselves", those WITHOUT SELF-EMPOWERMENT, attacking those "Who practice Independent-Thought, Who Think for Themselves", those WHO ARE SELF-EMPOWERED. You have people who make up their own mind, being attacked by people who let the Television do it for them. Which is what Television is really about, it is about "Telling us what to Fixate On", telling us what is Important Subject-Wise, and it is about "Telling us what to Think" in regards to particular subjects. The Media and News have "Sunken and Fallen" over the years.

What you have are a bunch of Idiot-Citizens being "Trained" to defend the Corrupt who are making a Mess of the World in the first place, and they are being trained to "Attack" the very people who would "Fix this World". It has gone from Corrupt-Politicians who have Created Corrupt-Programs, which attracts Corrupt-Agents, which then gets "Passed On" to Stupid-Citizens who are "Used for Violating the Rights of Others": and based on what they DO NOT KNOW. These Idiots get Used to Enforce a Corrupt-System. In my case in particular I was "Put in this Mobbing" because I knew of a Pedophile Cop, and a Racist 911-Dispatcher and Murder, and I knew of these

Two Cover-Ups. COINTELPRO got involved because of this, Framed me, and "Put me in this Mobbing" which none of these Idiot-Citizens knew, or knows, about. They simply "Believed everything they were Told", and supported very Corrupt-Agents who were supporting a Pedophile-Cop and a Murdering Racist Redneck in Health Services. THIS is what these Grey-Laws and Illegal-Programs are for, and this is what this "Terrorism Scare" is for: people have lost their minds. Out of Fear they are being Manipulated.

Only an Idiot does not see this as ILLEGAL IN AMERICA, and only an Idiot does not see this as "Preparation for a new Government-System"—Closet-Fascism. Only an Idiot does not see these Illegal Laws and Programs that are being Implemented, as the Creation and Implementation of Fascism. COINTELPRO, this "Mobbing", and Laws like the Patriot Act, are NOT about American Ideals or Democracy: they are a VIOLATION OF THIS. This is about "Democratic Ideals" is it? This is about, and because of, "Terrorism" is it?

When "Terrorists" have already CHANGED AND TRANSFORMED A DEMOCRACY INTO A NATION OF FEAR, BORROWING "TACTICS AND MEASURES" FROM A POLICE-STATE, THEN "TERRORISTS HAVE ALREADY WON". This is when the Ideals of America and Democracy SHOULD BE RAISED AS A STANDARD BEARER, THE CONSTITUTION RAISED AS OUR FLAG, AND "DEFIANCE BE GIVEN" TO ANY THAT WOULD SEEK TO DAMAGE OR MAIM OUR COUNTRY BY "CHANGING OUR COUNTRY FROM WITHIN", AND GETTING US TO "TRAMPLE OUR CONSTITUTION, FREEDOM, RIGHTS, AND IDEALS" OUT OF FEAR. This IS what being an American is about. This IS when being an American counts the most—those willing to give up our "Constitution, Rights, Ideals, and Freedom" for "Terrorists" can KISS MY ASS. Are you an American, or are you "More Afraid of Terrorists"? DON'T TREAD ON ME.

As far as I am concerned this "Talk of Terrorists" separates REAL AMERICANS from "WANNABEE AMERICANS", people I call BUFFOONICANS. People who are absolute Buffoons willing to "Throw Away the Constitution", the HEART AND SOUL of our Country, people who are willing to let "What makes us American" be "Thrown-Out", people who are willing to let the "Constitution be Trampled" for "Terrorists", for complete "Losers and Ass-Wipes". Where are your Heads? You are an American? As far as I am concerned you only live in America. The "Requirements" are greater than simply "Living in America" and "SAYING" you are an American, YOU ACTUALLY HAVE TO BE ONE. It doesn't take anything "To Say that You are One". You can not throw out the Heart and Soul of our Country, change it into something "LESSER" and claim to believe in America: you can not Throw-Out WHAT MAKES US AMERICA. "Saying you are an American" is one thing, BEING ONE, requires more.

All of this FEAR-MONGERING, which is EXACTLY WHAT IT IS, all done in the "Name of Terrorists", IS OUT OF CONTROL. "Terrorists" are "Fanatics and Extremists" willing to do "Anything for Their Cause" and now Government is acting the SAME WAY. All of this RESEMBLES, AND IS, THE RED-SCARE ALL OVER AGAIN. Now the Public is acting Fanatical as well, and in regards to Itself. Coincidence?

NOW ILLEGAL-LAWS AND PROGRAMS ARE BEING IMPLEMENTED WIDE-SCALE, WHICH TARGET AMERICANS. Out of Fear, and Fear-Mongering. The "People" have become "The Enemy". And now "Government is being put on a Pedestal". And now COINTELPRO and my "Situation/Mobbing" are being Used and Misused to Implement this, to "Put Government on a Pedestal", to make sure it is no longer "We The People", but "We the Government". The "Shell of Democracy" remains, while the inside has been "Eaten Away" and Destroyed for a bunch of Criminals, Corrupt-Politicians, in the name of "Terrorists".

This IS the Red-Scare all over again. This "Time-Period" we live in now, in the future when looking back on History, will be labeled the "Terrorism-Scare". And until people "Reclaim Their Brains", which they have lost because of a "Single-Large-Trauma", the Twin Towers and 9/11, we will continue to "De-Evolve" as a Country, and as Americans. America is "Running Backwards and In Reverse" because of it, America is being Reverse-Engineered because of it. The "Heart and Soul of America", the Constitution is being Over-Run by Corrupt-Government and people have virtually lost all of Their Rights. Behind Your Back, the Government now has "The Right" to do anything it wants, in regards to your life: it never used to be this way, or so bad. The Twin Towers, 9/11, was a Trauma, Induced and being Used, to Manipulate "The People and Masses", MK-ULTRA on a Global-Scale. The "Terrorism Scare", IS the Red Scare, all over again.

CHAPTER 12

The Path of Experience vs. the Path of Consciousness

You have a big hole in your roof. You have two ways of learning here: through Consciousness, Thinking, Conscious-Thought, or through Experience. The conscious route involves looking at the situation clearly, realizing there is a problem, that it needs to be taken care of right away, and then doing something about it. CONSCIOUSNESS, SELF-EMPOWERMENT, RESPONSIBILITY, AND ACTION are the formula here. If you do not do something about it, then "Experience" will result. Being conscious means you Think, Realize, and CHOOSE not to have "the Experience" that would happen if you did nothing. You choose something different, a different course, a different route.

In my case I like the conscious route, the "Path of Consciousness".

Now the other way of learning, the Masses "Way of Learning" for a very long time now, the dumb way, is called the "Path of Experience".

In this case you have a big hole in your roof but choose to do nothing about it, perhaps lying to yourself, "I'll get to it later", or perhaps procrastinating, deciding to do it next week, realizing you have no way of predicting the outcome of the weather until then. Or perhaps you don't really think about what can happen at all, and simply ignore it for the SAKE OF COVENIENCE

AND COMFORT: this is LAZINESS-OF-THE-MIND. This is called being IGNORANT, and it is still a matter of making a choice. In this case what happens, is that the predictable happens. It rains.

If we ignore problems they tend to get bigger until we are forced to deal with them. This is one of the ways "The Universe", the Divine, the Source, Allah, God, Evolves-Consciousness. This is a matter of Energy and what I call Human-Physics, it is really a matter of "Divine-Law". In essence this is EVOLUTION—the dumb, non-thinking, non-learning, die off. This is "Darwinism of Consciousness". Thus it is better for Mankind to deal with problems as they come up before it is "too late", before problems get so large that by the time we "realize they are there" (because we have been looking the other way), we are shocked by "the Experience". If we try to ignore a problem, however small in the beginning, it will continue to grow, until it calls our attention; unfortunately by the time it does we often times end up Shocked and Traumatized by the Experience. All because we chose the Path of Experience, allowing the problem and it's Energy to Build to shocking proportions.

As the problem is ignored the Energy Builds, it grows and gets worse, to call attention to itself. The longer you wait, the more complicated and difficult it gets to repair. This is a matter of the Law of Cause and Effect, a Universal Law, a Divine-Law. This is a matter of Balance. Thus the longer you wait, the more Energy is required to repair a problem. The more "Negative-Energy" that builds up, the more "Positive-Energy" it takes to repair a problem—again Balance, the Law of Cause and Effect.

This is what is happening in this World today, Mankind is beginning to Wake Up and realize the mess it is in. A large segment of Mankind is beginning to Wake Up, realize Responsibility is needed, and is wanting it's own Self-Empowerment back (thus taking it back from Government and Politicians), because THE PEOPLE see that their "Leaders" are not actually doing their jobs. As a result of people Waking Up, they are relying more on themselves to see the Truth of Things, and seeing more, including Manipulations of the Elite, Corrupt-Politicians, and the Problems of our World Today. Positive-Energy, Self-Empowered people, the Path of Consciousness, is beginning to respond and take effect, pushing back against the insanity of Corrupt-Politicians, the Elite, and their deliberately made problems: their Reverse-Engineering of Humanity and the World. As "Corruption spreads like a Disease", so does "Self-Empowerment" and "Doing the right Thing", except in this case we can think of this as being "White Blood Cells", "Anti-Bodies", a building up of our "Mental Immune System" to that which is Negative and Corrupt. In one very real sense, this is what I do: I try to bring up the problems with Humanity, so they can be dealt with before it is too late. I promote the Path of Consciousness. But back to my example of "Mr. Laziness" and rain.

Through bad decision making, you have chosen a Negative-Experience, whereas with Conscious-Intent you could have chosen Responsibility, and a "Good Experience" in comparison. So you end up with a lot of water pouring into your house, you end up totally stressed out, yelling and cursing while you get drenched inside of your own home, running around with a bucket, like some deranged Captain Ahab bailing out your house. But of course it did not have to be that way. And of course, strangely enough, you CHOSE IT: you chose to have this Experience.

Ultimately you learn not to do it this way again, and learn to do what you would have done in the first place if you had chosen the "Conscious Way". You learn to take care of the problem right away before the problem grows, you learn to take Responsibility as problems appear. Or as a lot of people are, half-assed and resistant to becoming more Conscious, they learn to be more Conscious in regards to a similar experience, a particular set of circumstances. Like say they only learn to take the Path of Consciousness when they have a hole in their roof again... Really "The Conscious Way" involves doing all things this way, taking care of ALL problems Consciously and Actively, but at least for some it is a start.

The Path of Experience is the stupid way, it makes more sense to fix problems when the minimal amount of Time (which can not be replaced) and Energy is needed to repair a problem. Unfortunately here on planet Earth, the Masses are "Drunken on Dodo-Berries", and prefer "Experience". I however do not.

Now on my home planet, I call this learning the hard way. The funny thing is, once this Idiot has this Experience, something strange happens to him. He learns, he becomes More Conscious about things. The next time he has a hole in his roof, he will choose to think first, choose the Path of Consciousness, not the Path of Experience, realizing that either way he will have to take care of the problem. THE IRONY IS, NO MATTER WHAT PATH YOU CHOOSE, YOU WILL IN THE END, CHOOSE THE PATH OF CONSCIOUSNESS. If Mr. Laziness really learns, he will apply this formula to all things in the future. So why even bother with the Path of Experience, it is stupid and a waste of Time and Energy. Lazy-Minded people choose the Path of Experience, but in the end learn to be More Conscious anyhow.

This is how Thought Evolves, this is how "The Universe" Evolves Consciousness. It is a Natural-Law, a Divine-Law, built into our Reality.

The Law of Cause and Effect can be found in our daily lives, it is all a part of what I call, Human-Physics: People and Energy, Power of the Mind.

Now what if this Mr. Laziness was our neighbor, or if we were an observer of this Lazy-Minded Individual with a hole in his roof. What would we do?

Why would someone want to help a person, who is not interested in fixing his own house in the first place? And what is the point of doing it for him, if all he does is expect someone else to do it for him, and gives up his Responsibility and the Learning that comes with it to others? Who would, and what Type of Person would, take this guy's Responsibility upon themselves and for what PURPOSE if they did? Under those circumstances almost all Normal People, would not help out. It is a matter of accepting Responsibility for the situation at hand, and trying to remedy it yourself that counts. For most people who would help you out, this is what counts. I think we all can agree on this. No one is going to simply come and fix it for you, but if they see you struggling and trying, many would and will help.

Well, how are we as Americans in regards to the Problems of our Country, and in regards to Problems with our Government, and Corruption? ARE WE TRYING TO DO ANYTHING, AND WHO DO WE EXPECT TO CHANGE THINGS, TO FIX THINGS, AND WHY? Are we expecting Corrupt-Politicians to "Fix" a Corrupt-System? How do you think they view the Masses, the Public? Probably as Stupid and Gullible, don't you think? And another important point and issue is, what are the "Motivations" of those "Willing to Fix-Things" when WE AS A PEOPLE, are not willing to do so ourselves? Many times anyone "Willing to Help Someone Too Lazy to Fix Their Own Problems" is going to have Ulterior-Motives, as there is little point otherwise. The Corrupt know what they can get away with in regards to the Public: they know that the Public, the Masses, are Lazy-Minded.

This is the Masses, Mankind's "Dysfunctional Weakness", it's lack of Self-Empowerment and Responsibility for itself. We "The People", Humanity, use this to our disadvantage in regards to electing Politicians and Leaders. Corruption is rampant causing ALL of the problems of our World today, yet we expect Corrupt-Politicians and Governments to "Fix Things". We throw our Responsibility and Self-Empowerment at Politicians, and this is what Corrupt-Politicians and the Elite understand and want. I have also provided an example of how this looks, this Mentality, Mankind's-Flaw, through the Path of Consciousness and the Path of Experience, how this looks from the "Outside-In". This "Path of Experience and Consciousness" is a reoccurring theme throughout this book, and for a reason: it is what Americans, the Masses, World, and Humanity are choosing, and they CONSISTENTLY "Choose the Path of Experience". WHAT DO YOU WANT? The World you live in, or the World YOU WANT TO LIVE IN? If you want the World you WANT to live in, then YOU are going to have to WORK FOR IT. This means SELF-EMPOWERMENT, TAKING RESPONSIBILITY, AND ACTION. This again means working for what YOU WANT, and not waiting for others to do it for you.

CHAPTER 13

The National-Debt Part One: 2011–2012, Not "If", but When and Why the American Economy will Collapse, and the World-Economy with it: A Reality Check for Everyone who Likes their Money

> "If you don't believe in your Country, and Sell-Out, how do you expect it to continue to go on and exist?"
>
> —MR. LEE

> "How about a War Against Corruption?"
>
> —MR. LEE

This is very basic, simple to understand. Seeing as how this is about peoples money, and to make a point, I imagine that people are a lot more questioning and analytical about information such as this. That is to say, I bet a lot more people are "Less Blind" when it comes to looking at information such as this. Because they feel MOTIVATED out of Self-Interest, and are therefore less Lazy-Minded when it comes to looking at Facts. People are more Realistic, Observant, Logical, Deductive, and more Reasoning, as a general rule, when it comes to their money. Having said that, as a Country, as a whole, it does not look that way.

The question here is, what happens to a Country, when it can not pay the Interest on the money it owes to other Countries? Never mind the money it owes, but the Interest, owed for the money "Borrowed".

As a country, America, the United States of America, owes an absolutely enormous amount of money, to other countries. America has an enormous debt, THE NATIONAL DEBT, that literally grows worse and worse by the minute. It is ridiculous to think that with the way Politicians and other Government Branches are performing and behaving, that this Debt will EVER be payed off. No one is taking care of it, no one is being Responsible. CORRUPTION is not only causing those who are Responsible for taking care of the Debt to turn a blind eye, but is in fact "THE CAUSE" of much of the money being "used up", and "Disappearing", thus INCREASING THE DEBT. In fact if you look, a lot of money is being STOLEN by Corrupt-Government-Employees, Corrupt-Politicians sharing the gains with other Government Employees and "Contractors", who hold key positions for helping this money "Disappear".

America now faces a CRISIS that most Americans do not even know about as a result of this "Disappearing Money" and the National Debt. So what is the deal? The fact of the matter is that the AMERICAN ECONOMY IS GOING TO CRASH IN THE NEXT YEAR OR TWO, probably in 2011, but 2012 at the latest. The Politicians know this, how could they not, but the American People do not. The American Economy is going to CRASH and the American Dollar is going to become WORTHLESS. This is based on very SIMPLE ECONOMIC PRINCIPLES.

The "Economic-Problems of 2008" was merely a "Profiling of the Masses", for an Economic-Crash that is about to happen. A "Field-Testing" and "Fine-Tuning" for the Governments, and in regards to the Public and People.

We as a Nation are approaching a point where the American Economy, and all of the money it makes in a year, will not even pay for the Interest owed on the National-Debt. We literally will not even be able to pay for the Interest, never mind the Debt itself. When this happens the Dollar will become WORTHLESS, the American Economy will CRASH, the World-Economy will spiral out of control and then CRASH, there will be a lot of "Instantly Poor People" as a result, and a lot of people will be angry. Especially Americans, but many others across the World as well, as THEIR ECONOMIES will also CRASH. This is in the process of happening NOW, and will be Deliberate when it does come to pass.

Now keep in mind that American Politicians, the American Government knows of this Impending Reality, and so do OTHER POLITICIANS AND

GOVERNMENTS ACROSS THE WORLD. Thus there is a MUTUAL AGREEMENT to remain silent about this, by World-Governments, a CONSPIRACY against The People of the World, against the Masses. No Government is speaking out about this coming Economy-Crash, no Government is doing anything about this to prepare it's people. NO ONE IS DOING OR SAYING ANYTHING TO FIX THIS.

But one thing is for sure, when the American Economy Crashes and the dollar becomes worthless, the Politicians will still have money, and still be rich. This is what we are paying them for?

Now why would the American Government, and other Governments, take this approach? Obviously their "Being Quiet" is "Mysterious", and obviously they must have SOMETHING PREPARED for all of the CIVIL-UNREST this will cause. These Governments know that people will be VERY angry. If they are saying nothing and doing nothing, then IT IS SOMETHING THEY WANT. I'll get to this later, but keep this in mind—the Truth of this statement is obvious. This "SECRET" of a coming Economy-Crash, is only a "SECRET" from the Masses, not a "SECRET" from other Governments. But back to American Corruption, Corruption within the American Government, that is "using up" all of America's money, all of the American People's money that is: TO THE POINT THAT THE AMERICAN ECONOMY WILL COLLAPSE.

As an example, a big one, on September 10, 2001 2.3 TRILLION DOLLARS was said to be "MISSING", "LOST", in "Unaccountable Transactions" as said by our "Representatives": 2.3 TRILLION DOLLARS was said to have "Disappeared", was said to be "MISSING" (and that is just for ONE YEAR). Which is of course impossible. It is in the pockets of Corrupt-Politicians and others within Government helping them to "Funnel the Money Away". One way in which this money is "Disappearing" is that Corrupt-Politicians are AWARDING CONTRACTS TO FRIENDS who own Companies working for Government and who are being EXTREMELY OVERPAID; no doubt as a result of "Lobbying" and "Friendship" (Remind anyone of 10,000 dollars for a hammer?). This money in turn gets "SHARED WITH THE POLITICIANS LATER", it is "MONEY LAUNDERING" for Corrupt-Politicians. Money is being taken from the American People, "Given to Others" in very large amounts (DELIBERATE OVER-PAYING), and then part of it is being "Given to Politicians" who end up later working for these "Overpaid Companies" which are owned by "Friends of Politicians". In one way, shape, or form these Corrupt-Politicians are receiving some of this money later. It is for all intents and purposes MONEY LAUNDERING, Illegal, and everyone is cut in on the action. The American People get ROBBED, and this is happening yearly. Every year it is estimated that one TRILLION DOLLARS "Disappears"—and I heard this on the Radio (Nov.2010) where all can hear it. It is all Corruption and it is causing the American People's money to "Disappear" at the cost to our Country, National Debt, and Future-Economy. At the cost to Democracy and Capitalism as well, as they will later try to sell us Socialism, and "Declare Capitalism to not Work"—which is a Farce, it works just fine, minus Corrupt-Politicians.

America is being DESTROYED by these Corrupt-Politicians and "Their Elite Friends" who head Corporations and Companies. They want a System that does not share the Power and Wealth with The People. Want "Proof"? How many people know that MANY LARGE COMPANIES DO NOT EVEN HAVE TO PAY TAXES? AND MANY THAT DO "PAY TAXES" GET A "TAX CUT" AND PAY FAR LESS THAN "THEIR FAIR SHARE". We are talking about Companies and Corporations which make MILLIONS AND BILLIONS OF DOLLARS, and many DO NOT even Pay Taxes, while others are getting Tax Cuts and pay far less than they should. Think this is an "Accident", a "Coincidence"? The "Tax Standard" for "Individual Americans" is far "Higher and Harder" than it is for Companies, why? These Elite have all the Power, they "Make the Laws", and are "Sharing the Wealth" with Themselves while "Robbing The People", that is why.

At the time that this 2.3 Trillion Dollars was "Declared Missing", it was estimated to be 8,000 dollars for EVERY man, woman, and child in America. This should give you some perspective. With that in mind is it any wonder that America has a National Debt? AND WHY DO WE HAVE ONE? When this much money "DISAPPEARS" it is obvious that CORRUPTION is the CAUSE of our National Debt, not America as a Nation and most certainly not Capitalism. America makes a very LARGE AMOUNT OF MONEY yearly, so why are we not paying-off the National-Debt, why in fact is it GROWING LARGER? When we make so much as a Country we should not be going broke. Obviously CORRUPTION is the issue, anything else is an EXCUSE.

This "MISSING MONEY" would have been found by now, if it were not for the fact that The People who are Responsible for it, are The People who STOLE IT, are The People who "Lost It", are The People who have "Given it to Friends" by OVER-PAYING THEM FOR CONTRACTS. This is really where a VERY LARGE AMOUNT of this money is going. If these Corrupt-Politicians were not The People who STOLE this money, which is what this is—STEALING, a big deal would have been made about finding this money, and the THIEVES CAUGHT. Which would of course expose a vast "Web of Corruption", and expose the Connections between Corrupt-Politicians and their "Friends with Contracts" working for Government and a host of other "Money Laundering Schemes". In other words these Politicians are not finding where the money has gone, because it has gone to THEMSELVES, and will lead to THEMSELVES. It is no wonder our Country is going down hill, we are being made BANKRUPT by Corruption and this is occurring every year. It is no wonder that CRIMINALS AND CRIME are drawn to Politics when this much money can be made SIMPLY BY STEALING. Especially when The People being stolen from are doing nothing about it. Give an inch take a mile.

And Americans are expecting Corrupt-Politicians to put everything back into place? THEY EXPECT POLITICIANS TO TAKE CARE OF "CORRUPTION ISSUES" WITHIN GOVERNMENT, OBVIOUSLY, OR ELSE THE AMERICAN PEOPLE WOULD BE FIXING IT THEMSELVES. How stupid, or Realistic is this? If you are going to be this STUPID AND CARELESS WITH YOUR MONEY then people are going to steal from you, regardless of Job-Title. This IS the World you live in, not the World you want to live in. And I bet you lock your doors at night.

In other words you may not get Jeffrey Dahmer's Viewpoint, or the Zodiac Killers psychology, but you lock the doors nonetheless BASED ON THEIR ACTIONS. Regardless of whether you understand these people or not, YOU BELIEVE, you see their ACTIONS and lock the doors and windows. Yet you can see the ACTIONS of these CORRUPT POLITICIANS, THESE ELITE, YET DO NOT BELIEVE, because of a JOB-TITLE. A "Label and Stereotype" of "Politician" like this automatically tells you about the "QUALITY" of the person you are dealing with. This is a Stereotype, a simplified form of thinking, a short-cut to thinking, an avoidance issue of "Actually Thinking", not a Reality. It is a stupid way for determining Truth. How damned stupid can anyone be? This is SLOPPY and Laziness of the Mind. Whatever it is it is no excuse for not seeing the Truth, and it is in fact an "Allowing of Negative-Energy" to build and grow, in this case Corruption. In essence the American People have signed themselves up for the Path of Experience, for learning the hard way. The other end of it is that Americans are being VERY LAZY-MINDED, and since there are "No Observable Affects", well then it must be that things are "not as bad as they seem". Stuff, information like my book here is simply "Hype". I have news for Lazy-Minded Jackasses like this, your way of thought here is about to catch up with you, and the "Affects" are about to become "VERY OBSERVABLE".

Because of the simplistic attitude of Americans, these Corrupt-Politicians and their "Friends", feel that they are "Entitled to do what they Want", feel that they are above all others, and so feel they are a "Class Unto Their Own": they feel that they are "The Elite". So I refer to them this way throughout my book. In REALITY these "people" are CROOKS, CRIMINALS, ORGANIZED-CRIME THROUGHOUT POLITICS, GOVERNMENT, AND INDUSTRY and any other name is a distraction from the Truth. When I speak of the "Elite", keep this definition in mind. This is not some "far away kook talk", this is a simple matter of CRIME and CRIMINALS, being drawn to EASY MONEY AND EASY POWER. It is EASY for these CRIMINALS because Americans and others across the World are too damn Lazy, Pathetic, and Apathetic, to actually do anything about these CRIMINALS.

Thus Criminals and Psychopaths, CRAVING POWER AND MONEY, are attracted to Politics, and when THAT is your FOCUS for taking a "Job", it certainly does not even begin to be about "Serving People" and "Representing People", this ought to be obvious. These "people" are about "Representing Themselves", you need only look at the FACTS to see that this is True. And if you have "doubts", or do not believe, then assume it is True for a moment, and then "Look with These Eyes"—you will see that everything makes sense this way, because it IS TRUE. But back to Corrupt-Politicians "Awarding Contracts and Giving Away Money".

In 2006 the House Armed Services Committee took a look at the Defense Departments Budget and it was stated that in 1999 2.3 TRILLION DOLLARS was "MISSING", and that in the year 2000, 1.1 TRILLION DOLLARS was "MISSING". Folks this is a lot of MISSING MONEY and it can only be "Missing Yearly" if it is being STOLEN YEARLY. This is not "Accidental Over Spending" this is STEALING, and "Criminal Neglect" as an excuse at this point in time, is a LIE. Quite simply this is not "Criminal Neglect" it is INTENTIONAL AND DELIBERATE—"Criminal Neglect" can be ruled out. There is no excuse.

Within TWO YEARS 3.4 TRILLION DOLLARS has "Disappeared", it is no wonder our Countries Economy is going to Crash, look at these figures for only two years, think about it and the fact that this is YEARLY, how could it not Crash? And America makes A LOT of money, so why is it going DOWN-HILL? The money is not being spent on America and Americans, it is being STOLEN AND WASTED on Corrupt-Politicians and "Their Friends". Folks our Country, America makes a lot of money yearly, so how can it be that "We are Running up a Debt"? This is not possible with RESPONSIBLE POLITICIANS IN CHARGE. Things can only be the way they are, America losing all of the money it makes in a year and then some, with Corrupt-Politicians running the Country. And then they have the nerve TO TELL AMERICANS THAT THEY NEED MORE MONEY FOR "DEFENSE". This is all an absolute LIE, FARCE, and SCAM.

This STEALING of the American People's money is going to COLLAPSE the American Economy. As an added point for you to think about, since Politicians know of this Impending Economic-Collapse, and know they are to blame, what do you think they will do when that time comes? Turn themselves in? Allow this "Web of Corruption" to be exposed? Perhaps they are "Awarding Contracts to Friends", overspending an INCREDIBLE AMOUNT IN PREPARATION for an angry America. In other words perhaps these Corrupt-Politicians are paying Friends to "Construct Measures" for an angry American People—like Prison Camps all across America which DO EXIST, as one example. But back to the National-Debt.

As it is we HAVE "DISAPPEARING MONEY" RUNNING UP THE NATIONAL-DEBT, and in VERY LARGE AMOUNTS. It is said that a TRILLION DOLLARS A YEAR DISAPPEARS, at the least—some years being more Exceptional than others. The National-Debt would be no where near where it is, if this were not so. If we had RESPONSIBLE POLITICIANS, with all of the money America makes (Capitalism does indeed work), we would NOT HAVE a National-Debt. And a STEREOTYPE, the "Reputation of Government", of "Red Tape", and Bureaucracy, is being used to sell "The Ineptitude of Government" as if this explains 2.3 trillion dollars missing, as opposed to Corrupt-Government-Employees. This would not normally be the approach of INNOCENT HARD WORKING GOVERNMENT EMPLOYEES (people such as this would not provide "cheap excuses" rather they would fix the problem), a Cheap-Lie and EXCUSE for "Missing Money". So what we have is Corrupt-Government-Employees stealing billions upon billions of dollars, TRILLIONS OF DOLLARS, and running up the National-Debt to THE POINT OF IT COLLAPSING OUR ECONOMY. Running up the National-Debt to the point of our Country being DECLARED BANKRUPT. And they know this. Is this AMERICAN, is this LEGAL?

Do you think these Corrupt-Politicians do not know this? They most certainly are aware of America's-Financial-Status, they most certainly are aware that America's Economy is going to CRASH when America can no longer pay the Interest it owes. They know all of this, yet are doing NOTHING about 2.3 trillion dollars missing, doing nothing about billions upon billions of dollars MISSING YEARLY, except giving BAD EXCUSES, CHEAP LIES. What do you think of these Politicians when it is put this way? Because this is the TRUTH of the matter. They know of "All of this Missing Money", know that they are Responsible for this "Missing Money", know that the American Economy is going to COLLAPSE because of it, AND ARE SAYING NOTHING ABOUT THE COMING AMERICAN-ECONOMY-CRASH, yet are turning around and PROVIDING CHEAP LIES AND EXCUSES FOR WHERE THE MONEY IS. To top it off, these Corrupt-Politicians have built Prison-Camps, in preparation for an angry American People—this IS THEIR IDEA OF RESPONSIBILITY. Obviously it will be the American Peoples fault for being angry, NOT the Corrupt-Politicians fault when the time comes. At least this is how these Politicians are dealing with it. Excuse me for saying so but you have to be an IMBECILE to believe otherwise about these Politicians and other Government Employees.

If they say something NOW about this Economy-Collapse the American People will be FURIOUS and want to pursue Justice NOW, and want to pursue these Corrupt-Politicians NOW: if the Economy-Collapses first and then Americans find out, it will be too late to really do anything. So many people will be in SHOCK that they will not be thinking Rationally enough to actually catch anyone. In fact people will be EMOTIONAL NOT RATIONAL, and so Corrupt-Politicians will get away in all of the CHAOS. These Politicians will have a "Legitimate Excuse" for Declaring Martial Law. Thus these Corrupt-Politicians are waiting for it to happen so they can "Get Away"—this is part of an "Escape-Strategy", a "Tactical-Retreat". To top it off, this American-Economy-Crash fits a LARGER-AGENDA AND LARGER-GOALS which these Elite want. In my opinion once this Economy-Collapse happens it will be too late for Americans, and the rest of the World.

You can forget it, it has everything to do with why they are saying nothing about the Imminent-Collapse of the American-Economy in the first place. AND KEEP IN MIND, THAT ASIDE FROM CORRUPT-POLITICIANS, there is no reason why America should not be one of the RICHEST NATIONS ON EARTH. If it were not for these Corrupt-Politicians it would be. What makes these Corrupt-Politicians so special? The American People do, WE DO, by turning a blind eye to all of this CORRUPTION. The American People do all the work, these CRIMINALS get all of the money. To top it off the American People's LAZINESS AND APATHY makes Politics and Government attractive, a "Honey-Pot" for CRIME AND PSYCHOPATHS. We can not take care of our National-Debt, when we have Corrupt-Politicians in charge of OUR MONEY. Our Government can not function properly, with dysfunctional people within it, with Corrupt-Politicians, Corrupt-Government-Employees in charge of our Country. Dysfunctional Government Employees means a Dysfunctional Government, take a good look at it because this is what it is these days.

At some point the American People are going to have to have a "WAR AGAINST CORRUPTION" and CLEAN UP THEIR Government and that time is quickly approaching. It is better that the American People begin doing it NOW, before it is too late, and while they have a chance of stopping this Economy-Crash, and everything following it, from happening. By my estimation the American-Economy is DESIGNED TO CRASH in September 2011.

Your Worthless Dollar: The Mainstream-Media is a Corrupt Media—The Proof is in the Silence: when the Time Comes We Will Be "Sold" the "Failings of Capitalism and America", Not the Truth Which is a Corrupt Government and Corrupt-Media

What most Americans, and people in other lands, do not know is that the American-Economy is about to CRASH, America is about to become BANKRUPT. This will happen when America can no longer pay the INTEREST owed on the money owed, on the money borrowed. This will happen when the Interest we owe, is more than our country makes in a year. Think about this. To top it off in early 2010 (I believe April) I heard on National Public Radio, that this figure was at an 86% point. Meaning that there is only another 14% to go before the Economy Crashes. When we reach 100% this means that we can no longer pay the Interest on the money we owe as a Nation. This was not a big story on the radio, it was in fact a few sentences, and in fact if you did not know about any of this, no doubt you had no idea of what was being spoken of in the first place. It was "Deliberately Short and Vague".

Where is the NEWS of this Impending Economic-Crash, where is all of the talk about stopping this? Think about this and the FACT that when we can no longer pay the Interest owed, America will be DECLARED BANKRUPT, DEFUNCT, and the American Economy will CRASH. Where is the News of this, why is the "Media" not doing it's job? Obviously the "Media" is CONTROLLED for there to be such a "lack of interest" over such a worthy and important story: one which affects every American. Obviously the Media, larger elements of it, is CORRUPT. The People in charge of it, are obviously CORRUPT, and in league or a part of, this Elite. This does not mean that every reporter is in on this, but it does mean that people in Key-Positions are. Why the American-Economy-Collapse is NOT an issue, makes this OBVIOUS. Anybody working for the Media arguing otherwise is an IDIOT (in denial or a Deliberate-Liar), and can look at the FACTS. But back to the subject which the Media is not covering…(maybe I'll do a chapter on Paris Hilton)

When we as a country can no longer pay the INTEREST we owe for money borrowed, the entire American-Financial-System will collapse. Everything you have stored away in terms of dollars will become absolutely worthless, as in "junk", fancy toilet paper. The American dollar will no longer be money, it will be worthless, and everyone around the World will totally drop it. Your stored savings will be worthless, your 401-k, your bank account, your stocks, your assets, you name it, it will all be JUNK. EVERYONE will be

INSTANTLY BROKE, as in penniless, and your dollars WORTHLESS. You will have as much money as a person in a "Third World Nation", except unlike them, you will not be able to produce your own food: most do not even know how. Americans across the country will be broke, everyone except The People who "Raided and Stole" from the American peoples coffers, because THEY quite frankly know this is going to happen (do you think there is a single banker or economist out there who does not have enough sense to know this?), and are prepared to keep themselves fat and happy regardless. We are dealing with a very simple, BASIC ECONOMIC REALITY Folks, one created by very CORRUPT people, who are more than prepared for this collapse to happen, they expect it to, and want it to. They know it will because they have DESIGNED it this way, and based on two principles: THEIR GREED AND THE AMERICAN PEOPLE'S IGNORANCE. This is YOU FOLKS, this is YOUR FUTURE.

So have you heard of this? Is it in the NEWS? Is this "Small Detail" even in the Mainstream Media anywhere? No? Why? Because Corrupt-Politicians are in charge of the American Government, and ALSO IN CHARGE OF THE MEDIA. Obviously and with TRILLIONS OF DOLLARS to go around, how hard can this be? For some reason these Corrupt-Politicians are not mentioning THE COMING ECONOMIC-COLLAPSE, and are in fact SUPPRESSING this information. Why? Because they WANT IT TO HAPPEN. Or else these Politicians would be taking Responsibility and Action and be doing something about it. Instead they are COMMITTING CONSPIRACY BY HIDING THIS from the American People, and Committing Conspiracy against the American People. To believe otherwise would be stupid. So much for American Government and Main-Stream Media which is not even beginning to do it's "Job". The Media is now "A Branch of Government".

These Corrupt-Politicians are protecting THEMSELVES and their own SELF-INTERESTS. They are Hiding the fact that they are stealing all of this money and getting away with it to the point of BANKRUPTING AMERICA. This is why many Large Companies who make millions and billions of Dollars a year, do NOT Pay Taxes, and why many others get Tax-Cuts. These Corrupt-Politicians have set it up so that their Elite-Friends heading these Companies, get to "Keep the Money" and no doubt share it with these Elite-Politicians later. It is all a SCAM, and clearly Companies and Corporations making millions and billions of dollars should be Paying Taxes. Instead only the "Average Citizen" is paying Taxes, and apparently Paying to "Keep Companies Alive" which have been Deliberately Run Into the Ground. This is Ridiculous.

Do you honestly believe that American Politicians have no idea of "America's Money Situation", that they have no idea of America's Economic Situation? No, THEY DO KNOW, but this SECRET is kept so that WE THE AMERICAN

PEOPLE DO NOT KNOW. When it does happen it is going to take most people in America and the World by SURPRISE, a WORLD-SHOCK, a WORLD-TRAUMA, which is part of the Plan. A Plan Designed by Corrupt-Politicians, Organized-Crime within Politics, a Plan Designed to ROB THE AMERICAN PEOPLE and to discredit Capitalism and America. This will all be a FARCE, LIE, AND SCAM, as it will all be about Corrupt-Politicians Manipulating this to happen, not about Capitalism which does work, and not about America which also works: both capable of working minus Corrupt-Politicians, and Corrupt-Heads of Industry and Companies (like the Media and Car Companies), that is.

Folks this is absurd, this is OUR MONEY, and it has been, and is being "Stolen" by the very people, Politicians, Government, and System, we have put into place. It is not that the Government and System we have does not work, it is that The PEOPLE WITHIN IT ARE CORRUPT and could care less about their elected positions, or "The People". They quite obviously only care about themselves to the Detriment of the Masses, which is how people who are psychotic are: they have no empathy, they only care about THEMSELVES. This however, is also the fault of the American People, turning a blind eye, Apathy.

As an added point, too many people are relying on Mainstream-Media to "Point Out Issues", and are allowing the Mainstream-Media to decide what is an issue, and what is not. Too many people are allowing the Media to do their thinking for them. Are we thinking for ourselves at this point, or being TOLD what to think, and being TOLD what is an issue or not? Again look at something as stupid as Water-Boarding, and whether it is "Torture" or not. I can sum this up in one word, "Duh!". And stuff like this is in the Media all of the time. Another good example, is the idea that "Cows are Contributors to Global-Warming". We are talking about "Methane and Cow-Farts" here Folks, how ridiculous is this as an issue? Cows are not the problem, PEOPLE obviously are. Look at the amount of pollution we cause, need we point a finger anywhere else? But I digress, back to the Public turning a blind eye.

The American People are also to blame: they are not being Realistic about, or Vigilant of their Representatives, they are not taking on their Responsibility as Americans. They are also "NOT CARING". This is also about the American People. OBVIOUSLY THE AMERICAN PEOPLE DO NOT CARE, if they are going to ALLOW this to happen. The American People are not taking RESPONSIBILITY for their Country, their Democracy, and the System they live within. They are not keeping it "Clean and Functional", like the environment as an example. If things are not kept "Clean and Functional", they deteriorate, and stop functioning like any environment. THIS IS LAZY

AND IRRESPONSIBLE and will lead to even more Robbery. And as you will experience, the American People will be duped yet again when a "New Currency" is adopted, unless Americans decide to do something about this Impending Economic-Crash.

Keep in mind that this American-Economy-Crash is going to affect the WHOLE WORLD. Where are the "News Stories" and "Reporting" on this? Who are we waiting for to take this up as an issue? The Mainstream-Media, the Government?

Americans and people across the World, the Masses, will be SHOCKED, the Politicians across the World will not be. It will be a Massive-Robbery, and a "Staged-Event", a FAKE-EVENT, a Manipulated-Event, to install Socialism in America and finally the World. This will be the beginning of installing a One-World-Economic-System, and I believe, with a "Second-Staged-Event" coming shortly after this one, an attempt to install a One-World-Government. After all, America "will no longer be able to stand on it's own" and "Capitalism" will have been "proven to not work".

This will be all we hear about in the Mainstream-Media which is also CORRUPT. When the time comes, it will all be about how Capitalism does not work and ADVERTISED AND SOLD all across the World this way. To get people, the Masses, to "BUY INTO THIS LIE" rather than question WHAT went wrong and WHY. This will be because of Corrupt Government and Corrupt-Media not doing their jobs as they are meant to—rather misusing their "JOB TITLE" for their own gains: which is of course what Corruption is all about. They will "ADVERTISE AND SELL" the "Failings of Capitalism" (which it won't be) to keep people from seeing the Truth of the matter, that this was allowed to happen, and that Politicians, Governments, and Mainstream-Media across the World kept their mouths shut about it all.

Instead they will try to get the Masses to "Buy Into" the idea that America does not work, and that "Capitalism does not work", which will take the BLAME AWAY from Corrupt-Politicians who Orchestrated it all in the first place. The same as was done with Communism when Soviet Russia broke apart. This will be ADVERTISED BY THE CORRUPT-MEDIA NON-STOP and eventually will be about how we Americans and America need to "Join the rest of the World" in a World-Economy and World-Government. Which will be the "Next Part of the Plan" that these Corrupt-Politicians try to put into place. These Corrupt-Politicians in America, and in other Governments across the World, want a One-World-System to CONSOLIDATE ALL OF THE MONEY, and TO CONSOLIDATE ALL OF THE CONTROL. They want this for THEMSELVES, not for the Masses, but seeing as how they are HIDING all of this information it should be obvious that their intentions are not good.

So shortly down the road Americans will be broke, AS IN HAVE NO MONEY, and at THAT POINT you will no doubt, FINALLY, be very Pissed Off with your "Elected Officials". But by then it will be too late. Why didn't you care sooner? And who doesn't care about their Country? You see it is a lot easier to SAY that "You do care about your Country", something that requires no effort, than it is to actually SHOW that you care by DOING SOMETHING ABOUT IT. If Americans are going to be this STUPID and LAZY about "Taking care of their Country", then this is going to happen. These Corrupt-Politicians certainly get this, that most Americans talk this way, but few actually do care enough to want to LOOK, or do anything about what is going on. How else do you think that these Politicians are getting away with so much CRIMINAL ACTIVITY? With so much ROBBERY? And to the point of running the most powerful Country in the World into the ground...And I might point out BANKRUPTING, what is supposed to be, and should be, one of the richest Countries in the World.

Do you actually think these Powerful Thieves, this GROUP OF ORGANIZED-CRIME WITHIN POLITICS AND INDUSTRY, these Elite, are so stupid that they do not have some clever way Planned out of this, some clever escape from the American people? Do you think they have no idea of the amount of anger there will be? The amount of RIOTING? Forget it. They do know, and they do have PLANS for dealing with this, for dealing with the American People, and for dealing with the rest of the World's Population, which will also be angry as THEIR ECONOMIES will Collapse as well. When the American Economy Collapses, the entire World-Economy will Collapse. All of it by DESIGN, by Corrupt-Politicians, the PROOF is in the silence across the World by ALL Governments in regards to the Collapse of the American-Economy and the effects it will have.

And when the time comes, the American People need to remain CALM and call for an INVESTIGATION of Corrupt-Government, not get Angry and Riot. This applies to other people in other lands as well. This is what the Elite are expecting, that everyone will react EMOTIONALLY and WITHOUT THINKING, which will give the American Government, which has been Hijacked by THIEVES, an idea of WHO IS ANGRY at it, WHO is "against it", and WHO the Government should TARGET and Lock-Up, or Destroy. This is what these Prison-camps, "Custodial Facilities" as the American Government calls them, all across America, are for. For locking up Angry Citizens, and anyone with a "Viewpoint" that is not "Pro-Corrupt-Government". These "Facilities" are for locking up people who want to go after these Corrupt-Politicians. When the time comes, Americans need to remain RATIONAL AND CALM, need to DECLARE A "WAR AGAINST CORRUPTION", and INSIST AND FIND these Corrupt-Politicians who let this happen, and who in fact CREATED THIS MESS. This means throwing out Politicians, NEW AND INSTANT ELECTIONS, putting Honest-People in place, and then PERSECUTING AND PROSECUTING THESE CORRUPT-POLITICIANS who are taking America Down, for their own Selfish-Gains.

A New Economic System

When the American-Economy crashes, it will be a THEFT, BY CORRUPT-POLITICIANS, all over again. You will get PENNIES FOR YOUR DOLLARS, when the Current-System crashes, the dollar becomes worthless, and a NEW-SYSTEM is instituted. This New-System will be based on a "New Currency", a replacement for your dollar, and for every dollar you will receive only pennies of this "New Currency"—so Americans will be ROBBED TWICE. The New-System will be centered around SOCIALISM, which is why the American Public now owns a large portion of Car-Companies due to "Financial Mishaps", and IS WHY MUCH OF WHAT OBAMA SPEAKS OF IS SOCIALISM; the Government, the Corrupt-Elite, already have a "Solution" planned. This is why the National Debt, AND THE IMPENDING ECONOMIC-CRASH IS NOT BEING ADDRESSED (in the form of speaking to the American Public), TAKEN CARE OF, OR FIXED. They want the Economy to Crash, so they can Implement PLANS for gaining MORE MONEY AND MORE CONTROL.

People across the World are not even being informed of this "Situation" by their Governments or Politicians, it is ridiculous. The Corrupt-Media across the World is not doing it's job either, which shows that it is Controlled by Government and by the Elite. The Masses across the World are obviously being kept in the dark.

Instead of informing the Masses they are laying down the "GROUNDWORK" for SOCIALISM, which is THEIR "SOLUTION". This is also why there is a lot of talk about "Health Care" and "Health Insurance" these days. So that this System, Socialism, is achieved more easily and smoothly; and so there is less to do for putting Socialism into place when the time comes. To get people used to the idea as well. When this happens you will get pennies for your dollars, and the Elite will make a HUGE PROFIT. They will have all of those "OTHER PENNIES"—they will make a FORTUNE all over again. They will have STOLEN MONEY to the point of Bankrupting, what should be, one of the richest nations on Earth, and then even afterwords, ROBBED the American People all over again. THEY WILL HAVE ROBBED AN ENTIRE COUNTRY, AND RIGHT IN FRONT OF EVERYONE, right in front of the entire World.

How is that for Stupid? Americans at that point will be the Biggest-Suckers on the planet, and very tempted to take Socialism right away out of

FEAR OF HAVING NO MONEY. It is a SCAM playing on "Survival-Issues" so that people will accept Socialism right away instead of questioning the Economy-Collapse and the Corrupt-Politicians RESPONSIBLE FOR IT: AMERICANS WILL NEED TO REJECT IT FROM THE BEGINNING and PURSUE THESE CROOKS INSTEAD. And then people wonder "Why" someone would do such a thing. This is the World you live in, not the World you want to live in. People have got to deal with Reality first, not their "Fears and Desires". If you as an American are NOT going to do your part in maintaining your Country, then this is going to happen, and already is. These Corrupt-Politicians could care less about you, they Control the Media and Government, and this ought to be obvious. You have only to look at America's Finances to see the Truth of this. They are too greedy to be able to hide themselves on this point.

Of course the Elite within Corporations, are involved as well in AMERICA'S DEMISE and is why we have had "Financial Mishaps" with Car-Companies, and there will be more to come. The "Heads" of these Companies receive so much in "bonuses" and "retirement-plans", that they are ROBBING the Companies of their own money and profits. This is by Design and is to cripple these Companies, and to make "Capitalism" look bad. Just like with America, they ARE DRAINING ALL OF THE MONEY AND RESOURCES FROM WITHIN, UNTIL THE COMPANIES "IMPLODE". Until the Companies become Bankrupt. The same FORMULA USED for Bankrupting America.

When the amount of money these "Heads of Companies" receive lacks common sense, LACKS BUSINESS-SENSE, lacks logic, and goes against THE SURVIVAL OF THE COMPANY ITSELF, it is obvious. It is TO PUT OWNERSHIP OF COMPANIES INTO THE HANDS OF "THE CITIZENS", who "Spend Money" trying to keep these Companies afloat—which is all a FARCE. We are talking about "Paying Debt" by Creating more Debt—does this even make sense? It is all about ROBBERY and Bankrupting America. It is all about putting Socialism into place—now the "Public" is beginning to own "Companies", and this will no doubt continue to go on until the Economy Crashes.

By Deliberate-Design Companies are becoming "Publicly Owned" and in such a way as to be "Acceptable to the Public", by Deliberately-Bankrupting Companies: which in the case of American-Car-Companies, this has occurred. The same thing is happening with the American Economy. In both cases the Financial-Resources are being DELIBERATELY DRAINED TO COLLAPSE the American Economy, and to collapse these Car Companies. It is the same formula, it is all a part of a Bigger-Plan. It is all a "Heist and Robbery". They are stealing money until America Implodes, and then when they offer a "New Currency" you will only get pennies for your dollars, they get all of the other pennies, you get ROBBED TWICE. And once

Socialism is in place it will be even EASIER to rob The People—at that point Government will OWN ALL. You on the other hand will own next to nothing, with little future in sight, as Capitalism will be NON-EXISTENT and the idea of "Getting Ahead in Life" will be a "lost dream". The "American Dream" will be finished. This is what happens when we turn our backs on CORRUPTION, and are too Lazy-Minded to even look honestly at what is going on in our Government. This is what happens for turning our backs on Reality, Reality being a Government and Media that is controlled by Corrupt-Politicians, Corrupt-People, across the World...

The Implementation of Socialism will be so that these Corrupt-Politicians own all of the money and power.

Watch, there will be more things which happen, "FAVORABLE TO PUBLIC OWNERSHIP", as this is part of a FORMULA. A Formula for Socialism and the Destruction of America from within. If you are familiar with history it sounds like COMMUNISM, doesn't it? I wonder why? And to think that when the Berlin Wall came down they said that "Communism does not work"...

What The People Should Do

My tip to the American People, and people in other parts of the World, is the following.

At that time, when it happens—the American-Economy Crashes AND the World-Economy with it, the American People, and others across the World, need to remain CALM, and surprise these jerks, BY CALLING FOR NEW AND INSTANT ELECTIONS, and throwing these Greedy-Crooks out. These Corrupt-Politicians, this Organized-Crime in Politics, these Dementoids are expecting an Emotional-Reaction, a NON-THINKING REACTION, not a Calm and Logical Approach to catching them. Thus if the Masses remain CALM this will throw a major monkey wrench into their PLANS. Having said that, when the American-Economy Crashes, and then the World's, these Elite have a SECOND-EVENT PLANNED to distract the Masses from a Logical-Calm-Approach in the first place. WHEN THIS SECOND-EVENT happens the Masses must still remain CALM AND FOCUSED and continue with their PURSUIT of these CRIMINALS. The Masses must remain focused on getting to the bottom of what happened, regardless of "World-Events and Traumas", regardless of Events Orchestrated by this Elite, for the very purpose of distracting the Masses from their PURSUIT OF JUSTICE.

It is that simple. Then the American People, and other people across the World within their respective lands, need to really investigate what has happened to ALL OF THE MONEY, and find out WHO has it, and HOW. EVERYONE CAN START WITH THE ISSUE OF "WHY WERE WE NOT INFORMED OF THIS COMING

ECONOMIC-CRASH?". People, newly elected under this crisis and because of it, who TRULY REPRESENT the American People, or The People of their land, at this time of crisis WILL be able to get this done. They WILL be able to find answers, and catch a lot of these people, because when the time comes a lot of people involved in this Corrupt-Nonsense will not be happy with the Reality of it all. No doubt there will be people who "sell-out" once they are caught, there is no loyalty among thieves...

When the World-Economy crashes people need to remain CALM, hold new elections, and throw out their Crooked-Representatives, who did nothing to speak out of this problem, or to remedy and divert it. Instead they will have CAUSED IT TO HAPPEN, PREPARED FOR IT TO HAPPEN, WILL HAVE ALLOWED IT TO HAPPEN, TO SHOCK AND TRAUMATIZE THE WORLD, for the sake of MORE EASILY Manipulating THE PEOPLE, THE MASSES, THE WORLD. For getting what they want. When a Big Shock like this happens it takes a while for people to think straight. Remain CALM AND RESPONSIBLE, and these Elite and their Plans will become screwed up.

When the time comes, insist for NEW REPRESENTATIVES, and INSIST FOR "A WAR AGAINST CORRUPTION".

Keep in mind that ALL Governments are involved in "Secrecy and Conspiracy" against their people in regards to this WORLD-ECONOMIC CRASH. This means ALL Governments have a VESTED INTEREST in what will transpire. This is CONSPIRACY, and it exists out in the open—the coming Economic-Crash is an ECONOMIC REALITY. They will ALL GAIN something from this happening: MORE MONEY AND MORE CONTROL. Obviously a One-World-System, a similar Economic-System across the board is one of the goals. A New World Order, One-World-Government with a One-World-Economic-System in place. Their Plan is to have and Control ALL, while the Masses have and Control NOTHING...

The essence of it all, at the heart of all of this Manipulating and Scheming, this THIEVERY, is this: in order to implement this New-System of Socialism within America and the World, the economy has to collapse, everyone will have to lose everything. Capitalism will be "Declared to not Work", the way it was "Declared that Communism does not Work" when Soviet Russia went defunct. This will be to convince the World that Capitalism does not work, and that it is not a "worthy pursuit", which is a LIE: America and Capitalism do work, they are only being SABOTAGED to look as if they do not. When the Standard Bearer for Capitalism "Crashes and Burns", Socialism will be sold to the World. This is to convince the World that Socialism is the "Only System" that will work, Capitalism "Being Proven" to be a "Farce" like "Communism and Soviet Russia": which is ironic because they will actually be pushing for a System that is very similar to the Old-Soviet-System. Obviously there is something wrong with this picture.

Again it comes down to Organized-Crime in Politics and Industry, Psychopaths, Power-Mongers, who want Money, Power, and Control, at all costs. They could care less about the Masses, and from "Their View-Point", the Masses could care less about themselves as they never seem to do anything about these obviously Corrupt-Politicians, this Elite, in the first place.

This "Economic Crash" will supposedly be "Proof" that CAPITALISM does not work (a lie), and WILL BE ADVERTISED THIS WAY IN THE MEDIA. America will get the blame for it all when the World-Economy Crashes, and then Americans will basically be offered Socialism, AS A CASH INCENTIVE, which these Elite already have picked out and on the table. This "Cash Incentive" is called "SURVIVAL", and will be what motivates people to "Buy into Socialism": as I have said it is all a Manipulation, and if YOU LOOK, if you play detective, you can see that all of the signs and evidence are there. What I see, I see clearly whether you like it or not. And what I see will happen unless enough people begin Speaking-Out and Taking-Action. Speaking-Out will increase the Awareness of the Masses and put a spot-light on these "Politicians" and their "Plans". As a general rule these "people" do not like to follow through with their Plans when they feel enough people are watching. Thus if a large segment of the Masses is alerted and begins OPENLY QUESTIONING THEIR Government AND "POLITICIANS" ABOUT THESE POINTS, these Plans can be averted. People across the World need to do this, as this is not only about the American Government, and American Elite.

In short, when the time comes the Masses will need to wage "A War Against Corruption".

So with that in mind, BE AWARE AND INFORMED, BE SELF-EMPOWERED, BE RESPONSIBLE, BE ACTIVE, SPEAK OUT TO OTHERS, AND DO YOUR PART TO TURN THINGS AROUND. When the time comes DO NOT LET YOURSELVES BE DISTRACTED BY "WORLD-EVENTS", BY WORLD-SHOCKS, BY WORLD-TRAUMAS. A Conscious-People, an Aware-People, a Questioning-People, a Vigilant-People, is a Free People.

CHAPTER 14

Why Such a Good Country and Powerful Country is Going Down-Hill: America the "Standard-Bearer of the World" is Being Deliberately Reverse-Engineered into the Ground

Well like I've already said Corruption is eating the Country up from the inside, it's the Final-Stage of any Country or Culture. If you don't believe it look at history. Having said that America is being deliberately destroyed from the inside, by Corrupt, and by anyone's definition, Selfish-People. You can add Psychotic, Psychopath, as well, as what other "Personality Type" could resort to such means for such gains, and at cost to the Country and World.

Perhaps I'm getting ahead of myself here, and if you find such a thing hard to believe please try to bear with me a few more pages and chapters. I at least think most can and do agree Corruption is a problem that creates more problems, however many no doubt will disagree as to HOW CORRUPT the System is. And many no doubt will disagree as to how deliberate Corrupt-Politicians are being in destroying America. I don't doubt that there are those who find it hard to believe that America is being "Deliberately Destroyed",

REVERSED-ENGINEERED, as a Culture and Country, as a beginning point for REVERSE-ENGINEERING THE WORLD. However, I think I can make some valid points to this though, if you read on a bit further. I also was not a believer in such an idea as this, at first. I had heard such ideas before but considered it "Cliche" and "Kook-Talk". What have you got to loose, aside from a few minutes of time? It is not as if I do not understand this Viewpoint. But again why America? I'll tell you why; there are people in America who don't want to raise the standard of the World across the board, that would create GREATER-EQUALITY and give ALL, the Masses, GREATER-POWER. Democracy is a Distribution of Power amongst the Masses. America used to be a Standard Bearer for the World, now America has gone down hill. These Corrupt-Politicians want more Power and Control, and want to Consolidate Power and Control, rather than spread it out amongst the Masses through Democracy. Rather the idea is to destroy America, deliberately, and take the Whole WORLD down with it. Which in the end, as these Corrupt-Politicians have planned, will lead to One World Government, the Ultimate-Consolidation-of-Power. It is of course a "Power-Issue", something most of us can't understand without having rubbed shoulders with this type of person (which I have—the Government Spooks in my life), that's what all of this nonsense in the World is about.

Remember, Democracy is a SHARING OF POWER amongst ALL, and a Dictatorship is a CONSOLIDATING OF POWER at "THE TOP": it is where a MINORITY has ALL of the Power.

So many leaders, CORRUPT-POLITICIANS, and Governments of the World are concerned with more Power NOT The People they represent—things in this world are DELIBERATELY REVERSED. COUNTRIES AND CULTURES ARE BEING REVERSE-ENGINEERED, taken down a downward-spiral. I'll be giving a couple of examples shortly, along with my experiences with these type of people, Power-Mongers. I will also attempt to help the readers understand the "Psychology" of this type of person, due to my long years of suffering and analyzing, sick and demented Government Spooks, who have been wreaking havoc in my life for 17 years now, and specifically the last eight years. These people are again, total Power-Mongers, just on a smaller scale than what I call the "Elite". The "Elite" being those who see everyone else as being "The little People", the Masses a stepping stone to Power, and underneath their "Elite feet".

Just like most people, the Public, the Masses, I also didn't believe, or get this "Type of Person", or their psychology. Unfortunately I had no choice but to do it, to "look at and understand" this type of person, in order to try and put an end to my "Situation"—which I will get to eventually. I have spent years experiencing the "Power-Monger-Type", analyzing, and sticking my mind into their "Sewer-Mentality" to get a better understanding: and quite frankly it has been an uncomfortable mental journey. It is NO WONDER no one understands these "Types of People" beyond a certain point: it is VERY UNCOMFORTABLE to put ones mind there. Most peoples energy, their mindset and personality, is incapable of going "There". I on the other hand have had no choice—so you can learn from my doing so, and take my word for it, instead. I hope to put my Experiences and Knowledge to good use and hope to increase the Awareness of this Sick Type of Person, the "Power Monger", which translates down to Government Agents, Fascism and Cults, Psychopaths, Corrupt-Politicians, the Elite.

CHAPTER 15

The Mentality of the Elite: A Down-to-Earth Look and Explanation of these "People": A Look at these Elite from a "Laymans Perspective"

If we are ever going to understand the World we live in, it's problems, how they come about and why, we are going to have to understand the "People in Charge", their Mentality. We should at least try, and if not we can simply "Look at Actions"—like my previous example in this book of the Zodiac Killer and others. This is a "tricky one", and has everything to do with why people living in this world can not see or understand what is really going on in front of their faces and why. Most of this has to do with the vast difference in Mentality, mindset and brainwaves, between ordinary people—the majority of the World, and the "People in Charge"—the Elite. These people are what I call Power-Mongers, and have a very different Mentality than the average person.

Now when we speak of the "People in Charge" we are not referring to all Politicians across the spectrum, of all Politicians across the World, or even just Politicians (but for the sake of being obvious lets stay with Politicians for now). We are speaking of a great many of them though, Corrupt-Politicians, across the board, across the World, who like any ORGANIZED CRIME, like any CRIMINAL-ORGANIZATION are working together for their own Agendas, own Power, own Wealth. If we are going to understand this Elite, my word for the Corrupt and Powerful united together, we are going to have to start with the basics.

The first thing to understand, REGARDLESS OF TITLE (Appearances-Stereotypes), is that these "people" are CRIMINALS. These people are Criminals within the POLITICAL SYSTEM, and as a result wield great Power and Influence. They wield Power and Influence over the Masses. As a result they are able to Buy-Out, Influence, Threaten, and Own damn near anyone they want in any sector or walk of life. Their "Playground and Breeding Ground" is the Political-Food-Chain, which is of course where all of the Power, and as a result (If nothing else due to insider information and Connections) all of the money is at. Money being another way to obtain more Power and Influence.

As an equivalent, think of Mob-Families. These people want to move up within Societies Ranks, the Social and Power Structure, for more Power and Influence, holding "Respectable Job-Titles" all the while being Corrupt and working for ulterior, behind-the-scene Motives and Agendas. All of which involve acquiring more Wealth, and in the end more Power. What is the difference? There is none, this just makes it easier to "see" these Elite.

Now seeing as how the "Playground and Breeding Ground" is the Political Field, this makes it hard, very difficult for Honest Politicians to thrive. They are automatically entering a "Playing Field" that is slanted against them; the "Real Players" both hidden and seen are against them, and even the "Rules", as a result, are hidden. Honest Politicians are entering a darkened room, where all is not as obvious as it seems. Again they are entering a room where the "Players" are against them. Corrupt Politicians with massive amounts of Power and Wealth do not want exposure, or "their rules" changed, the Playing Field cleaned up. This, from their Viewpoint, is Counter Productive. This automatically would result in the eventual down-fall of their Power, Influence, Wealth, and the "Playing Field" itself. This would result in a Playing Field and Game, as it is meant to be played, with the actual rules in place, as opposed to a "Game" which exists only for the sake of appearances. It would in fact result in a Political System as it is meant to be, where the Politicians actually represent The People, NOT their own SELF-INTERESTS.

In short it is that many Honest Politicians are on a "Team" where their own "Team-Members" are against them. Within the "Team" there is actually "Another Team".

This is one of the problems. Good Politicians do their job as it is meant to be, representing The People; Corrupt Politicians represent THEMSELVES. Corrupt Politicians put themselves first and then The People, they are Self-Gratifying-Ego-Maniacs whose world centers around Power and Wealth: Wealth being just another means for obtaining more Power. Any "Representing The People" is SECONDARY for the Corrupt, the Elite. First THEM, then the Public. They have to maintain appearances for the Public, in order to stay where they are. Having said that you can bet that much (if not all) of what they present to the Masses, for the Public, is SECONDARY to their own Goals and Agendas, they are just making things work for themselves. Thus what they present to the Public is SECONDARY, an Illusion, and therefore not what it seems to be. There is ALWAYS going to be a SECONDARY LEVEL, which is in fact the FIRST LEVEL: that involving their own SELF-INTERESTS. This is the REAL LEVEL, the first one, and is what remains Hidden from the Public Eye. It has to remain Hidden, the Public would object to having Criminals in charge.

Again a way of reaching an understanding, an equivalent of this, is to simply think of things in terms of known Organized Crime, the Mob. Imagine that they work themselves into the Political System, that actual members of the Mob are now Politicians, but in actuality Mob Bosses. How would they do things? What would the Mob do? Obviously they would maintain appearances for the Public and Masses, while secretly milking the System for more Power and Wealth. On the surface it would be one thing, while in Reality it would be another. They would no doubt be "misrepresenting" The People for their own SELFISH ENDS. Any "Representing The People" would no doubt be duo-fold, any laws being passed for the Masses no doubt being part of a Bigger Scheme, a step towards an Agenda of their own. "Small Steps" toward their own Agendas, being taken slowly and progressively, as a way of remaining "Invisible and not Obvious". "Small Steps" for their own Agendas, involving the acquiring of more Power, more Wealth, while appearance-wise still looking good for the Masses.

Now at some point in order to cheat the Masses, within a "Free-System"—like a Democracy or many a "Western Nation", of their Freedom and Self-Power, you are going to have to arrange "CIRCUMSTANCES" so that you can bring the "Invisible" Power-Monger-Level to the forefront. This happens when the Power-Mongers, the Elite, are ready to bring THEIR LEVEL out into the open, and are ready to substitute THEIR LEVEL for the level that is meant to be. A lot of ARRANGED SCENARIOS AND FALSE EVENTS happen and have happened, and will happen as a result of this. They are trying to push their Paradigms into place, their Level, for the SET-UP of Total-Control, which REQUIRES a Dictatorship.

Hitler used this and many others. It is Standard-Operational-Procedure for the Power-Monger-Types, because it is what works on the Masses. This is a matter of Human-Psychology on a "Collective-Scale". The Masses are not contemplative enough, to avoid falling into this TRAP, which is why many refer to the Masses as Sheep: including myself on occasion. In order to avoid falling into this Trap, the Masses have got to be more reliant on their own thinking, and less reliant on others to do it for them. Especially when the Masses are "Reliant on Criminals" to do it for them. INDEPENDENT-THINKING IS WHAT COUNTS, not "Externalized Thinking", being "Thought For", told what to think, by someone else. As long as you are Reliant on others for Thought and TRUTH, you will always be Manipulated; which is what the Mainstream-Media is used for, for Manipulating the Masses. My example of Water-Boarding being a good point—it should not even be an issue at all, never mind in the Media. In short this is the entire downfall of Humanity. People have got to think for themselves and be Independent with their own minds: Self-Empowered and Self-Thinking. This starts and begins on an Individual Level, it can start on no other. A COLLECTIVE of Individual Thinkers, is a smarter, more Conscious, and more Free-People. Manipulation becomes a more difficult avenue of approach for Power-Mongers when this is the type of people they have to deal with. They are forced into honesty with a Conscious-People, and therefore will have less Power. This is why so much in the Media is a Farce and Lie, to dull our senses and to Deliberately Mislead us, to keep people "Less Conscious".

So the Elite at some point have to "bridge paradigms", have to bring THEIR LEVEL into the fore-front in order to take over and CONQUER the Masses. It is all about Power, and to have Total-Power they have to make a transition, from being "Hidden" to out in the open. The Level of Power they want can not be obtained otherwise. They do not want to "spring the trap" too early, to do so would result in large scale, across the board, Resistance. The shock of "Western Style Freedom" to Dictatorship would result in significant losses on their end, in the end P=B-C (again the profits equals the benefits minus the costs). This MUST be done slowly, and has been done slowly, to negate the difference in SYSTEMS, and LEVELS. This is to Negate Shock and Backlash from the Masses. It is all a Manipulation. This entire Transition of Systems from the hidden to more obvious, is currently going on. Until New World Order arrives, and thus Dictatorship, the Elite are still trying to be "Invisible" to the Masses, all the while their desired Invisible-System is breaking the surface, and becoming extremely apparent for the Free-Thinkers, and Self-Empowered of Society and the World.

But back to my earlier point of "Small Steps" for their own Agendas, while still looking good, "Appearance Wise", for the Masses.

The motivation for this IS obvious, to hold their positions, their Job-Titles. Of course the Ultimate-Goal would be to bring the entire System to a point where they no longer have to hide, can exist out in the open, without ever having to worry about loosing their Power, thus able to do what they want. If you look throughout the course of history, this has happened many a time, these Systems are called Dictatorships—where the FEW rule the Masses. There are many ways of getting to this type of System, but if you are dealing with countries where Freedom, Equality, The People, are the basics of the lands—like all "Westernized Countries", then moving slowly and progressively is the best way to go.

From a "Power-Monger" point of view this would be the most intelligent path to follow, with the LEAST AMOUNT OF BACKLASH FROM THE PUBLIC, THE MASSES. Thus slowly REVERSE-ENGINEERING "Freedom Oriented Governments" into TOTAL-CONTROL Governments. Now seeing as how these Corrupt Politicians are united across the World in their Criminal Enterprises, like different Mob Families, they are working towards a uniting of these different branches, into one Centralized-Family. Regardless of any "In-Fighting" amongst various "Families and Branches", these people have agreed to a basic goal, and have united for it. In this case, this is already being done and implemented, it is called New World Order.

New World Order is the uniting of these different Political Systems, with seemingly different Politicians, under One Banner, One Flag (in reality, the "first level", the hidden reality <the selfish interests of Power Mongers> and the Hidden Players, are already united, it is just about bringing it out into the open). This is the Centralizing of Power and Government on a Global Scale, which is what any Dictatorship is about. Dictatorships, Fascist-States are about the CENTRALIZING OF POWER, taking it from the "MANY", the Masses, and putting it in the hands of the FEW. Thus New World Order is a plan for "Global Dictatorship", the taking away of Power from the Masses on a Global Scale.

IT IS THE OPPOSITE, THE REVERSED-FORM OF DEMOCRACY.

Western-Style Nations, are being Deliberately REVERSE-ENGINEERED, for the sake of creating a Dictatorship.

It is just that in this case these Corrupt Elite, want to turn the ENTIRE WORLD into a Dictatorship, not a single country. Once these New World Order countries are united, Economic-Pressure alone will get all others to comply. Now keep in mind that behind the scenes these Corrupt Politicians across the World are working together for their own selfish egotistical desires, like any group of Organized Criminals. They are united by their Criminality, they are Organized Crime. They are Organized Crime in the Political Field, thus the scale is different. This also has to do with the "Mentality" of these Corrupt Politicians, the Elite, this World Organized Crime Family.

This World Organized Crime Family is far different from any Mob Family, or Families, they are not as simple. They are like a particular religion, say Christianity as an example, with all of it's many different off-shoots, branches, variations. Or like a large Corporation with separate arms or branches, with specific goals and functions. As an example of this point think of a car company that has separate branches, dealing with particular types, models, of cars. Just keep in mind that there are deeper and different levels with, and within, this Demented-Group of Nitwits.

Back to New World Order for a few moments.

If you are going to have New World Order, uniting all of these different Political Systems under One Banner, then you are going to have to level the Playing Field, creating One Political System for all. Tricky right? No it is simple, and thanks to the Government Spooks in my life, I have come to understand this Power-Monger Viewpoint, their style, techniques, and "flavor".

Uniting all these different Political Systems is going to be much easier than it seems. It is called Economic Collapse, and no doubt will begin in America, from there spreading across the World. So far we have had a taste of this (2008), as the Elite, this World Organized Crime Family (WOCF), fine tunes its plans for doing so.

Lets start with America. America is (was) the great Standard-Bearer of the World. A good example of Freedom, Democracy, and Capitalism: as it is sold around the World. Now America, technically speaking, has the responsibility of raising the Worlds-Standards across the board, bringing greater equality to all, and empowering the Masses across the World. It is, or was (this is becoming less and less these days—by Orchestrated Design) seen this way across the World. So by bringing America down, from the inside, you can bring the World down. However this Corrupt Political Playing Field, with its Corrupt Politicians are about anything but "Distributing Equality". Distributing Equality comes down to DISTRIBUTING POWER. With Equality comes a higher standard of living, less polarity in this World, a spreading out and dispersing of "Power", as it is shared with all—again a higher standard of living comes down to a sharing of Power. Democracy IS a sharing of Power, dispersing it amongst the Masses; NOT a hoarding of Power at the top of a Political Food Chain. A Dictatorship IS THE OPPOSITE of a Democracy. People automatically wanting Power, are against the Masses, and in favor of a Dictatorship, not Democracy.

Democracy is an equal distribution of Power amongst the MANY, not a hoarding of Power amongst the FEW. Like say Politicians, the Rich, and Corporations and Businesses. Any hoarding of Power at the Top, making a System "Top-Heavy", is indicative of a Dictatorship, a Fascist-State: anyone who has studied history knows this. Hoarding Power at the top, Centralizing it there, is contrary to a Democracy, contrary to any "Freedom Oriented Society". It is contrary to what most Westernized-Systems and people, American or European, have come to expect. It is a danger sign to anyone who likes their freedom, this hoarding of Power at the Top. Thus America and its Ideals, have got to come down, be lowered, and is being done so, by the Elite. Quite Deliberately. Take a good look at America, do you really believe that Politicians could actually be so Inept that America is going down so quickly, deteriorating so rapidly, that it could be anything but Deliberate?

America and what it stands for is counter productive for these Elite, these Power-Mongers.

If you are a Corrupt Politician, a Power-Monger, in order to have Power you can not share it; by doing so you narrow the gap between the Power you have and others. This is counter productive from a Power-Monger perspective. Automatically you can find that the ONE SYSTEM in which Corrupt-Politicians, Power-Mongers, can thrive in and therefore the most desirable, is a Dictatorship, a Fascist-State. This is a System in which Power is hoarded at the top of the Political-System, which is what Power-Mongers, by definition, are about. Thus at some point as Power-Mongers within a "Non-Dictatorship-System", gain more and more in Power and Influence, the System they hide behind is going to have to COLLAPSE and be reformed to accommodate the "Hidden System". This is a natural "evolution" of the circumstances, it is actually DE-EVOLUTION of the System being hidden behind. The REAL LEVEL, the one the Corrupt do not want you to see, the one of their "Self-Serving Egotistical Intent", begins to grow and over shadow the very System it is hiding behind. In the end this is what the Corrupt want any-how. In the end Power-Mongers want, and exist best within, a Dictatorship.

Can you imagine Hitler, Stalin, Idi Amin, Pol Pot, existing within a Democracy, or any other "Western Style/ European Style" type of Government for very long? In Secret to top it off? Of course not, these kinds of people, the Corrupt and Psychopathic, do not thrive under these conditions: they thrive under a Dictatorship. At some point they would become obvious because their ACTIONS would speak for themselves, their Actions would over-shadow their words. Kind of like what is happening now within many a "Western Nation", across the World. If we look at ACTIONS these days, we can see what is going on. Western-Style Nations are indeed being Reverse-Engineered to "FAIL".

The problem here, with doing this within "Western-Nations", is that not everyone is "asleep". Thus as this "Hidden-System" begins to "Break-the-Barrier", begins to "Emerge and Surface", many people take note of it and begin to Speak-Out. Which is what Healthy-Minded-People who love their Country do—like say "Patriots". Power-Mongers do not want this, it is Counter-Productive to have people in Society, Cultures, Lands, speaking out about the Emergence of such a System, Counter-Productive to have "Dissidents" within the land, so THEY BECOME TARGETED. This has a lot to do with people being in "My Situation", Mobbing as it is commonly referred to (Localized Fascism), and has a lot to do with things such as "The No-Fly-List". Thus we do not have a lot of "Instant and Potential Terrorists" but we do have a lot of people speaking out about Government, and criticizing it, specifically things which are a result of Corruption. All of this is to CENTRALIZE POWER, and to TARGET INDIVIDUALS who are "IN THE WAY". This is about TARGETING PEOPLE who are "Anti-Corruption", "Anti-Dictatorship", people who are Pro-Freedom. People who love their Land. This also makes these people into "Suspects", without them actually creating a crime. IT IS ABOUT A MENTALITY, AND TARGETING PEOPLE WITH A MENTALITY THAT IS IN THE WAY OF THIS ELITE.

But back to America.

What to do? It is again simple; you take America, as a World Standard Bearer, and bring it to it's knees. The purposes and intent are many fold, all of it "Good Logic" if you are a Warped Power-Monger, a Psychopath and Dementoid, in the first place. Remember these people only care about THEMSELVES, THEIR life, THEIR Power and Money, THEIR health and well being, not the Masses.

First of all Capitalism due to the collapse of the dollar and America the country, will be claimed to not work, to be Dysfunctional. Just like with the Soviet Union, Communism was declared to not work, to be Dysfunctional. Sounds familiar right? NONE OF THIS IS ACCIDENTAL IN REGARDS TO AMERICA, it is a Formula. In the case of America, America is being Deliberately Sabotaged from within by its own Corrupt-Politicians, to not work: it is being REVERSE-ENGINEERED to Collapse on itself.

Corruption as anyone who has studied history knows, is the "Last Legs" of any Society, Country. Corruption is a Societal-Disease that eats up Populations and Countries from within, until they implode. Like a large tree eaten from within by termites, it looks good on the outside, until all of the inside and underlying structure, causes it to collapse, at which time it is apparent to all, the damage done. At any rate it is ironic because Capitalism does work if Corrupt Rich and Powerful People (the Elite), heading Companies and inside of Government, are not Deliberately Sabotaging it so it does not work. But this is one reason why America is being Targeted and brought down: TO "PROVE" CAPITALISM DOES NOT WORK.

When the Standard-Bearer of the World, and shining example of Democracy and Capitalism falls, it will cause a tailspin pulling down the entire World. Like a watch, one broken piece affects all. Economic chaos will ensue, and a whole lot of mess with it. The Masses will be angry, to say the least. What will be the "Solution"? WELL SOCIALISM OF COURSE. This is why much of what President Obama is talking about is Socialism. This will be the SOLUTION OFFERED for a PROBLEM DELIBERATELY MADE. The Ground-Work, through Obama, is already being laid down for this, it is a definite sign of what these Elite have planned. Already the "Public Ownership of Companies" is turning into a Theme, which sounds a lot like Communism doesn't it?...

The MAJORITY of people living in this world are already poor, when the Economic System Collapses across the World, it will be more so, and in the "Western Nations" as well. People will be clamoring for solutions. The "Solution" will be Socialism, Capitalism will be declared to not work, America as a Democracy will look a disgrace, and Socialism will be on the World Table as the "Fix". Shared wealth at that point, when everyone is POOR, will seem indeed like a very good idea. This NEW SYSTEM will gain a lot of automatic support from the MAJORITY of people across the World, THE MAJORITY OF THE WORLD, from the Masses in every land, REGARDLESS OF THE TYPE OF GOVERNMENT. Everyone will be "United in their Poverty".

The problem, Man-Made, Elite-Made, will be GLOBAL, the "Solution" GLOBAL. Countries will be turned inside out. The "Solution Offered" will cut across International Boundaries, and act as a CASH INCENTIVE, a Survival-Incentive (playing on fear), for the Masses to go along with it. Again, the Masses in every land will be united in their Global-Poverty, thus a "Solution" of Global-Socialism will be in effect a CASH-INCENTIVE for the Masses, and a Trick of the Elite to get one step closer to the type of "System" they really want, which is a Dictatorship (They may even settle for Socialism, hiding behind the facade of it, because it will not be that far off from the System they want regardless). In Reality, this Ploy to "Sell Socialism" and offer it as a CASH-INCENTIVE AND SURVIVAL-INCENTIVE, will be the GREATEST CON-GAME AND ROBBERY OF ALL TIME; while at the same time totally removing any Power left in the hands of The People, and giving it to a CENTRALIZED GOVERNMENT, a New World Order.

America and Democracy will be declared "A Sham and Disgrace", people will want their Money and SURVIVAL, and will be all too ready to throw out "Democracy" for the sake of their own asses, and due to their BEING BLIND to what is really going on. Capitalism AND Democracy will both go out the window in America AND around the World. In theory, if all goes as planned for these Corrupt-Nit-Wits. I speak of other peoples Plans and what CAN happen, what MIGHT happen, not necessarily what WILL happen.

Having said that these Elite have many Plans and counter-measures in place, THE REAL THING TO PREVENT THEIR SUCCESS IS THE PEOPLE, Consciousness, Speaking Out, Being Active, and most importantly of all: being SELF-EMPOWERED. What I speak of is a very clear Game-Plan of the Elite, it is not the only Game-Plan. There are many variables involved as far as what choices the Elite make. For example there are different groups within the Elite (different Crime Families, to use the Mob metaphor, or simply "different branches") with conflicting Agendas who are fighting for Control and their own Agendas, they are fighting for Power. A few other "larger variables" are: the choices the Masses make as a whole based in part on its Consciousness, the Elite will make choices based on what the Masses SEE and what the Masses DO, the Elite are aiming for a 2012 date for reasons I will get to later (these reasons alone separate them from Mob Type Families and makes them more complicated to understand and see), and so on and so forth: there are many variables that will effect the final outcome, but those are a few of the bigger ones. Having said that New World Order has been set up as a Definite-Event. How they get there and what they do, based in part on the Masses, is what is questionable. Plans for a New World Order are not. A lot of time, money, energy, Manipulating, and Control is going into the creation of this, too much for it to be a "Possible Scenario". New World Order, a World Dictatorship, is a Planned-Event. Having said that I am still optimistic, AND STILL BELIEVE that the Masses can prevent this by simply being CONSCIOUS, ACTIVE, AND SELF-EMPOWERED. However simply believing the best will change nothing unless The People become this and do this: again SELF-EMPOWERMENT is key and TAKING RESPONSIBILITY AND ACTION IS KEY.

Further more the Elite have picked 2012 as the Set-Date for reasons they CAN NOT CONTROL, and if a "Monkey Wrench" is thrown into their Plans by the Masses, their Whole-Calender of Plans, can be thrown off. I get to Their 2012 Deadline, the reasons why, in my second book, yet to be printed.

To top it off, the Elite have one more step, for AFTER New World Order is established. This "Next Step" is also being "Broadcasted" and has a lot of time and energy put into it. I have a single Chapter devoted to this subject, and if you are Religious and/or familiar with Religious Prophecy of Various Faiths, as you will see the "Next Step" is what many Prophets throughout the ages have referred to. I give a brief summary of this "Next Step" below.

The next logical demented step, "Solution, will be to have a single SPOKES PERSON for this New World Order. A "Real Miracle Maker" in terms of "Helping" the World. This gives an even further singular view and vision for the World. It narrows peoples, the Masses vision down to a laser-beam-Viewpoint and thus no peripheral or hindsight vision. The Elite, considering many options and having many choices, is seriously considering this. In fact they have it PLANNED. This "Next Step" depends on the "Success" of the ones before it, and exists as an "Option" for now. It involves putting All Hope onto one person.

Having a singular Spokes-Person negates the idea of National Pride, and National Differences, thus negating SEPARATION and further "Uniting the Masses". National Pride and National Differences will no doubt remain even after a "Uniting through Global-Government", and is one more thing the Elite will not want, because at that point it will be Counter-Productive for their purposes (ironically "divide and conquer" has worked very well for them for hundreds of years). In short the Masses will not be divided by National Lines (Government-System lines, Government ideology already removed by One-World-Government), but united UNDER A PARTICULAR PERSON, someone who appears to be "Humanitarian", a "Fixer of Things", someone who everyone is impressed with, and likes to like. The "Poster Boy" and "Spokesman" of the Elite, the Pied-Piper of the Masses. The "Proposer of Ideas of the Elite", a puppet. Someone "To Love and Admire".

He will offer "Solutions", already concocted, for problems also previously made and put into effect, so that the Corrupt can Centralize Authority even further and into one person: the basis of ANY DICTATORSHIP. All of this Step-by-Step, Planned out and made by Corrupt People, the Elite, this "World Organized Crime Family". At this point, at these Final Stages, a few years into it, things will eventually AND TOTALLY FALL APART. This whole "Fake Messiah" thing will totally unravel and bite everyone in the ass: mostly the Masses, the Elite have their Underground Disney Land Bases to go to...

The Bible and Koran, under different names refer to this figure, along with many other prophets throughout the last few thousand years, as a Fake Prophet, a total ANTI-HUMAN, NON-HUMANITARIAN; the Anti-Christ, Dajjal, as I choose to call him simply because these names are so common, not due to beliefs on my end. So, this is my explanation of the Elite, hopefully in a way that people can understand, an explanation of what they are doing, and an explanation of what they want to create and where they are going with their Plans.

CHAPTER 16

Flying the Friendly Skies—the Restricting and Shutting Down of the Airways: the "No-Fly-List"—People Have Not Changed, the Government Has

In the future you can forget about flying anywhere. You already see the signs, see this is happening, again it's being done slowly to Desensitize and Acclimate People to this Warped Reality. If it's done too quickly people will object in numbers, and Publicly. That is why it is being done the way it is being done, because these sick people at the top of the Food-Chain are smarter than that. Society at large and Humanity, apparently are not. You are the frog in the pot being slowly cooked. You are Bugs Bunny in the large soup-pot asking for some soap.

So how is it being done and what is it based on?

Well first of all there is a heavier Police-Presence at many airports to Acclimate you to "Martial Law", a "Military-State". Second of all there is a "NO-FLY-LIST" allowing the American Government to prevent people coming and going from America. It was said to contain a list of a few thousand names, people who for whatever reason were on this list. However in 2006 the television show "60 Minutes" got a hold of the list and it had somewhere around 44,000 names on it.

Does anyone honestly believe there are 44,000 Terrorists, or potential Terrorists running around America? All of a sudden since 9/11 these people have sprung up overnight? This is absurd, it is much more likely that the list is being kept for DIFFERENT REASONS, and that the "Criterion of being a Suspect" has nothing to do with being a "Terrorist suspect" or a "Security Threat". The numbers point to this. This list has been "Developed and Added To" over time for use in the future, when Flying will be heavily regulated.

I think really this List is merely growing day by day and is a List of people who are Freedom-Oriented and Outspoken against an obviously Corrupt-System, which by definition would be run by Corrupt-People. By definition these "Targeted People" would also be in the way of Corrupt-People's-Plans. In all likelihood, this List is to control movement and to monitor these people the same way the Soviet Russia used to do—this is about "DISSIDENTS". Again this is looking more and more like a prequel to "Martial Law" which of course will lead to a DICTATORSHIP. Once the American Government, run by and controlled by Corrupt-Politicians, institutes Martial Law, it will never take it back. "Martial Law" will in and of itself be a "Prelude to Dictatorship", and the end of "Democracy" in America. No doubt an American People angry at Government for an Economic-Collapse, will suffice. As you may have noticed, the American Government has yet to inform it's Citizens of this coming Economic-Collapse, never mind that "Other Governments" are not speaking of this. This is called Conspiracy.

The Plan is for these Criminals to eventually step out into the light of day (This is one idea at any rate, it depends on what choices they make—other Plans involve them "Hiding Behind a Spokes-Person"), once they've achieved Total Power. In the meantime these Criminals in Government continue to hire Criminals to do the work underneath them. This is the reason I have these Corrupt Government Agents in my life, they are Criminals working for Criminals, in one sense. Each layer of crooked Government Employees has it's Power and Abuse-of-Power issues and they are a perfect example of someone Misusing and Abusing "Job-Title" for their own personal ends. We are talking about Power-Mongers, people who want and Crave Power, and who can not have enough of it, and who Enjoy Misusing and Abusing it. These Spooks in my life right now, are very much like the Elite in one sense.

These people are very different than the average person, keep in mind we are talking about CRIMINALS, Criminals with "Badges and Titles of Authority". People put in charge for their "Personal Defects", their Psychotic-Tendencies. People without empathy. People who torture without reason other than their sick need for "Fulfillment"; like any rapist this "Fulfillment" is externalized and thus has to be re-fed over and over for the "Emotional High". They are morally and ethically disgusting, repugnant. Lowlifes and quite frankly, evil. Their Mentality and Viewpoint of life is hard for the average person to understand. When I say "Evil" I don't mean a cliche, I mean this literally. I have experienced things you can not imagine at the hands of these people, these Spooks.

As a result I have been forced to stick my head into their Sewer-Mentality. I understand this Type of Person much more clearly, and way beyond the average person, but certainly can not relate to it. It took me many years to even believe that these people were "Real" in regards to their Mentality and what they were doing in my life. This Type of Mentality was too hard to grasp all at once, regardless of the nasty experiences I was going through. I will get to this later, it's extremely important to know, and gives me a very good understanding of this Type of "Mentality and Psychology", of these Elite. Because of my "Situation" what many refer to as "Mobbing", what is actually "Localized Fascism". But I digress.

Getting back to the "No-Fly-List".

Another thing/point, is all of the searching going on at airports. Again this is to Desensitize and Acclimate people, to get people used to having less rights, and in the end to accept this "NEW NORMAL": this WARPED REALITY that the Elite are creating for us.

It's CORRUPT NOT NORMAL. It's to Acclimate people to "Abuse of Authority". When it happens to you, you may feel differently. This abuse, these "Safety Measures" are only going to get worse. Give an inch take a mile, and The People at the top want to take all. Still think all this "Policing and Security" is for your benefit? You might want to ask those who have been KILLED by "SECURITY" at the airport—which has happened and will happen again. How safe does that make you feel? How secure? "WHO" is the threat to the American People at these Airports? Maybe those "Security People" need to be on the "LIST" because quite clearly THEY ARE A THREAT and disruption to The People at the airport. But you won't see that one happening. In fact I bet none of these "SECURITY PEOPLE" received more than a reprimand, after all an atmosphere of Blind-Compliance to "AUTHORITY" is what is being sought after by Government, it's the real reason that none of these murderers with "Authority and Badges" ever gets in trouble. Murder is Murder and if you go to the Airport to catch a plane and then end up DEAD from "SECURITY", then someone needs to be CHARGED WITH MURDER. But this is not happening. The Government wants and is encouraging "Law Enforcement" to see the Public as "Second Class" and beneath them. Not people to be Protected, but people to be Controlled. This is also the attitude of this Elite and people in Cults in regards to the Masses. If "Security" is killing people, Citizens, and NOT BEING HELD ACCOUNTABLE, then what is it really there for? Again this is about CONTROLLING PEOPLE, scaring people into Compliance, it is NOT ABOUT PROTECTING PEOPLE.

THE NEW MOTTO OF LAW-ENFORCEMENT IS: TO SERVE AND PROTECT GOVERNMENT.

Keep in mind that FASCISM IS WHEN GOVERNMENT BECOMES CULT. This is about "Blind-Obedience" and Compliance to Authority, no matter what.

Don't believe it? Have you noticed there is a rise in Violent-Abuse by Authority Figures? Have you noticed that the police across the Country are becoming more BRAZEN in where they are Violating People and to what extent? In other words these Corrupt Police are now committing CRIMES against the Public, Crimes against Humanity, in full-view of the Public. INCLUDING MURDER. And unless they are videotaped they are getting away with it. At most it is usually a reprimand, a slap on the wrist: of course the Police Department involved always stands up for it's criminal employee, even in cases where murder has been committed. Again for the sake of Appearances and Looks. Why not just do the right thing and put their employee in jail when he or she seriously breaks the laws? THIS would look best to the Public. If an employee in any other walk of life commits a crime, what happens? They of course go to jail. If I work at Mc-Donalds and kill a customer, I do not get to go free because I work at Mc-Donalds. Only in "Law Enforcement" is "Fast Food Justice" served to it's Employees. Given my "Life Circumstances", as I get to at the end of this Book, I would know.

So how do you feel about all of this? What does your gut tell you in regards to all of this? Well let's take one more look at numbers, shall we, the No-Fly-List made for "OUR PROTECTION".

In 2008 the American Civil Liberties Union stated the "LIST" was now over one million people. Maybe you think 44,000 people is possible as a suspected Security Threat (how is beyond me) but ONE MILLION? Come on Folks! How blind can people be? How deaf? Wake up!

The People flying have not changed (nor have The People in our country), one million threats to Air and National Security did not suddenly "pop-up" overnight, THE GOVERNMENT CHANGED!

This growing list of people on the No-Fly-List is a list of people who Criminals-in-Power feel threatened by, not The People. I don't feel threatened by The People at the airport (only by "Security" actually) and on the plane, or any where else in the country for that matter, but by The People in uniform in heavier and heavier presence, heavily armed, and CAPABLE of getting away with Murder. In general if you can not be held accountable for Murder, I tend to be worried. So what's the deal? Don't be emotionally yanked around and Manipulated by these Criminals so that you are putting up with what they want, and set up for something later. Take a look at the argument I present and think about it. Believe what you want but 44,000 people is a Total Farce, ONE MILLION people is totally ridiculous. What has changed, The People or the Government, and where is the Government headed? YOU DECIDE.

As one last added note here. Almost all of The People in Guantanamo Bay are suspects, and suspects only. Again I've discussed the reasons why and why they've never had a fair day in court; it's about PROPAGANDA and you don't see people in power giving a damn about the innocent. If you are a "Suspect" and on the No-Fly-List, perhaps the next thing you should be worried about is Prison Camp. Think about it, one million "Suspects" in Prison Camp. And as a matter of fact, as I cover later, there are Unused-Prisons across America which the Government has acknowledged exist. So think about this, after all you are a SUSPECT if you are on the No-Fly-List... Think about it and decide the kind of World you want to live in.

CHAPTER 17

The Internet Shutdown in 2012: The Net Has Brought Unity of The People; It's Got to Go: and the Weeding out of Alternative Thinking and Ideas; So Much for the Free-Flow of Ideas

Well this one is pretty easy to figure out once you understand how these people think. A DIVIDED PEOPLE IS A CONQUERED PEOPLE, the Internet has brought UNITY to the Whole World. Therefore it has to go, has to be Totally-Controlled so that Unity is not a problem for those wanting to Control The People. Apparently this is already Planned.

Fortunately I came across some kids (in their twenties I'm guessing) on "You Tube" who came across someone on the inside of this nasty Plan for shutting the Web, the Internet, down. Although there will still be a Web, it won't even begin to resemble what we have today. It will be Totally-Controlled and all alternative information and free thinking will be Eliminated. A cable T.V. system will be implemented, this is of course to shut down the free flow and free exchange of ideas. Go to "You Tube" while you still can and look up "2012: The Year The Internet Ends", you'll find the information you need there from a collective of young adults, from Belgium, and yes it is in English.

I wondered how the Internet would be shut down, here I have found it is going to be heavily controlled to the point of being worthless. As an added note here the Government Spooks I am dealing with have something on our computer so that certain information can not be found again or accessed, it's a filter of some sort, you can expect this in the future as well World-Wide. People will have to come up with innovative ways of talking about subjects so they are not blocked and cut off. Or flat out banned from the Internet. Look at China on this one, except I live in Europe...

But back to the Plan for Controlling the Net.

The idea is that there will be a basic fee for the Internet, enabling you to access certain sites etc. Like cable T.V. you will have a list of basics you can access for a general fee. However if you want to look at something else, somewhere else, you'll have to pay for this. Like an extra channel on cable T.V., say like HBO. I sure as hell don't have money for this kind of system. So a lot less people will be going to a lot less sites. As a result of less viewers, a lot of sites will lose their financial backing and eventually shut down. It's Total Control, so much will be wiped off the Internet.

The entire Internet will be reduced to "pop culture".

This will totally break down the free flow of information and eliminate all sorts of information, almost from the beginning. The "little guy" will have no chance to exist on the Internet, and people like me will only be able to view junk-for-the-mind, mainstream-pop-crap. Which I won't, if I want junk-for-the-mind, I have television. Having said that I watch almost no television, I have a DVD player and usually watch something on DVD or nothing at all.

The idea is Total Control, and to destroy the UNITY AND AWARENESS that has come about World-Wide between People of Different Lands, as a result of the Internet. The idea is to take away the one TOOL the Masses have for "Interacting and Uniting across the World". To destroy an excellent form of Communications for the Masses. This will happen in 2012. There is a reason why these Elite have chosen 2012 (notice they'll be crashing the World-Economy around 2012 also), but I'll get to the subject of 2012 later, let's stay with the Internet for now. So what do you think? Plans have been made to change the Internet in 2012 (along with other big World-Changes in 2012), the Internet is going to suck, and it will be changed for all the wrong reasons. Again you decide. In my opinion the Internet belongs to The People, the Masses, and when the time comes people should be extremely vocal. On the other hand "When the time Comes", people will have much bigger worries as there are many other BIG-PLANS for 2012 prepared by the Elite.

CHAPTER 18

Religion: A Divided People is a Conquered People

In regards to Religion people need to put Humanity first, HUMANITARIANISM FIRST, and then the rest will naturally follow. A cleaner, more caring, more Honest World will come into being. All as a result of putting Humanitarianism first. HUMANITARIANISM IS THE ROOT OF ALL RELIGIONS, was the ORIGINAL INTENT of Religion, to Unite Humanity and bring it One Level Up, one step closer to the Source, to what many call God, Allah, the Divine. Even if you don't believe in the Divine, I think you can see the wisdom of putting Humanitarianism first, indeed maybe even more so. Further more it includes all, including Atheists.

You will notice that now Religions have become secular, dogmatic, Corrupted. Religions have become Clubs, TEAMS, "Exclusive-Membership" is required. This is opposite of the original intent of ALL RELIGIONS. Notice a pattern here, "Exclusiveness", "Club-Ship", Divide and Conquer. Now (well for a long time—but Religion is being deliberately Manipulated again) people of different faiths are more POLARIZED, more Separated, and instead of reaching out to Common Ground as the Faiths were meant to do, they are more Divided and Hostile, attacking and even killing one another at times. Again it is Divide and Conquer. A Fanatic by any other name is still a Fanatic, whether they hide themselves, "Cloak Themselves" in Religion or not. The majority of people following their own particular faith would agree with this. Whether they be Christian or Muslim or any other faith; I am just picking the two being most "Polarized for Manipulation" in these times.

This "Division of Religions" is in fact to push people away from God, and into Hatred. A "Divided People is a Conquered People" and in this case people who are supposed to be focused and rooted in "God" are being Deliberately "Steered Away from God", and very much of this being done by a Corrupt-Media that is being Used and Misused to Incite and Inflame "Religious Division". God, the Divine, Allah is supposed to be THE FOCUS of Religion, NOT HATRED, DIVISION, AND DIFFERENCES. This is being "Perpetrated and Orchestrated" by people who could care less about "God, the Divine, Allah", but who see "Goodness" and "Good People" as BEING IN THE WAY. Thus they are trying to get "Good People" who are "Rooted in God" to fight amongst themselves rather than be against these "Corrupt-People" within Governments Orchestrating it all: The Elite of course.

Thus it is all Divide and Conquer, and the "People Falling for This Ploy" are not at all actually that "Rooted in God", or they would not be so easily "Swayed and Suckered". They would not be "Fighting One Another", or "Blaming One Another", as the "Real Culprits" of Terrorism are NOT actually Religious. Thus "People of Religion" are being Used and Divided: again the Root of All Religions is Humanitarianism.

Who amongst you would kill or attack another in front of the person of which your Religion was founded on, sheerly because someone else does not follow your Faith, your Religion? Folks this is backwards, and absurd. As an example, would you kill in front of Jesus, Krishna, Buddha, or Mohammed? These people were sent by the Source, God, the Divine, TO UNITE PEOPLE AS ONE and bring them back to God. To bring back Humanitarianism to a race that has sunken and fallen, that has lost True Vision for itself. They all came from the same Source, from the Divine, yet each Religion acts as if it is "THE ONE". Which is a Lie. This is Short-Sighted, Ignorant, Lazy, and totally on the "Surface". Totally backwards as well, and Totally by Design. This is where "People have Hijacked Religion" and taken it away from "God, Allah, the Divine", and taken it away from it's Original-Intent. The "Real Intent" of RELIGION IS WITHIN US not on "The Surface". You Do Not have to go to a Church or Temple to be with God.

If Buddha, Mohammed, Jesus, and Krishna were in a room, do you really believe they would fight and attack one another to see who is the greatest, the "Most Correct"? Do you think they would fight over "Who is Closest to God"? This is absurd, anyone doing so IS FURTHEST REMOVED FROM GOD—you would expect this of "Kooky Cults", not people of SOUND MIND. It is ridiculous to even think it. God, the Source, the Divine, sent these great people of our race—the Human Race, through different time periods and places to Unite Mankind, NOT divide it. Certain members of Mankind, Power-Mongers, the Elite, have used Religion to divide Mankind, and a lot of ignorant people have gone along with these Reversed-Intentions: this is REVERSE-ENGINEERING OF RELIGION. Say what you like but the intent of the Divine, the Source, is clear. This is why all Religions are roads which end at the same Destination. Anyone preaching or believing otherwise has taken a detour, fallen into a trap, and is serving someone else's goals—not the Divine's, God's, Allah's, the Source's. All Religions are "Rivers leading to God".

These "Various Prophets and Enlightened Beings" have been sent across various regions, been born into certain areas, throughout time, to UNITE PEOPLE ACROSS THE WORLD UNDER ONE FLAG AND BANNER: that called "God, the Divine, Allah". GOD by any other name is still GOD. Has everyone lost their minds? OBVIOUSLY RELIGION IS BEING MISUSED.

The Elite have existed for thousands of years keeping their Backwards-Mentality and Misused-Power alive: Suppressing, Repressing, and INFILTRATING anything and anyone that poses a threat to them. Thus Religion has been Deliberately Misused, Abused, Falsified, and Mis-Taught, FROM THE INSIDE OUT. This is how this Elite and Power-Mongers operate, they are CORRUPTERS AND INFILTRATORS of what is "Good and Pure". Anything that is "Good and Pure" being an "Obstacle" and "In Their Way". Evil attacks "Goodness" and if "Goodness" does not fight back then "Evil Triumphs". Thus "Evil" for a very long time now, and in Present-Current-Time, is trying to "Divide and Conquer Goodness" through Religion. In many ways Religion is being "Divided" through "East and West", "Christianity and Islam"—many people who CLAIM to be followers of these Faiths, are being Divided, Manipulated, and Allowing themselves to be. These Religions are being Deliberately-Polarized.

Are you really practitioners of your Faiths? If you are not "Practicing that which is Good", Humanitarianism, which is the root of all Religions, then you are NOT. Religions should be practicing "COMMON GROUND" not "Divide and Conquer", which was and is, the Original-Intent of Religion—to "Unify People" and take them "Collectively Closer" to the Divine. Religion for some time now has lost its original meaning and purpose for the Masses. IN ORDER FOR EVIL TO FLOURISH, GOOD MUST PERISH, this is why. Thus Religions have been Corrupted and "Watered-Down" to serve other people's Interests and Goals. People of Faiths should not be "Oriented by an External Form", by a "Type of Religion", but rather oriented by "Goodness, Actions, and Compassion". They should be looking from "Person to Person", not "Faith to Faith". In the end people should be "United in Goodness" and anyone stepping outside of this, "Killing in the name of God, Jesus, Allah, whoever", should be recognized as NOT BEING RELIGIOUS, they should be recognized as MERELY CLAIMING TO BE. In this way Religions can not be "Divided and Conquered" against each other, and this also happens to be the Truth. Anyone "Killing in the name of God" is not Religious, they are merely FANATICS, claiming to be Religious. "Fanatics" are a class unto their own, they are "Psychotic and Stupid" not representing any Faith or Religion. That anyone even turns such a thing into "Religious Issues" is beyond me.

Just to make things clear I don't think Religion or people following Religion are bad. I think the majority of Religious People are good and try to do good, but these people are the Keepers and Shepherds of their own Faith and Religion. I think that those who are good need to start speaking up, and in keeping with the tenants of their Religion, make sure their Religion is PROPERLY REPRESENTED. They also need to make sure, and work at keeping "Common Ground" with other Religions. They all need to make sure that their Faith and Religion is not Used and Misused for "Divide and Conquer" of the Masses, Goodness, and Religions.

Those who are good need to make sure their Religion is not Misrepresented for the sake of DIVIDE AND CONQUER. Which is what the Elite of our World are trying to do today. Those who are Religious and Good need to make sure that Tolerance, Acceptance, Peace, Humanitarianism, and Unity are the dominant characteristics of their Faith, and make sure that those doing evil in the "Name of their Religion", are put in proper context. No one committing atrocities out of "Religious Ideology" is a True-Believer...People who do so are against Mankind, against God, Allah, the Source, the Divine. People who "Kill in the name of the Divine" are merely wearing a "Label of Religion", they are not actually Religious. This applies to all Religions and Followers, whether they be Christians, Muslims, what have you. Religious People practice Humanitarianism which is the ROOT OF ALL RELIGION, people who "Kill in the name of the Divine", who "Kill in the name of a particular Religion", are not Religious, they are merely FANATICS hiding under Religion. And no matter what, "A FANATIC by any other name, is still a FANATIC".

Polarizing The People: East and West: The "Misuse of Religion": Who is the "Fanatic"?

I wrote this after hearing a lot of Ignorant People speak out against Muslims around the anniversary date of 9/11. I could not believe the amount of Ignorance that was being Spoken, you would think that people would have more Brains—or Not. People like this are "Feeling what to Think", not actually "Thinking First", and a lot are simply "Going with the Flow" because at least on this particular subject, being Ignorant is "Fashionable and Acceptable" with many. It is Stupid.

The "Terrorists", Middle Easterners and supposed followers of Islam, were "Picked" to foment BIGOTRY which makes people very easy to Control, and a host of other reasons. There are very many reasons which I will not get into, and many details involved in such a SHAM (It's similar to "Wag the Dog", the movie) but suffice to say, anyone really of the Religion of Islam, WOULD NOT EVEN BEGIN TO CONSIDER SUCH AN ATTACK. It is a Deliberate Ploy to Divide the World, AND ESPECIALLY AMERICANS, in terms of "West and East", and Religion, like "Christians and Muslims". It is a Deliberate playing on Stereotypes. IT IS TO POLARIZE PEOPLE, to Create Extremes Amongst the Masses, FOR GREATER CONTROL, and for Manipulations of The People down the road. You will note that America has much interest in the Middle East, and the population of America is much more "Supportive of the Government" in regards to the East after 9/11, these days.

Some points should be "Cleared Up" here. If you are TRULY A MUSLIM OR CHRISTIAN, you do not do such a thing. Christians, fanatical ones, also commit murder, and other deviant things. Doctors who perform abortions are murdered by Christians, people "Wearing Religion". What's the logic here? Save the babies, but kill them when they are older? DO YOU THINK Jesus would go running around with a gun shooting people he did not agree with? It is stupid. What about Mohammed? Needless to say such people are not actually Christian, merely "Wearing It" as an excuse for their own WARPED MENTALITY.

These people are FANATICS, using Religion as an EXCUSE, and no matter how you slice it, A FANATIC UNDER ANY OTHER NAME IS STILL A FANATIC. RELIGION IS NOT THE PROBLEM. So lets see things for what they are here and Not Be Divided. A CLEAR LABELING, DEFINITION IS REQUIRED HERE: People committing Terrorism are not "Religious", THEY ARE EXTREMISTS, FANATICS. People should not be divided over Religious-Grounds, RATHER UNITED over a common enemy, who is playing on "Religion", AND THEREFORE ATTACKING RELIGIONS.

RELIGIONS ARE NOT THE PROBLEM, FANATICS ARE. EXTREMISTS ARE. I am no more "Afraid" of a Muslim than I am a Christian. Which is to say that I am NOT afraid of either, I see things for how they really are. Fanatics are the cause of these problems, NOT Religions. This has to do with INDIVIDUALS, Warped-Sick People, NOT RELIGIONS.

You have to look past the STEREOTYPE, the "Job-Title of Religion", have to look deeper, to see the Truth of the matter. DON'T BE LAZY-MINDED. Fanatics are responsible, NOT Religion. People need to stop buying into Stereotypes which are "Being Sold" through the Media, a Media run and controlled by the Elite. These Elite are "Playing on Stereotypes" through the Media, and trying to get people "To Feel what to Think", which is a Diversion from actually Thinking in the first place. These Politicians and their "Other Friends" are Power-Mongers, interested in Manipulating Religions for their "Own Ends". Many of the Corrupt-Politicians are only too happy to "Play on this Division of People and Religions". Look at what was announced on the news on the tenth of September, and the "Timing of 9/11", that in and of itself should say something.

People have to look past the STEREOTYPE, the "JOB-TITLE of POLITICIAN", people have to look deeper, to see the Truth of the matter. The Public has to look at the "Actual People" and Their Actions, not a Job-Title. Look at all of the missing money, at least a Trillion Dollars a year is "Missing", look at how the Tobacco Industry started, now the "Food-Chemical-Industry", Chem-Trails, Secret-Technologies used against an Unsuspecting Public, the "Misuse of the Media" to Promote-Dumb-Ideas and to "Program the Public": these are THIEVES, and PSYCHOPATHS, NOT POLITICIANS. They are their own FANATICS, wearing JOB-TITLE, for their own Selfish Ends. Corruption is Corruption, and just because you are a decent person it does NOT mean that These People are. Just because you do not get their "Mentality" it does not mean they do not exist. It is because of THIS kind of thinking, that these people are able to get away with so much.

People allowing themselves to be MANIPULATED over "This Issue", of Religion, are giving in to an "Emotional Viewpoint" that has nothing to do with REALITY. They are giving in to a Stereotype being "SOLD" on the Television and News, a Stereotype being a "Shortcut for Thinking", for Lazy-Minded-People. This is what a Stereotype is, from an intelligent Viewpoint, and a Truthful Viewpoint. People out there are trying to MANIPULATE how the Masses think, through their EMOTIONS. Are you against Muslims as a result of 9/11? If you think this way YOU are giving in to Bigotry, Hate, and "Personal Issues" on Your End. STEREOTYPES. That you can not see other Bigger Issues here, is Lazy-Minded, and means you do not want to look deeper and find out the Truth of the matter. What is stopping a lot of people here, from looking deeper, is their EMOTIONS. For many this is simply a "Wanting to Look Down on Others", a giving in to an Emotional-Viewpoint, and then "Selecting Information" from this Viewpoint, without practicing Deduction, Logic, and Reasoning. BECAUSE PEOPLE LIKE "THE FEELING"—and it is being Encouraged by the Media.

Whatever the deal is, if you are "Hating Someone" based on Their Religion, then you most certainly are not following Your Own.

You are "Going with the Flow" because there are "Others out There" with a "Similar Viewpoint", and this "Viewpoint", THIS SCHISM AND MOVEMENT, is being encouraged by the Media and people within Corrupt-Government. It has TURNED INTO A MOVEMENT which this Elite are Manipulating for their own Self-Gain. It "Feels Good" to look down on others with disdain, a "Selected Group"—doesn't it? In short you are NOT THINKING rather you are FEELING FIRST, and to people like me, YOU look like a Fanatic. I could care less about your Religion, YOU look like someone Without a Brain, without a Clue, and are willing to be worked into such a "Viewpoint", and an EMOTIONAL ONE at that. YOU ARE BEING WORKED UP INTO ATTACKING OTHERS. You are Emotional first, "Thinking" second, and you "Think" from this Emotional Viewpoint. YOU SHOULD BE MORE HONEST WITH YOURSELF, MORE QUESTIONING, AND MORE LOGICAL. Fanatics are responsible and are the problem, NOT RELIGION. The question is are you a "Fanatic"? I hope this clears things up for some people...Anytime you want, we can "Water-Board You" and find out the Truth of the matter. Blindness and Apathy is for Idiots: Lazy-Minded-People, people incapable of the Responsibility for the upkeep of their own Country. Buffoonicans.

This is the World you live in, not the one you PRETEND to live in.

CHAPTER 19

The World-Economy Part Two

Well early in 2010, listening to News on the radio, I heard an "Investor from the East" being interviewed. He had clear "Insider Information" and this was not "Conveyed Obviously" but was "Conveyed Deliberately", and in regards to the Future-World-Economy-Crash. He said that America and China would Remain Strong, that the "Dollar would not Crash", because of America's Connection with China. A "Deal is obviously Worked Out". America is one large "Entity/Country", and therefore a deal worked out with one Large Country like America, makes sense, as opposed to many "Different Countries of Europe" in which there are many "Variables Involved". The idea here is "The Safest Most Stable Choice". It easier to "Work out a Deal" with "One Large Country" rather than many small ones.

He said that the Euro, the European Currency, and EUROPE, WOULD CRASH INSTEAD. And this makes sense, as they do not actually have enough Gold, as I understand it, to "Back up their Currency". This guy obviously had inside information, and I have watched knowingly.

Now Europe has had nothing But Problems, since starting this Euro. The Economies of various European Countries have been on the verge of crashing repeatedly. No doubt most would never have had this problem if they had not "Gone in with the Euro". Yet it remains in place, Stupid or Deliberate?

Now I have also written a previous lengthy chapter about all of this as well: in regards to the "Dollar Crashing". Is it all irrelevant then? No, there are many ideas and factors within this chapter which are still very valid, the information does not begin to be irrelevant. THERE WILL STILL BE A "WORLD-ECONOMY-CRASH", and no doubt the "Dollar and America" will "Plummet". Hard times are still ahead, and America may still be the one that causes this all.

As another point by "Speaking of the Future" we have the "Power to Change It". The "Reverse of This" used by the Corrupt, is that by "Hiding Information about the Future" we can "Change It", thus as you can see there is indeed Truth to my statement. I believe that enough "People Speaking Out" have "Changed a few Things" in regards to this Economy-Crash. Having said that this remains to be seen.

There are several things going on here, and I write about what is most probable to "Help Make Sure" these "Things do not come to Pass". But again there will be a World-Economy-Crash, it WILL BE a mess, and there will be many problems and Manipulations as a result. This World-Economy-Crash will still be the Beginning of All, in terms of Elite-Plans. So my FIRST CHAPTER REMAINS.

Having said that if you are paying attention, the News is all about, these days (Jan. 2011), the Dollar, and China's Economy and Currency—this is all a sign. Obviously "Deals have been Made" behind the scenes, in which America and China will survive this World-Economy-Crash, by "Supporting each Other". This is not hard to imagine with "China Expanding", and America being a land of "Many Consumers". What will "Come to Pass"? I still believe that the my previous chapter on this subject can stand on it's own....

CHAPTER 20

Hidden Technology for Hidden Government Programs: Electromagnetic-Technology/ Exotic-Technology: My "Introduction" to Exotic-Technology: A Laymans Explanation of it: Its Exotic-Misuses for Governments and the Elite

> "Imagine. To peer across the world and know the enemy's secrets. To place our thoughts into the minds of your leaders. Make your teachers teach the true version of history, your soldiers attack on our command. We'll be everywhere at once, more powerful than a whisper, invading your dreams, thinking your thoughts for you while you sleep. We will change you, Dr. Jones, all of you, from the inside. We will turn you into us. And the best part? You won't even know it's happening."
>
> —IRINA SPALKO, CATE BLANCHETT'S CHARACTER FROM "INDIANA JONES AND THE KINGDOM OF THE CRYSTAL SKULL"

Articles and Stories in Mainstream Media Which "Point To" This Technology: Smaller Pieces of a "Larger Technology"—Some Brief Examples From Mainstream News Which Point to This

I might point out here that articles in the mainstream press have come out (2009) involving scientists now being able to "View Peoples Dreams", Dream-States, and being able to "Send Images" into peoples Dream-States. This is in the Public-Sector, and although The People are hooked up to a machine via electrodes, wires and such, quite obviously the technology is being developed and is there. Also in the Public-Sector (September 2010) technology has also been developed to translate unspoken thoughts, brain waves, into words which can be read on a monitor. This Technology also involves "Recording Brainwaves". Thus with this Technology people can "Read the Thoughts" of someone who is paralyzed or unable to communicate for whatever reasons, and they can "Record the Thoughts" of someone. All of this is in the Public-Sector, and both are "Breakthroughs in Technology" which are each a "Piece of the Puzzle" in regards to the type of Exotic-Technology I deal with.

Another example from the Public-Sector is that in 2008, I heard Paul Harvey mention a Technology that helps people "Sleep Well". I haven't my notes here, but he said it involved the use of Electromagnetic-Energy, and could give the experiencer a perfect nights sleep, and within a relatively short sleep time. "Compressed Sleep" as I call it. Paul Harvey mentioned that perhaps this "Exotic-Techno" would be used for "Sleep-Clinics". So this "Sleep Machine" points to Exotic-Techno which uses Electromagnetic-Energy. What Paul Harvey mentioned was specialized, and related to a particular "Theme of Sleep". In other words what Paul mentions "Focuses merely on Sleep", what these Agents in my life have, can do hundreds of things, and it is also Electromagnetic Related, in terms of the Energy it Uses and Manipulates. What Paul mentions here, like the other examples I give, are merely "Smaller Pieces", with Less-Capabilities, then what Government has.

Another thing is that the American Military has fighter jets, which are "Thought Responsive", reacting to a pilots thoughts—again one more aspect I deal with in regards to this Exotic-Technology. And which shows that the Military is ahead of the Public-Sector, obviously. Having said that the Public-Sector is light-years away from the Government. These Technologies I mention above are not as advanced as what I deal with, for now with these you have to be "directly hooked up" as I understand it, to these Technologies, but they do show what is possible. Having said that the Exotic-Technology I deal with does much more, these Public-Sector-Technologies are "Smaller Individual Pieces" of a "Bigger Puzzle", something a lot "Bigger".

Keep in mind that Government is far ahead of the "Public-Sector" in terms of Technology. It is said by some that for every year of the Public-Sector (the Private-Sector is also ahead and independently hired by Government—for the sake of cheating laws), the Military-Sector advances fifty years. As this "Gap in Knowledge, Science, and Technology" advances and progresses between the Governments and Public, such "Inventions and Exotic-Technologies" when revealed by people like me, seem "Far-Fetched and Implausible", however they are not. This is the result of a "KNOWLEDGE GAP", a deliberate one, A TACTIC, so that Governments can get away with MISUSING such Technologies and Manipulating Individuals and the Masses, in the first place.

If people do not know of such Exotic-Technologies, like I did not, then people can not Identify the Problem, protect themselves against it, or do anything about it. You can not fight what you do not even know is there. An "Invisible Enemy" is the most dangerous and YOU CAN NOT FIGHT WHAT YOU DO NOT KNOW EXISTS.

Look at how this Exotic-Technology is being used in my life in the following pages, and ask yourself if a "Violation of my Constitutional Rights" even begins to be The Issue here—we are talking about TORTURE, and Other Issues as well which exceed anything expected in a "Democracy". Guantanamo Bay is a moot point with this Exotic-Technology as is "Water Boarding" and a host of other "Torture Techniques"—as this Exotic-Technology enables one to "Listen to Thoughts" in addition to being able to "Send Thoughts". As long as these Technologies are HIDDEN and unknown of, they can be Misused. It is blatantly obvious isn't it? We are talking about Spy-Stuff, Espionage-Technologies, being Used and Misused on Citizens, The People. Clearly Governments see "Their own People" as a threat, as they are being "Targeted by these Technologies"—other Governments already know of this Technology. How were you expecting them to get away with it, by being out in the open? Governments know of these Technologies, The People, the Masses, DO NOT, and by Deliberate Design. THIS KNOWLEDGE GAP IS FOR MISUSING THESE EXOTIC-TECHNOLOGIES, and to some extent for another Plan of the Elite, one I refer to as "New Atlantis". In all cases these Exotic-Technologies are for Misuse.

Misuse and Violation is why these Exotic-Technologies remain Hidden from the Masses.

As a note to my readers, and further point in this regard, a lot of the Terminology which I use comes from a Former N.S.A Whistle-Blower, John St Clair Akwei, a man with a Brain, Intelligence, and Courage, who Filed a Lawsuit against the National Security Agency and ended up being "Harassed and Hindered" with this Technology: just like Me I might point out.

The Science of this Electromagnetic-Technology

Currently, as many people in this World view this Technology it is all "Hocus-Pocus" it is all "Unbelievable"—having said that it is ONLY TECHNOLOGY. For those of you "Doubting This" read on, this is where "Those Doubting" this Exotic-Technology, look less brilliant on their end, as I assure you, you are. This is a matter of "The Gods Must Be Crazy" in regards to people like you. I have seen too many people "Ridiculing the existence of this Technology" and personally experienced too much Torture and Danger, to allow this Level of Ignorance to go on. I do what I can to make this presentable and understandable; in my opinion if you want to be ignorant after reading this, then knock yourself out. It might make you feel more comfortable to believe your own Viewpoint in regards to this subject, but you certainly won't change Reality with it.

The point of all of this is to Educate People and to let them know of this Dangerous-Technology, and what it is being used for. Once you realize it's potential to do good, you realize it's potential to do evil—and when you see what it is being Hidden for, you realize that everyone has something to worry about. My life is one very good example of this, as well as The Clown is a good example of people who should be the Last On Earth to wield such Technology. When you look at how Governments are and behave today it should be obvious anyhow that this Exotic-Technology is Dangerous, and will only be Misused. The fact that this Technology is Hidden, speaks of this. Governments see this only in terms of Power, not in terms of the "Good it could do for the Masses, the Good it could do for All". Governments are quite aware of the "Good it could do for the World", and have only chosen the "Good it can do for Themselves"—if you understand the World you live in, then this should not be difficult to understand. People are in Government for Power, and will do what they can to Maintain Their Power, this Exotic-Technology being perfect for doing so. Until we address problems like this, this will continue to be the World we live in, as opposed to the World we WANT to live in. So lets begin.

To begin with we are dealing with Electronic Equipment, Computers, with other elements involved which are able to Manipulate Electromagnetic

Energy, Electromagnetic Frequency—we are talking about Advanced Technology, but the Science of it is there for all to see. Each thing that has an "Electrical Current", Humans included, can be affected. Anything alive with electrical currents gives off an "Electric Field", a "Bio-Electric Field". "Things" such as Humans therefore have an Electromagnetic Field around themselves, and unless I am mistaken this is what many refer to as an "Aura". Now we as Humans "Think" and "Thought" is Electromagnetic Energy, the "Quality of One's Thought" dictating the "Quality of One's Energy", this is very basic and easy to understand. This is where the idea of referring to someone as having "Bad Vibes" comes from, or at least you get what I mean. So we as Humans have an Electromagnetic-Field around us, an Energy-Field, a "Bio-Electric Field", and we have "Thoughts" which emanate from this "Field". In other words we have an Electromagnetic-Field around us, and we emanate, radiate, Electromagnetic "Waves", "Vibes", our Thoughts. Thought is Electromagnetic-Energy, anyone disagreeing with this, has absolutely no idea of what they are speaking of.

Now we are talking about Energy, and as such, that which is SCIENTIFICALLY OBSERVABLE. To top it off we are not even talking about "Unknown Energies" or "Unknown Theories and Principles", this is all very simple and basic to understand. Now seeing as how we are talking about Science and "Known Energy", we are also talking about that which can be measured Scientifically. This Energy can be studied, observed, measured, and with time "Controlled" once enough research has been done and the Technology created to do so. Which has already taken place.

Technology has been created which can control Electromagnetic-Energy, which can control Electromagnetic-Frequency. Having said that Humans have an Electromagnetic-Field, Humans give off Electromagnetic "Frequencies and Waves", which is what we call "Thoughts", Thinking, and therefore Humans can be Affected, Manipulated, and Controlled with this Technology as well. If someone can Manipulate another person's Electromagnetic-Field, can Stimulate a person's Brain with this Energy, then someone can Manipulate another person's "Physical and Mental Well-Being". It is all simple logic. By Manipulating one's Electromagnetic-Field we can Control Them, Manipulate Them, both Mentally and Physically. So this Exotic-Technology is not a stretch of the imagination. Not at all. It is merely a matter of Science and Technology.

This Electromagnetic-Technology enables people with it, to "Hear the Thought of Others" and to "Remote Control" others.

And if you think about how Governments are into Power and Control, and Weapons for Warfare, you have to be stupid to think that they have not developed this Technology. Of course they have, do you think the Americans, American Government, would refuse to research and develop this Technology based on "Moral and Ethical grounds" yet let the Russians develop it instead, forget it. So this Technology exists and is becoming more accepted and more used by Government Agencies. Within Government at least it is more commonly known of. I base this on my own experiences of being in two different Countries where this Technology has been used on me, and by Two Different Agencies.

Now with time, the amount of research and development is going to become more and more detailed and refined. Which it has. Lets look at this with Logic. So we have a Machine, Technology, which can Manipulate Electromagnetic Energy, and Humans are "Composed of This". So we have the basics for Manipulating and Controlling People. Now with time we do more and more research and acquire more knowledge. Everything becomes more refined and developed. For instance we begin to figure out "Specific Frequencies", "Specific-Finger-Prints", Specific-Energy-Prints, for various parts of the body. Keep in mind that all organic things give off an Electromagnetic Energy, and so when looking at a Human-Body we can begin to break it down into Specific-Organs and body parts. For instance a heart has a "Specific-Finger-Print", a Specific-Frequency, as well does a BRAIN. Now since we are talking about this Exotic-Technology we will deal with that involving the Brain, obviously the Government has "Mapped Out all Frequencies and Signatures" in regards to the Human-Body, but the Brain is what is relevant here.

So they then begin "Mapping Out" different "Brain Sections", parts of the brain, and functions. They also begin experimenting. If we do this, to this part of the brain, raise the Electromagnetic-Frequency, then this happens. We know that this part of the brain is responsible for this type of behavior so if we decrease Electromagnetic-Energy here, lets see what this does, and so on and so forth. Along the way they make a lot of New Discoveries, finding out that this part of the brain is responsible for this here, and if we do this then we end up with this "Mental Disorder here", etc. Basically they observe what Brain-Parts, what portions of the brain, affects certain Mental-Conditions, and they "Map This", all of this information, and all of the different possibilities, and put it into ONE COMPUTER-DATA-BASE. Thus they learn how to "Create Mental Disorders" in the process of experimenting, and learn what the "Signatures", the "Finger-Prints" are for these. Eventually they "Map out all Potentials" and all "Interactions" of Electromagnetic-Energy on various portions of the brain, and they put it all into one Machine, one computer capable of making all of this. They also do this in regards to Electromagnetic-Energy on other parts of the body, but for now this is irrelevant.

Now as they have developed this much, a machine and computer capable of interacting and Manipulating and Controlling Humans, they decide they want more. Whats the next step? A machine with more "Fine Details" of course, more capable of Manipulating People and Controlling Them, one which can be used for Manipulating People on a very fine and PERSONAL LEVEL. So the next step is that they need something capable of recording "Thoughts, Emotions, Brainwaves, Energy", and doing so in regards to people specifically. They have been recording all of their observations obviously, but they need something for doing this with specific people, people who are interacted with, and Manipulated by this Technology. So they do, which again makes this even more Dangerous, as the intents of all of this is certainly not for good.

And I might point out here that in, I believe November of 2010, it was announced that they now have Technology in the Public-Sector which can do this. Which can "Record Thoughts" and translate them into "Readable Text". In other words in the Public-Sector they now have Technology which can "Translate Unspoken Thoughts" into readable text. This means "Mapping out the Human Vocabulary" and giving a specific "Signature", Finger-Print, numerical-value, to each word: each word having a "Specific-Vibe and Print". So no one but an idiot in a case of Hard-Core-Denial can tell me that such Technology is not possible. Hell, I deal with Idiots at the end of this Machine, this Exotic-Technology, and they even believe that it exists….But back to business.

So in the Government's desire for more Power and Control they now "Upgrade their Machine", this Exotic-Technology. They develop this Exotic-Technology so it is also capable of Recording Thoughts, Feelings, Emotions, even Physical Ailments. All that involves the Brain and Body, they are able to record: parts of your body also having a "Specific-Signature". And what they can "Record" they can also "Send" I might point out. So at the end here we have a Machine, Exotic-Technology, capable of many a thing, and all based on Electromagnetic-Energy, Frequency. And in fact they have developed this Exotic-Technology even further. But lets look at the Bigger-Picture now, and interaction with People, now that we have "Engineered this Exotic-Techno".

So what does this all mean?

Well for starters this means that each Individual Person has their own "Energy-Print", "Finger-Print". Seeing as how the space between the Earth's upper atmosphere and the Earth herself is all one Electromagnetic-Field this means that with the right Technology and with your "Energy-Print", the Government can "Locate You" when it likes, as it likes, for whatever reasons it wants. The Earth's Electromagnetic-Field acts as the "Medium" for which they can locate you. They simply "Tune-In" to your Specific-Frequency, Energy-Print, within this Larger-Electromagnetic-Field. They "Dial your Number" and the "Connection is Made".

The next thing you have to realize is that YOUR THOUGHTS are Electromagnetic Energy. If you have a Machine, Exotic-Technology, which can Detect Electromagnetic-Energy, then you have a Machine which can "Detect Thoughts". As I have already pointed out, the Public-Sector now has Technology which can translate unspoken thoughts into "Words", so does the Government. So the Government with this Exotic-Technology can "Tune-Into-You", your Energy-Print, like a Homing-Device, and then "Listen in on your Thoughts". They can "Read your Thoughts". Each "Thought" and "Each Word" having it's own particular "Signature" it's own distinguishing "Mark/Characteristic" separating it from others—just like Finger-Prints. Now all of this information is "Stored" in a Government-Data-Base, and at this point has been incorporated into the Technology itself. Thus each of these portable Exotic-Technologies comes with "Training and Manuals" as to the "Different Frequencies and Signatures" for affecting various parts of the Human-Brain and Human-Body. All of this simply based on Electromagnetic Energy and Frequencies. "Signature-Prints", "Identity-Markers", for different body-parts, and functions, and the results of Electromagnetic-Energy on these.

The next step of course is that if they can "Read your Thoughts", then they can "Send Thoughts" as well. No doubt originally this all started as the "Ultimate Spy Technology", which it no longer is, judging from my life. If you have this Techno and are a Spy, this means that you can "Send Messages" and "Receive Messages" without being detected. Now of course this means that the Government can "Listen and Spy on your Thoughts" and also Manipulate your thoughts by "Sending Thoughts". If you do not know yourself so well, are not Conscious, then you can be easily Manipulated here as any thought in your head may be "Construed of as your Own". Your brain, your mind, like an apartment or house, can be "Wire-Tapped".

Except it gets worse. This Exotic-Techno has been used to "Map out all brain Functions" and so your Brain can be literally "Stimulated", or the OPPOSITE, for whatever purpose they want. Specific parts of your Brain can be "Stimulated" for a particular "Desired Response"—whether it be to make you hungry, tired, angry, happy, horny, sad, you name it—and all of this can be done on various "SETTINGS". They have an "INTENSITY LEVEL" which they can and do use; in other words HOW Hungry you get, or HOW Tired you get, or HOW Horny you get, or HOW Angry you get, is all determined with an INTENSITY LEVEL, the amount of "Stimulation or De-Stimulation" a part of your Brain receives. It obviously depends on "How much Energy" they use to "Stimulate or De-Stimulate" your Brain. So in addition to "Receiving Thoughts" you may also have "Feelings which are not your Own"—and the "Intensity of these Feelings" can be affected. If the two are put together, a "Thought with a Feeling", then you have a very Convincing and Powerful-Means for Manipulating someone. Which is how I originally caught this Exotic-Techno being used in my life. And as has been done with me and others around me, since then. How this Technology has been used in my life is ILLEGAL, SICK, AND DISGUSTING. To top it off The People using it on me belong in Prison, yet are allowed to use it because of a "Government Badge". There is something wrong with this picture, never mind that The Clown's Agency has given this Techno to a Psychopath—someone who would use this on her own Superiors just to "Get Ahead" as an added point. Brilliant.

So Thoughts and Feelings can be Manipulated with this Exotic-Techno, and Total-Criminals are allowed to Use it, Abuse it, Misuse it. The bottom line is that Government can listen to your thoughts, Wire-Tap your brain, and Remote-Control you, as well as any Corrupt-Government-Agents who run around on their own doing as they damn well please.

Now more points in terms of "Use". In regards to "Thoughts and Sending Thoughts" they can go right to that part of your Brain responsible for sound, the Auditory-Cortex of your Brain. By doing this they "Skip your Ears" and therefore you "Hear Without Sound". You can hear "Speech, Voice(s), any kind of Sound", without any External-Stimuli. For an enemy, doing this, would and could "Incapacitate Them". For someone who is "Unknowing of what is going On", they would think they are "Schizophrenic", and no doubt really go crazy as "Thoughts with Emotions" get "Pushed Into their Energy-Field" as their Brain is Stimulated from this Techno, this "External Source". Now as there is an Audio Component so there is a Visual Component. They can also Stimulate the "Visual Cortex" of your Brain, bypassing your eyes and without any External-Stimuli, and therefore "Send Images" as well. Thus with your eyes closed you can "See Images" not your own. In this case as with all the other things, the Reverse of this is True as well. They can see "Visual Images" which you "Pull Up", they can see "Visual Memories", and quite obviously they can Record these just as they can record Emotional States, and See these, on a Monitor, a Computer Screen. Having said that they can also have someone sit down, "Plug Into" this Technology, and

"Experience your Emotional States". Again based on observation, I have noted a lot of this along the way.

One point of doing any of this, would be to render an Individual, to make them incapable of, Independent Thought. When such things are going in "Your Head", "Thinking Clearly" and on your own, becomes impossible.

Now in regards to Emotional States, Thoughts, Sounds, Memories, Moods, Mental-Images, and probably a few others I have forgotten, they can RECORD ALL OF THIS, and EVEN PLAY IT BACK. Thus you may think a "Memory" has "Popped into your Head" and in Reality it can be Spooks with Technology simply trying to find out more about that "Particular Memory". They can even play back a "Mix and Match" of "Emotions and Feelings" to go with a particular Memory, Thought, what have you. By the same token they can BLOCK MEMORIES, and even ERASE THEM as you "Recall One". Having said that they can Block-Emotions as well, which for obvious reasons is unhealthy—by doing so you experience the world, like a robot. You know what your feelings "Should Be", you just can not "Feel Them": I have had this "experience before", was threatened to have it "Left on me Permanently" and it is "Unpleasant" to say the least. Again there are many possibilities.

Another thing they can do is affect your Speech. As in what you say and how you say it. This involves "Several Variables", as they will have a "Script" prepared, of what you will say, and even how you say it, if they choose. This involves somehow "Turning Down Your Will-Power" and "What you want to Say", and "Turning-Up their Will-Power" and "What they want to Say". This involves "Will-Power", "Turning it Up and Down", like two separate knobs on a stereo, one turned up and one turned down, a "Prepared Script" of what they "Want you to Say", and a "Giving you a Feeling" that it is "Yours"—I believe this is done, by "Attaching your own Personal Finger-Print and Feeling" to the "Whole-Dialogue". In other words this "Prepared Dialogue" is "Flavored with your Energy" and a "Feeling of You". Obviously this is Deceptive.

I have seen this done daily, around my house, and even seen The Clown do this to a Government Agent of this Country, a Field-Agent, because The Clown wanted to show me what she could get away with, and was "Disgruntled" with this particular Agent for having "Failed Whorbat Repeatedly". This woman was laughing at her own behavior and dialogue, because it was obviously not herself—of course she took it as being herself, not Techno, but I think you get what I mean. Around my flat, this occurs quite a bit, so it was easy to recognize, never mind that I know this Field-Agent pretty well, and know what she is like.

They can affect ALL of your Motor-Control Functions, that part of your Brain responsible for Motor-Control—your interaction with the environment around you. Thus all Physical-Action can be controlled, including "Speech". As a result your "Coordination", your physical functioning can be affected. As an example they almost made me "Steer Into Someone", as in I almost "Ran into Them" while walking by them in a store, because these Spooks told this Idiot I do not like her, and this was to "Convince Her" that I was "Being Aggressive" with her. I of course do not like this Idiot at all, this Idiot has done things for Whorbat that were absolutely Ignorant, but that I would "Deliberately get Physical" is a joke. Having said this, no doubt this Ignorant Jackass, believed it all and took this "Remote Controlled Tweaking" of my Coordination as True, and no doubt this "Verified for Her" what she had heard. Instead "She Verified" that she is "An Ignorant Jackass" all over again to me. To top it off, this person thinks they are "Cool and Know All" as a result of these Spooks. This is how it is for a lot of these Citizen-Idiots.

And again, with ALL OF THESE CAPABILITIES, they can "Mix Abilities" for more "Convincing, Complicated and Confusing Effects". The possibilities are almost limitless, I think at this point I have experienced hundreds of "Mixes". Keep in mind they can do more than "One Thing" at a time.

Remote-Controlling People and "Scripting Your Life"

Every thought, every physical action you make, every reaction you have to physical stimuli, every reaction you have to sound and visual stimuli, every "Accident" you have, literally everything you do in your life has a "Signature", that which can be seen, measured, and labeled, and then later "Changed" with this Exotic-Technology. They can "BLOCK PARTICULAR SIGNATURE PRINTS". They can "Block your own Prints", and therefore "Thoughts and Actions", while "Inserting THEIR OWN". They can "SUBSTITUTE your actions with Their Own", based on this. They also have an "Intensity Setting" in which they can decide "How Hard" their "Chosen Behavior" is "Forced On You"—this involves a "Turning Down" of your "Will-Power" while "Turning Up" their own "Machined Will-Power". You become a "Different Person" as your Behavior, Patterns, Speech, Emotions, Actions, get Re-Written and Erased by someone at the end of this Exotic-Technology. You can Literally be RE-DESIGNED INTO SOMEONE YOU ARE NOT, and it is disgusting. Especially when The People at the end of this Exotic-Techno are LOWLIFES and COMPLETE CRIMINALS. You literally do not get to be yourself, you end up having to "Remember who you Were". Talk about disgusting, and this is where

I and my Family have been for sometime now, and in some ways, things have even gotten worse.

Your life becomes SUBSTITUTED by someone else's idea of "How you should Be", or as is more likely the case "How THEY WANT you to Be". This is called "Scripting" and eventually your Daily-Life is "Lived on Remote-Control" as you are "Motivated and Impulsed" into choices and behaviors that are Influenced, Decided, Controlled, by someone else. As are many of your Emotional-States as well. "Events" in your life become "Scripted", a Pre-Written Event picked by someone else, where even your "Moods and Emotions" are decided before hand by Pre-Programmed Automated Technology, and all of this "Pre-Programming" is done by Government-Agents. In essence YOU the Target are "Hooked up to a Machine", Artificial-Intelligence-Technology, which then keeps you running by way of "Remote-Control" on and according to, a "Script" which has been Programmed into this Technology, along with certain "Computer Programs". Specific-Computer-Programs, for certain "Events". You become an "Un-Willing Actor" in a play.

Your Daily-Life is literally run according to a "Script" that has been "Pre-Written" and "Pre-Formulated" by Government Spooks, days in advance (at least), and then "Loaded onto a Computer-Program", which is then played on that particular day it was Designed for.

In this way your entire life can be Manipulated and Controlled on a Daily-Basis as your Brain is "Stimulated in a Variety of Ways" to achieve particular Responses, Behavior, Emotional States, and even your "Physical-Health" is chosen in this manner. They being also capable of deciding whether you are "Ill or Not", and deciding quite literally what "Physical Ailment" you will have. Simply by affecting a "Particular Organ" which has it's own "Signature-Print", and then "Lowering or Raising" it's Energy-Field. Or they can take a "Signature-Print" of a "Particular Illness", and "Send You" this particular "Energy and Vibe"—and can affect how fast it works, by the "Intensity", the Energy-Level.

If you live with other people like myself, then they get "Scripted" as well, and many times the "Scripting" is unpleasant as "Disharmony" and "Fighting" can be a common theme. And at this point in my life, is a Theme, as The Clown for all intents and purposes is a Power-Monger and Rapist on this and many other points. We are talking about an individual who is literally psychopathic, and an Agency that is no doubt literally clueless about the Truth of this statement.

Again the possibilities for this Exotic-Technology is huge in terms of what they can do. Again each "Thought, Action, Object" has a particular

"Signature", a particular "Energy-Print", and with time the Data-Base for Government, it's Knowledge for Manipulation of People with this Techno has grown and grown. The "Capabilities of this Techno" have grown and expanded. They can do all sorts of things, and many possibilities exist involving the Brain, People, their Emotions, and Their Physical and Emotional Well-Being. It is all seemingly limitless and with Governments keeping this Exotic-Techno secret, and with Corrupt-Agents using this Technology, it isn't getting any better out there. Now that I have given some of the basics and hopefully presented a Logical-Premise for the existence of this Exotic-Techno, on to other things and eventually back into other "Areas of my Life" in which I am interfered with this Techno. We will now "switch gears".

Knowledge and Power without Responsibility and Wisdom

Quite obviously as a result of this Techno the American Government is aware of such things as "Telepathy" and "Astral-States", has Technology to interact on this level. Due to this Technology and the physics involved in even making it, the American Government is aware of many things now, never before understood by Science. They understand a lot of "Esoteric Knowledge", a lot of "New Age Knowledge", and Ancient-Knowledge such as the Hindu belief in things like "Chakras", Astral-States, Auras, etc., that a good number of people do not even believe in, and they can observe it SCIENTIFICALLY. The Government knows these things are REAL AND EXIST and are able to Manipulate such things now, based on their Scientific Observation and Techno. A lot of information provided in Ancient-Texts refers to this knowledge. What was once "Kooky Religious Talk", "Kooky New-Age-Talk", and "Simply a matter of Belief", is no longer. A lot of this is NOW SCIENCE, Hidden-Science, and Kept Secret from the Public. Think of the amount of Mental Growth that would occur for the Masses if such Knowledge were available.

In short the American Government has acquired vast amounts of knowledge and insights as a result, without developing the Discipline and Responsibility that usually comes with this learning. IT IS SCIENCE WITHOUT WISDOM, IT IS SPIRITUAL KNOWLEDGE WITHOUT WISDOM. It is indeed like many of the myths of Atlantis and its downfall—Corruption and the Irresponsible Use of Power and Technology. This vast amount of knowledge applies to Physical-Ailments and Psychology as well, as they can "View Such Problems" from the "Inside-Out", can look at Physical-Ailments and see them in terms of "Energy" and can "See and Hear" Mental-Disorders in much of the same way. This Exotic-Technology can be used in many

Fields for many purposes, and has the Capability of Totally Changing the World for the better. But look at the fact that it is Hidden, and then read on and see what it is used for instead. I have a major problem with a lot of people in regards to how this Techno is being "Used", starting with the Feds in America, and then on to here in Europe with The Clown and Her Buffoons. But back to Science.

So as a result of this Techno they can "See and Cure" all Physical-Ailments and no doubt all Mental-Ailments as well. I'll give some examples below of this "Technology and Cures".

As a result of this Techno they now have a vast amount of Knowledge on things such as Mental Disorders like "Dementia", as they can view it from the "Inside Out". Of course because of this "Scientific Understanding" and because of the Technology which enables them to understand on these finer-levels they are also able to "GIVE SUCH DISORDERS" as well. Actually the Spooks here in Europe, in ADDITION to giving me "Dementia" (Usually when I go out at 5 a.m. to deliver newspapers—then in their "generosity" they "Take it Off"), have more or less explained how it works by doing so, and they know from being able to view a persons mind from the "Inside-Out". I know what this is like from experience.

They can also view things in terms of something akin to Kirlian Energy, or Auras. They can view Energy-Fields of people and understand the "Differences in one's Energy-Field" and what this means. An "Aura" being merely one's Electromagnetic-Field, and this Techno deals in and with Electromagnetic Energy. This Machine, this Techno has enabled the Government to understand Mental-Ailments, Mental Illness, in a very unique way, along with being able to cure these things by "Re-Balancing Energy" by "Re-Balancing ones Energy-Field". Thus in my case they "Reverse the Process" and are able to "Give such Ailments" as well. Having said that with this Techno they are able to "Mix and Match" things as well. In short this Techno has proven the existence of many a Religious-Belief, provided many an explanation for many a Mental-Disorder, along with being able to provide a cure for every Physical-Ailment out there.

Of course they can "Reverse all Processes", all things they have "Recorded and Observed" with this Techno.

With this Exotic-Technology they are able to give ANY Mental-Ailment or Physical-Ailment as well. Each thing has it's own "Signature-Print", including "Illnesses", and they are able to "Mimic" and Duplicate such things. Each person's Energy-Field, "Bio-Electric Field", has it's own unique "Fingerprint" as well, and so once "Recorded" you can be "Found" presumably anywhere in the World and "Interfered With"—one common way being with "Illness". They can "Plug Into" your "Signature Energy", your Energy-Field, your "Bio-Electric Field", and then transmit Electromagnetic Energy directly to you, doing what they want and please. They can take other "Recorded Energies" and transmit them to you, like any Physical or Mental-Ailment—again any Ailment also having it's own "Signature-Print".

I, MY KID, MY FAMILY, have suffered an enormous amount just in terms of the "Physical Ailments"—and it is Torture when it is happening to you. In fact currently, I and my child, have "Pneumonia" compliments of The Clown (Nov. 2010). Both of us have been sick for weeks, my kid missing school and I missing work. With my kid "They did this Gradually" although I knew what was going on, and in my case they did it "Instantly and very Painfully". In my case, and to give you an idea of how this is used for Torture in other ways (and in "mixed ways"), it felt like someone had lit a cigarette lighter in my lungs. I was extremely angry to say the least, and still am, but The Clown's Agency will be glad to hear that "I did not React or Speak Out"—thus I did "Not Fail an End-Test". This is what happens when Agencies do not Regulate or Monitor their Employees, and Corrupt-Filth is allowed to use this Technology.

As you can see this Techno has "Opened up new Doors" in regards to the "Scientific Applications". As things now stand this Techno is Hidden, and not used for anything good. It has given the Governments across the World a deeper understanding in regards to Physical Health, Mental Health, and many a Religious Belief and Spiritual-Belief. A far deeper understanding of things has been achieved, Science in many Fields has been "Expanded Knowledge-Wise", yet the Responsibility and Wisdom to go with this Knowledge has not been achieved. Even from what you have read so far this should be obvious.

This Techno works along the lines of ENERGY, FREQUENCY, "VIBRATION". It uses Electromagnetic Energy and is able to Alter, Manipulate, Control, and Measure such Energy. For what many call Meta-Physics, Spiritual-Physics, Physics on a deeper, more subtle, more refined level. In fact with this Techno, through the use of Electromagnetic Energy they can HEAL ALL ILLNESS. Think about this, take a look at what it is used for in my life throughout this book, and think about the GOOD this Exotic-Techno could do in this World. This Technology is "Misused for Evil", not "Used for Good".

THIS IS WHAT YOUR GOVERNMENT HAS, and others across the World, what it is Misusing it for, and is "Who" it allows to Misuse this Technology. Imagine "Hannibal" from the movies with this Techno—THIS is what I deal with. This is CRIMINAL, and to say that "My Rights have been Violated" is an understatement. Torture is Torture and arguing that it is not because "The means of Delivery" is somehow "Different", is ridiculous and ignorant. I and my Family suffer, and due to this "Miraculous Technology". But back to "Being able to heal all ailments and Illnesses".

The Governments of the World can heal ALL Ailments, there is not a single Ailment in the world that can not be cured as a result of this Techno. This includes Cancer, AIDS, Ebola, what have you. Yet this Techno is being used only for screwing around with people like me...This Technology would wipe out ENTIRE INDUSTRIES, but it is no reason for Hiding It. It is easy to see that this Exotic-Technology is being MISUSED AND ABUSED for Power, it is being used for all the wrong reasons, and being "Used in My Life" by Criminals, Power-Mongers, Rapists, people who enjoy Torture and the Misuse of Power.

Entering the Dream-State: What Happens when You Can "Read Someone's Mind" and they Fall Asleep: From what I Have Observed: a Difference Between the Astral-State and Dream-State and how this Exotic-Technology is Misused for these

The difference between the Astral-State and the Dream-State.

The difference is that the Dream-State involves the subconscious and its "Mental Landscape", the Psyche, and is a personal thing depending on each person. It is the painters canvas, our conscious-mind within this landscape, is the brush and paint. Thus we are able to look deeper within ourselves at our "Subconscious", our "Mental Landscape", consciously, and interact with it consciously, and bring these deeper-symbolic-insights back up to our Surface Mind upon awakening, for growth. Dreams involve CONSCIOUSNESS.

In America, the Feds with this Techno, liked to "Use the Astral State", here in Europe, the Spooks I deal with like to "Use the Dream State".

So have you ever wondered what happens if you can "Read someone's Mind" and then that person falls asleep? These Spooks with this Exotic-Technology simply "Follow you into your Dreams".

These Spooks can "Hijack the Mental Landscape", the Imagery, Control It, Shape It, and Create It, in the image they want. They follow your CONSCIOUSNESS into the Dream-State. Your dreams get "Hijacked" and you dream what THEY CHOOSE. This can be "Traumatic" as the normal safeties your mind has in place for bad dreams, are "Over-Ridden" by these Spooks as they so desire. They can also "Remote-Control" your actions within the Dream-State, and "Talk to you within the Dream-State". Thus with them a dream can be Psychologically-Traumatizing in terms of the "Content", and you are unable to awake from it. They keep you "Locked Into" your sleep, "Sleep Paralysis". When you do wake up, you can literally Wake Up Traumatized.

I awoke once from one of these (a nasty disgusting one), here in Europe, when I was "Not supposed To", and was able to make some interesting insights as to what was going on, some of the processes going on, on their end. There were two barriers: a DREAM BARRIER which I had to break through (like Jim Carrey's character in the "Truman Show"), and an ASTRAL BARRIER which I had to break through, in order to return into my body. Once in my body there was yet again another barrier: Sleep Paralysis—as I could not move. While in this state they tried "Putting me back to Sleep", which I would not do. Eventually I forced my eyes open, and twitched my fingers, and then they eventually stopped. I was not supposed to "Break Free" of my "Spook chosen Dream" but I did. I remained awake and analyzed the experience.

This puts a "Monkey wrench into some of my Ideas" as to the difference of an Astral-State and Dream-State. I admittedly do not have all of the answers here. Once I left my dream I still had to return to my body: thus where does dreaming occur? I assumed within ones own mind. So perhaps it occurred in my own private Astral-State, an "Astral-Closet", with its own unique "Astral-Fingerprint". There are several possibilities here, and a few notable differences as well. I hope to narrow it down here, for both you my reader, and myself...So lets take a look and see.

Dreaming involves "CONSCIOUSNESS". It is expansive, extends outward, and "encompasses" many things, straddling several things at once. It is energy expanding outward. Now an "Astral-Body", how you travel in the Astral-State, is enclosed, encapsulated, "bubbled". It is a Self-Contained-Unit and experience, similar to what we experience in the physical world. You are not necessarily in an Astral-Body, Astral-State, when you Dream, but when you are in an Astral-Body, quite obviously you do experience "Consciousness". There are overlapping points here, and when you toss in Exotic-Technology it is more complicated. I do not have all the answers but things are getting narrowed down here to a finer point.

Another possibility, in retrospect, which I consider the Most Accurate, is that I am "Grabbed-Astrally", put in some sort of "Astral-Cage", "Astral-Room", where all of this occurs thus I am not experiencing my own "Processes" here "Normally" (like any of this weirdness is "normal"). An experimental-head-trauma-shop, "Hogan's Alley" on an "Astral-Plane", an "Astral-Cage" for creating victims: and yes they profile you intensely on this level. However there is a lot to be said about this, profiling on this level can be stupid, there is a lot of room for error. This is an artificially created "Dream-Scape", an "Artificial-Room", an "Astral-Cage" where "Dreams are Simulated", as your "Consciousness" is with you in an Astral-Body. So I am "Astrally-Kidnapped", then my Dreams, Consciousness, is affected and Hijacked as well—there are TWO controls in place, TWO processes going on here. Thus once I broke free I had to break through a "Dream-Barrier" FIRST which is CONSCIOUSNESS-RELATED, then break through an "Astral-Barrier" (insinuating an "Astral-Kidnapping" of sorts) before returning to my body.

I consider this to be it in retrospect. In each case the "Barriers" were invisible walls, like glass, points which I simply could not go beyond. Except I did, as what I was experiencing motivated me at the time to do so. What "Motivated my Breaking Free" was an incredibly sick dream—compliments of The Clown, and of course aside from it being Psychotic on their end, it was "Sexually Oriented" as well—a favorite theme of The Clown and other Sick-Agents. This also gives you an idea of the amount of suffering one has as well, as sleep is no longer a "Restful Experience". You are also "Cut-Off" from your own Sub-Conscious and the "Growth" that comes with it. It does not feel normal, the "Sleep" is of an entirely different quality, it is as if you have "Half-Sleep" all of the time, figuratively speaking. There is no "Learning", no Self-Insights. It is as if you are always "Missing Something". And you are, you are cut off from a deeper-side, your own Subconscious, and the Growth that comes with it. You are unable to access this. I have been experiencing this for years (at least 5 years and every night) and unfortunately my child as well (also at least 5 years—also every night)—which in the case of my kid is stupid as it can damage a lot of her Mental Growth, learning about herself. Of course this Power-Monger and Rapist, The Clown, could obviously care less.

A couple of quick points. We are like "Ghosts in the Machine", think of it this way, "WE LIVE INSIDE OF OUR BODY". We are ghosts within our body. Astral-States, Astral-Bodies, enable us to go to actual-places, on another level. Life is but a dream.

The Astral-State is an actual level of reality, one in which everyone can go to consciously although most do so only during sleep—and usually do not remember. An Astral-Body is for an Astral-Level. What is commonly referred to as an Astral-State is for accessing another Level of Reality, an Astral-Plane, "Public-Grounds" for the Collective, for all people. Ones energy level, vibrations, affects what level you are on here astrally speaking. Thus the idea of the "Mansion with many Rooms". Thus who and what you encounter depends on if they have a similar "Frequency", Vibration—although there are exceptions to this. When in this state we can go where we want (and "will"—as in with intent), like ghosts. Thus the idea of "Astral-Travel" and accessing "Other Realms", Other Worlds. In this manner we can also access the Physical-Realm, Physical-World, thus the idea of traveling "Astrally—in Ghost Form", viewing the world in which we live in physically.

It is a meeting ground for others, lost relatives what have you. In terms of the Matrix, the movie, the white room where Neo and Morpheus meet is an Astral-State. When Neo sleeps and dreams, he is in a Dream-State. The Astral-State is for the Collective, the Dream-State is an individual thing—having to do with ones own "Consciousness". A lot of people meeting deceased relatives and friends in a "Dream-State" are often experiencing two things involving an Astral-State: a Particular Mental-Landscape on an Astral-Plane, and a "Visit" from someone in an "Astral-State". They knew you and can plug into your "Vibe" and visit you. You quite literally "Leave your Body" and "Travel" to a particular "Astral-Level", an "Astral-Room", which has an Existence of it's own—again the idea of a "Mansion with many Rooms". However many write off such an experience as only "A Dream". Thus you meet "Deceased" loved ones, at a particular place, where you can both interact, and they are "Tuned into your Vibe", your "Particular Finger-Print of Energy".

Thus an idea of a Government-Astral-Marker, an ASTRAL-HOMING-DEVICE is very plausible, and something I experienced in America with the Feds. Your "Vibe", your "Energy", your "Frequency", is an "Energy-Print" on an "Astral-Level", an "Astral-Level" in turn perhaps being merely one large Electromagnetic-Field. As the Earth's Field around it, is one Electromagnetic-Field, the space between the Earth's upper atmosphere, and the Earth itself—the Schumann Resonance. Thus Government, with your "Particular Energy-Print", that "Identifies you as a Whole" can find you anywhere on Earth, because this Techno operates on Principles of Electromagnetic-Energy, has your "Electromagnetic-Print", and we all live within one big Electromagnetic-Field. It is easy to understand, and makes things like "Searching for Osama Bin Laden" a moot point. As does this Techno, make Guantanamo Bay a "Moot Point".

Anyone within Government, and Agencies, with a problem with my Book, the Knowledge, and my understanding, can take it up with The Clown and Her Buffoons first—you have THEM to Thank. Never mind the stupidity, and Farce of this Government Program that you as Agencies have put me in. But back to business: Astral-Markers, Astral-Homing-Devices, and Techno.

This makes sense, there are still laws of physics, spiritual laws and principles which must be followed here, Meta-Physics. Thus I am guessing, and feel comfortable in doing so, that this is what the Feds had put on me in America, as I am about to get to.

In regards to further understanding "Dreams, Consciousness, Astral-Bodies,and Other Related Topics" I highly recommend the "Seth Books" by Jane Roberts. Quite frankly there are no better books than these, the approach is based on Logic and Science, and if you need an "Endorsement" at one point the American Government came to her asking for "Help". Obviously they were "Keeping Tabs" on her and considered her legitimate.

Unknown to Me at the Time: My First Experiences with this Techno

In 1997, four years into my "Situation", while living at Virginia Beach, Virginia, I began having strange experiences. These experiences were strange and involved Government Agents coming into my Dream-State. Having said that I was ignorant of the possibilities and had no idea of Exotic-Technology so I wrote off these "Strange Experiences" as something other than they were. Due to many experiences of my own involving out-of-body experiences, remote viewing, meditation, and my own personal philosophy, I realized this was more than dreaming, that someone was coming to me Astrally.

Now we kept "Meeting" in a room like the "White Room" in the movie "Matrix", while I slept, an Astral-Room and an Astral-State. Many times while we sleep we "Leave our Bodies" and "Travel", and so this was the same thing. The difference is, that I obviously was being "Grabbed and Kidnapped", and "Brought to this Room". I did not want to be "Meeting" these people, it was against my will, and I seemed to have "No Control" over these "Meetings". I am normally someone who does not have bad dreams, and who can in fact "Control the Dreams". So I would go to sleep, but "Someone was Waiting", and none to pleasant. I just did not believe that it was Government Agents—at the time, and even now for many, who would?

Now these experiences involved Agents, always three of them, always a mix of men and woman. The experiences were unsettling and disturbing, as were The People I was encountering, who were absolute Lowlifes and Filthy—we are talking about Totally-Corrupt and Seedy-People here. At the time I had no idea of this Electromagnetic-Technology, so could not believe that Agents were coming to me, Astrally, while I slept (Even though I was well aware of the Agency involved because they told me—"Innuendos"), but these "Experiences were Memorable" to say the least, which is a "Common Tell" of this Techno. As it is, I never forgot these.

The Agents coming to me were antagonizing me, I was "Locked into this Astral-Room", in fact could not even move my arms or legs, and could not break free. They were doing this, keeping me held in this State, so that I could not "Break Free" or Wake-Up. They were explaining things about the science of numbers (something akin to meta-physics), the Truth within Ancient-Texts (for instance the covenant of the ark can be reverse engineered from such texts), and of all things Atlantis (one was claiming to be "Astrally Projected" from Atlantis). Talk about weird, however I was living only blocks away from the Edgar Cayce Center at the time which I liked to frequent with my entourage of Spooks in tow, and no doubt a lot of these themes came from there just to screw around with my head. To give me the idea that these were "Just Dreams", a Cover-Story—which obviously I did not think. I had no idea of what was really going on but knew that what I was experiencing was not simply a matter of "Dreaming". As an added point the guy coming from "Atlantis" was wearing glasses, so much for Atlantis being technologically advanced, capable of healing all body ailments, and capable of "Astral Travel" if that is the case. Now to top it all off these people by anyone's standards were Arrogant, Power-Oriented, Ego-Oriented, Cold, and flat out "Evil"-Power Mongers. They were and are, Disgusting and Repugnant.

Since I did not know of this Technology at the time, I assumed that some other force was at work here, some Evil-Astral-Force brought into my life simply by contact with these Spook Jerks, and whatever unseen forces may have been present in "Their Own Lives". In short our own Energy attracts similar Energy, unless you are in my shoes—but for whatever reasons my circumstances were and are meant to be. I thought something attracted to Their Energy was invading my own.

These "Sessions" went on several times and were unpleasant and involved "Sex-Stuff" as well (Typical of the real Lowlife Agents, and is a favorite theme of The Clown), the last one I remember, the guy with glasses was squeezing my groin and did so three times while I was "Paralyzed in this Astral-Room" (a reference to their "Esoteric-Explanation of Numbers"). I could not move: I woke up from this "Dream" laying on the bed wondering what was going on, and felt this again WHILE AWAKE, the squeezing of my "Crotch". It was as if someone were doing it with their hand. Talk about creepy. I knew something extremely weird was going on here and I never forgot these experiences—obviously not just "Dreams" if what I am "Dreaming About" is occurring while I am awake. Of course an absolutely stupid psychologist, ignorant with and about Technology, no doubt would have a "Clever Explanation" for all of this.

Now we jump to1998 and 1999, before my trip to India (to Herakhan among other places). In retrospect at this point the Feds clearly were interacting with me via a "Strange Technology", it is just that I would not find out the Truth of all of it until here in Europe, due to The Clown. I once again, after years of "Silence" in terms of these experiences, began having them AGAIN while living at my parents house in a different state, to save money for my trip to India (Babaji)—which I eventually made. Having said that these Feds made sure I was followed across seas to India as well. Apparently if I had just left the country, these Agents would not be able to pursue me, so they arranged that they could by Misusing this Exotic-Technology, for Framing Me. Having said that the Feds Framed Me in the first place which is what created all of this "Mobbing"—and for a Pedophile-Cop Cover-Up, and a Murder Cover-Up, which I knew about obviously. They'll be glad to hear that I have it all sorted out now, right down to the fine details, have some of it printed here in this book for the World to see, and that I am more than slightly Pissed Off. But back to being "Revisited" by Corrupt-Agents with Exotic-Techno.

While living at my parents house and saving up money for my trip to India, Agents came once again to me, this time PUTTING SOMETHING ON ME IN THE ASTRAL-STATE, A HARNESS OR BELT of sorts, while I remained immobilized. It reminded me of my experiences at Virginia Beach except in this case this was not in an Astral-Room but in an Astral-State, as in a "Ghost State". I could literally see the Physical World around me, as if I were a ghost, but could not move: somehow they had me Immobilized, Frozen, in the Astral-State, and this particular "State" is how people "Astrally Travel". Again a mix of men and woman, again three, and while I could not move I began to do an Indian Mantra to ward off evil, "Vibrationally Speaking" this is supposed to work, however it did nothing, and the pig-of-a-woman laughed as they did this, at the fact I was doing a Mantra, that it was not working, and that I could do nothing about it. After they put this "Harness/ Belt" on me, I woke up, and I was disgruntled to say the least.

This was an ASTRAL-MARKER, an Astral-Tracer, an Astral-Homing-Device, and years later here in Europe, I would find out that the Government has technology to put an Astral-Marker, an ASTRAL-HOMING-DEVICE on anyone they want and are able to find them ANYWHERE IN THE WORLD. This was released in A NEWSPAPER ARTICLE which was also on the Internet (2006), a woman was "The Inventor" (no doubt the Government was first), and from some South-of-the-Equator Island-Nation if I am not mistaken. I am going on memory on this one (the Spooks have a "Web-Blocker" on our computer: blocking particular subjects), but it reminded me of my experience in America, the one I am describing, and perhaps this is what was put on me, I still have no idea but it is a SAFE BET. So the experience was strange, and whatever they put on me was to serve their own demented purposes. In retrospect I can bet that this is what it was, especially since I was going to India. In retrospect, this is "Why" they did this, and this is what it was.

Aside from these Feds Misusing this Techno during a phone call, before I went to India, these are most of the experiences I remember. During this phone call, they "Machined", "Remoted" my voice "Flat", so that my "Sarcasm was not There", and so that I sounded "Serious Instead", when in fact I had "Planned the Call" hours in advance, had INTENDED TO BE SARCASTIC, and knew that my parents phone was "Tapped". The Feds watching this as I "Planned, Prepared, and Looked Forward to This", and no doubt "Irritated" or perhaps "Feeling Clever", decided to throw a monkey wrench into the whole thing. I was going to "Rub their Accusations" in their face with Sarcasm, because I had not done anything I had been accused of, they had "Shoved it down my Throat", "Deliberately Stigmatized Me" with it, IGNORED ALL FACTS TO THE CONTRARY OF THEIR ALLEGATIONS (Deliberately as I found out Here), and Misused Everyone with this, never mind I was "There" for the "Things" I was "Accused Of"—their Allegations and Deliberate Lies.

So I called someone up, spoke to them, and instead of the "Long Awaited Sarcasm", and my "Giving it back to Them", I got a "Flat Voice" and was "Totally-Shocked" in the moment to say the least. It took a moment to recover, but of course it was too late. I never forgot this moment, it was "Unreal, Weird, and Shocking", and I could not believe it—not ever actually. It never "Made Sense", and I simply could not "Explain It". Thanks to The Clown, now I know why. This is YOUR TAX DOLLARS at work. So these Lowlifes, Misusing this Techno, did this, and I ended up "Pursued to India", then back to America for a few months, then to Europe. Three continents, and three countries, within the space of a year and a few months.

American Tax-Dollars at work, and Government Technology at work. For a Pedophile-Cop and a Racist 911 Dispatcher who Delayed an Ambulance for a Kid because of his skin color: and the resulting Cover-Ups which I knew about. This is why my Mobbing began, and twenty years later, is about. And now I have a Corrupt-Psychopath to deal with as a result: this Exotic-Fruitcake I refer to as The Clown, and who "Holds Me In" this Mobbing based on Personal Reasons, all the while Torturing Me with this Techno based on her own Frustrations....

How I Found Out about this Electromagnetic-Technology in the Beginning—Europe June 2006: The "Exotic-Anal-Retentive-Clown", and her Deliberate "Revelation of this Techno"

I moved to Europe in 2000. There have been two different people in charge of this "Mobbing" since coming here to Europe. A man who was here until 2002, and this woman I call The Clown—due to her Egotistical-Bungling-Behavior. Well as it turns out, this last man In Charge had this Techno, as did the Feds in America, but with either of these "groups of people" I did not know of this Techno. This all changed with The Clown.

As part of The Clown's Profile, being a Psychopath, she has several Characteristic Traits which eventually lead to me finding out about this Techno. Being Excessive, an Inability to Perceive Consequences, Self-Aggrandizement, no Empathy, and of course a Total Lack of Self-Control and Impulsiveness. To put it mildly, she loves Misusing this Exotic-Techno, and began doing it from the beginning. Having said that she continues to Misuse it on me and my Family, and quite a few others as well, as I have come to learn. As an example, she has Delayed my Book, two years with it.

As there are things which "Qualify someone for being in Mobbing", so there are "Qualifiers for being Free" of it. And in 2002, shortly after The Clown showed up, I figured out that it comes down to about one months time. Which seemed to irk her, but if you know anything about Psychopaths, they have an "Instant Anger" problem anyhow, it does not "take much" to irk one. So once The Clown was here, I figured out that there is a chance "To be Free" every month or so, and of course began paying attention to these "End Dates". I like anyone, looked forward to my Freedom. I began "Noting the Patterns" of this "Mobbing" and began sorting it out, how it works. Eventually I began reaching a point where I could be Free. So naturally The Clown being a Corrupt-Psychopath, she began Deliberately "Holding Me In". She began doing so, in fact, with Exotic-Technology. This was all during the first few years.

It did not take long for me to figure out that something was wrong, and that something weird was going on. In fact I always seemed to have "Problems" from out of the blue, on THE LAST DAY. Naturally as a result, I was "Held In", "Kept in this Mobbing". After all I "Failed the Last Day" and "Last Test", right? Not at all. You can no more avoid the affects of this Technology than you can a bullet from a gun, and if people who have this Techno, decide you are going to "Fail", then you will. But at the time I did not know of this Techno, like anyone, the "Thought of It" just never occurred to me. I had never heard of such a thing, and if I had, no doubt I would have "Seriously Questioned It". It is hard to believe. Or at least for me, it was.

But as it is, I quickly noticed things were not right. A "Common Tell and Sign" of this Techno being used in your life, is Un-Characteristic Behavior, Shocking Moments, in which something you do, say, or feel, is "Totally Different" from how you were. You get "Machined", or "Remoted" as I often call this, and afterwords you are left "In Shock or Surprise" at what has occurred. Your mind is left "Reeling" at what "You have Done" as it makes no sense, and is NOT YOU. Thus "Shock" and "Your Mind Reeling" are "Common Tells". Well, I began having these quite a bit, and specifically always at the "Last Day".

So for the next two years I began experiencing "Weird Coincidences" and "Weird Behavior" ALWAYS at the "End Test", and I came to realize in several ways that this was "Weird". That this always happened the "Last Day" and that many times I was not "Stressed" was one thing. In fact many a time my "Attitude, Mood, and Spirits" were high, in fine form, as I knew I would be Free. Except then there would be a "Total-Reversal", something would happen, and then there would be "Unusual and Excessive Anger", or something else, and for absolutely stupid things—as one example. Afterwords I would be left in "Shock" and wonder "How this Happened". Naturally it was The Clown. Many a time me and my Family would go from "Being Free", to "Traumatized" by behavior not our own, all within minutes. Again this woman is Sick and it was all being done by her.

On occasion this Clown would change her methods, Tactics with this Techno. Another example being that I would "Forget the Last Day" on the "Last Day" and do something stupid, that would of course "Get Me Held In", and of course these "Stupid Things" were affected with Techno as well. If you are in this "Mobbing" you no more "Forget the Last Day" than a man in prison forgets his day of release. So again as The Clown experimented with me, by way of this Techno, I was left "Wondering and Reeling" many times, in many different ways. All of this noted by me for the "Strangeness" it was, and noted by me, as NOT BEING ME.

That this KEPT HAPPENING REPEATEDLY was another thing I noted.

The "Things" which "Happened" were not based on my "Own State of Mind" at the time, and left my "Mind Reeling" as to how this was even possible. There was no "Logical Explanation" and I most certainly realized there was an "External Influence" affecting me. I developed a "Theory of Evil Spirits" or perhaps "Negative-Aliens" (I was grasping at straws here), and there is a lot of Dialogue with my girlfriend about this, so there is a "Record of The Clown's Cheating"—unless they have destroyed it. I mention this for The Clown's Agency, as they will receive a copy of this Book. Seeing as how the behavior was NOT MINE, it had to be "Someone Else's" and obviously this came down to an External-Source and Influence. Well eventually I "Caught this External Influence" as it tried to affect me. This happened one night in particular, and I hope this Clown's Agency pays attention here, as this Clown tried to put me in prison.

And for what? A Pedophile-Cop and Racist Redneck? You call this Justice? You call this "The Rules, Standards, and Protocols of Mobbing"? You just "Hand-Out this Technology to People" and don't even MONITOR, REGULATE, OR SUPERVISE? Have you lost your god damn minds? This is "Professional"? Obviously Psychopaths are Incapable of "Tolerating very Much", and again this gets down to The Clown's "Psychotic Temperament", but that said, I did not "Hire The Clown".

"I Catch This Techno Being Used On Me"

So one night, The Clown decided to "Machine Me", to "Remote-Control" a "Fight between me and my Girlfriend"—a very violent situation between me and my girlfriend I might add. I "Caught a Thought" "Pushed Into my Head" followed by a sudden "Surge of Extreme Anger" "Pushed-Into Me". I telephoned my parents to "Diffuse the Situation", and even spoke to my mother, telling her of "My Experience"—I literally said to her that something "Pushed a Thought In", which then "Pushed in Extreme Rage" right after the "Thought which was Pushed In", and that this came "From Outside of Me". For my readers it was that obvious, that this is what occurred—it was so obvious I said this to my mother that night without caring how this "Looked or Sounded". So there is a phone record of this for The Clown's Agency. As it is, I "Narrowly Diverted a Disaster" and this was a "Final Test Day" (nothing violent happened, I telephoned my parents). Obviously I was "Held In" and for Behavior NOT MY OWN. And as all of the other times The Clown CHEATED and I was Held-In with this Technology. How does one "Pass" when these are "The Options"?

Of course this also shows how Dangerous The Clown and her Buffoons are, and shows how Governments are Misusing this Techno. When I eventually get to how my "Mobbing Started" you will also see I was Framed in the first place, so none of this should be going on, never mind The Clown almost putting me in prison. This Techno is being Hidden for Misuse, and being Misused against The People.

I believe that this "Night in Particular" was Whorbat's (The Clown) response to her own growing Frustration with me. Some where around this time, I had "Passed this Mobbing" for the first time, and Whorbat had

"Held Me In" regardless, and for Personal Reasons. If "You Pass, You Pass", but with Whorbat it was not this way, The Clown was Pissed-Off instead. She "Held Me In" FOR NO REASON when I "Passed" this first time, and this was a Big Mistake on The Clown's end: as I knew at that point that she was Corrupt, and had no intention of "Ending This". Then shortly down the road she tried to get my girlfriend seriously injured, and me in prison—of course like anyone I did not suspect this "Exterior Influence" was these Spooks.

From a Warped-Psychopath-Viewpoint, I had "Beaten Her" which Pissed Her Off—and yes this Imbecile does think this way. This "Mobbing and Her Job" are not about "Mobbing and a Job", they are about HER which is why she likes to Show-Off, thus it was not a matter of "He Passed", it was a matter of "He Beat Me". This is how She is, this is about "Her Beating Others" and "Winning", not about Doing a Job, and getting a Job Done. EVERYTHING this woman does is always about "Her" and is always "Personal", and THIS IS HOW SHE VIEWS HER JOB. Her Job, is for "Serving Her", She does not "Serve a Job or Agency". She is not about "Following anyone's Rules", or "Rules and Standards of Mobbing", it is all about HER and "Her Rules" which are "Preferences based on Warped Fulfillment". This is a result of her being a Psychopath and "Bubbled" within herself.

We are not talking about a Normal Person with "Normal Brainwaves" here. We are talking about someone who behaves like a spoiled four year old kid, quick to anger, and who gets angry over stupid things—ESPECIALLY NOT GETTING HER OWN WAY. At these beginning years, like a "Rapist" she wanted that my "Will was Broken" and that I would be left behind, this "Mobbing Done", me left in a "Mess". Her "Preference and Desire", Personal not Professional, all about HER, not Her Job, which she only uses for Her Own Reasons—obviously. So once I began reaching a point of being able to "Pass this Mobbing" she used this Techno, unknown to me and others initially, to "Hold Me In", until she could get this "Personal Fulfillment". Imagine "Hannibal" from the movies, and you start to get the idea.

This is what happens when a Psychopath is given Power and a Job-Title of Authority. This is of course why she has chosen this Job, and why now that she has this Job, AND a Title of Rank, she does as she pleases: she can act with Impunity in regards to others, which of course she does. Psychopaths are "Bubbled" and are all about themselves, thus she Uses and Manipulates her Agency for her own Self-Fulfillment. And as I have noted, this is easy for her to do, as obviously there is little contact and "Oversight" between Her and Her Agency. Quite frankly she is Manipulating Everyone, Her Agency, the Citizens, and of course the Agency of this Country, who knows of my "Mobbing" yet does not even know such simple facts such as I was Framed in America, and that I have been "Held In" continuously by this Corrupt Clown. To top it off, these Idiots have given this Clown, a Psychopath, this Electromagnetic-Technology in addition to a Title of Rank. Neither of these Agencies has a "Clue" as to what is going on in my life, and going on around here. Neither has an idea as to "Who or What" this woman is. I DO, this Clown knows I do, her Buffoons are absolutely Filthy with the things they have done for her, and it has been "Unpleasant" to say the least. Never mind that these Corrupt-Idiots have Delayed this Book, for TWO YEARS, and with this Techno, and because of THEIR OWN REASONS, NOT AN AGENCY'S.

I find this ironic as this Clown has been the sloppiest of all people In Charge, yet no one seems to be able to "See Her", or catch her. So much for Government Agencies, it really makes you wonder how good any of them are at what they do. But I digress.

Now I will "Jump ahead Here", and as this "Totally Fits" I will add it here, since this SHOWS what The Clown was trying to do with me. Another example of the "Same Thing", and from as recent as 2010. If you are this Clown's Agency, or the Agency of this country, pay attention here.

In 2010, I believe late spring, The Clown began giving "Knife Innuendos", these Spooks began "Broadcasting a Future Move", and they did this over a period of many days to "Heighten the Suspense" and to cause Psychological-Stress. Obviously this was to "Intimidate Me" and to convince me I would be "Attacked by someone with a Knife". It was "Noted on my End" as I have already had the experience once before, in America, in which I fought off someone who tried robbing me with a knife—of course The Clown was playing on my past here. Naturally I took these "Innuendos" seriously, but I felt "Ready", and if you know anything about Spooks, they like to do this—to "Broadcast and Show Off" their "Future Moves". So what happened?

After a week of Innuendos, there was a violent knife attack down the street, and around the corner, from where we live. Imagine that.

This involved a young man stabbing his girlfriend repeatedly with a knife in their car. There were helicopters and police all over the place looking for this guy. What happened? Obviously the same thing this Clown tried with me and my girlfriend this one night back in 2004–05. Obviously these Lowlife Spooks did this, with Techno, they knew this was going to happen.

This was the CHOICE OF THESE SPOOKS, not his or hers. They were affected with Techno, no doubt "Got in a Fight" which was "Scripted" and "Started" with Techno, and the young man was "Remoted" Excessive Anger and Rage. The Clown was no doubt hoping to "Scare Me" into shutting up, and stopping this Book, instead this is in my Book. As it is She is Misusing this Techno on Citizens of this country, and I am sure that this countries Government Agency is unaware of what The Clown has done, and has been doing. For that matter I am sure The Clown's Agency is unaware of this as well. No doubt Her Own Agency will have several problems with all of this, with this Clown. For starters there is no "Ending this Mobbing" when it is done this way—that is because it is about Deliberately "Holding Me In". The next problem they will have being that The Clown is running around "Killing People" she does not like, with Exotic-Technology, and in a Foreign Land—and no doubt without "Authorization". "Personal not Professional". This Clown and her Buffoons, Totally-Corrupt, act as their "Own Agency". This is some of the "Proof", and THIS is the kind of thing this Clown tried doing to me and my girlfriend that night. THIS IS A "RULE OF THIS MOBBING"? Now back to this night between me and my girlfriend.

As heavy as this night was, "Life went On", but I clearly realized that things were not only "Not Normal", but that this "External Influence" was DANGEROUS as well. In the meantime both I and my girlfriend were "Slowly being changed Here" as well, at this point in time with this Techno—unknown to either one of us. Our "Personalities and Behavior" were being "Re-Designed", to "Fit someone else's Desires and Ideas" of how we should be. None of it for the better of course. How else could it be with Corrupt Spooks involved?

Well then in June 2006, these Spooks came out of the closet and said they had this Exotic-Technology. Typical Clown Fashion: Showing-Off. This was also Her Cheap-Excuse for "Raising the Level of Misuse" of this Techno.

Again another mistake of The Clown, for multiple reasons, as I will get to.

The first thing I did was point out that I was "Held-In" the night of this phone call to my parents—which became obvious right away. I pointed this out to my parents and girlfriend. This was a result of these Spooks screwing up—they got CAUGHT like the Idiots they are. I pointed out the obviousness of this and these Spooks had to Lie to both: which somehow they got away with doing.

Now this "Revealing of Techno" became used as a Cheap-Excuse for "Raising the Misuse of Techno" (It already was being Misused), Her trying to put me in Prison, having failed. If you are looking for "The Logic of This", there is none, it is Typical of Whorbat and Her Warped Mind. Like "Blaming others for her Mistakes", a common thing Whorbat does, and again a Trait of Psychopaths. She has constantly Blamed Others working for her who have made mistakes, and then in Secret, by Misuse of Her Job, attacked these "Others" for "Screwing Up". Sometimes with this Techno.

In reality, a lot of these "Mistakes made by Others" were a result of Whorbat not having "Thought Ahead" well enough, and not having her "People prepared well Enough". These "Mistakes" are in fact Whorbat's Own, yet she Blames Others, and then "Attacks Them". Having said that no doubt that was the "Logic" for Raising the Stakes with me, in regards to this Exotic-Techno. The Clown "Blamed Me" for "Figuring Out" there was this Techno, as I obviously "Recognized an External-Influence". Which is STUPID TWICE, as She was so "Excessive" with it, a complete Retard and Imbecile could recognize it, and then She is the one "Who Revealed this Techno to me and Others". We are "Talking about Warped-Princess-Logic" here, the "Logic" of a Psychopath, Typical of The Clown and Psychopaths.

At the time I could not figure it out, it was if she was "Revealing this Techno", but then "Attacking Me" for knowing too much—ie. "Secret Classified Information". Which made no sense all over again as Whorbat Revealed this Techno to a few people in my Life.

To me, to others around me I knew, and I believe at one point, even to a small portion of the Public involved in my Mobbing. Which would be absolutely Typical of Whorbat's "Showing-Off" and Self-Aggrandizement, her "Showing-Off" and being in the "Limelight". This was revealed specifically to parents and teachers of the kindergarden where my kid was going at the time. Peoples reactions were giddy and nauseating, you could tell people were excited by the knowledge of this Exotic-Techno, it is just that they had, and have, absolutely no idea what this is being used for. It was "Wondrous and Amazing" like a "New Show-Casing of a Car" by G.M. or Ford at a Car-Show. What a bunch of Idiots, no doubt Whorbat was enjoying this "Showing Off" of her "Wondrous Toy". And in regards to this "Show" at the Kindergarden, it may have been just that—in the end, for this particular group of people, they may simply have been asked and Told to "Act" this way. BUT I SERIOUSLY DOUBT IT. Her Agency can certainly check. In regards to others, closer in my life—there is no doubt that they know of this Techno, and as a result of Whorbat.

Now as I said this "Revealing of Techno" was a Mistake in Multiple Ways.

Once I knew of it, I began "Rewinding on my Life". Eventually I began looking through my Past to see what "Other Possibilities" there were, and I found a ton of them. Of course this explained my "Unusual Behavior" and "Failing Final-Tests" for a few years in a row prior to this. And as I did a "Rewind" I realized that this explained some very Weird Experiences I had had in America. "Weird and Unusual" experiences which I did not forget because they were so "Weird and Unusual"—a "Calling Card" and "Tell" of this Technology being used in one's life. Quickly I realized that these Feds in America had the same Exotic-Techno, had been using it on me, and in fact used it on me to "Frame Me" for following me overseas.

In fact I found "Weird Situations and Experiences" which were obviously the Feds with this Techno, and I kept finding more. In fact, although later, I realized "Why they acquired this Techno" in the first place. Because they were worried I had figured out all of the ways in which they had Framed Me, and how they had Misused and Tricked two of my girlfriends—and one in particular, never mind how they Deliberately Misused some of the Regular-Citizens. They were worried I had "Figured everything Out", including their "Motivations", and so worried about their own asses, acquired this Techno to find out WHAT I KNEW. They did this in 1997, and it began at Virginia Beach, Virginia.

Another thing I realized is that they Misused this Techno, used "A Feeling Blocker", to be able to Frame Me, "Appearances not Reality", so that they could have "Authorization to Follow Me Over Seas" as I prepared in 1999 to go to India. I had "Planned on being Sarcastic" during a phone conversation with someone to "Rub the subject of what `I Allegedly Did' according to Them" in the Feds face, as I knew our phone was Tapped. Instead my "Voice came out Flat", without any "Sarcasm", which "Shocked Me" as an understatement; especially since I had been Planning on doing it for a few hours previous to the call. It was my way of "Rubbing it back in their Faces" after all of the Non-Listening, Heavy Lies, Traumas, and Attacks I had endured. But the Feds with this Techno knew this, and "Misused Techno" to get this to "Backfire on Me". As it is, anyone with half a brain should have known I knew the phone was Tapped. But this has always all been about "Appearances not Reality", no one at the time knew of this Techno, and the Feds obviously succeeded. I will get to this "Blocking of Feelings" later, and "Scripted Speech/Dialogue", it is less hard to understand or believe than most may think. And this also meant that what the Feds had accused me of, they knew to be False, as they had this Techno for "Viewing my Thoughts and Memories"—this was not simply "Mistaken on their End". It is a bit of a moot point though, as they obviously Framed Me, and in Multiple ways.

And it is because of this Clown that I was able to "Rewind on my Life" and find out that in fact the Government Agency in America had been using this Exotic-Technology, quite a bit (Listening to my Thought Daily), and at times in Extreme Ways for "Manipulating and Lying". Thanks to Her, The Clown, a lot of my experiences in America now make a lot more sense. And again thanks to The Clown other people around here in my life know of this Exotic-Technology—as they have said as much to these people.

THE NEXT LEVEL OF TECHNO AND THE "NEXT LEVEL OF DISCOVERY"

So the "Next Level" with this Techno, of course being More Extreme. This was no longer about "Passing Mobbing" as I had already "Passed", this was about "Keeping Me In Deliberately" and trying to "Break Me" with this Techno, among other things. This was no longer about "Stress-Tests", this was now about "Surviving Torture", and it still is as I write this in 2011. All of it Personal, not Professional, and Whorbat was hoping and waiting for "Her Day to Come"—her day of having "Totally Broken Me". Her "Preference and Standard" as opposed to doing her Job, and Her Agency's Standard—"Her Rules" based on her Sick and Warped Mentality. She just could not "Let Go" of WHAT SHE WANTED, and could not put Her Agency, and the "Rules, Protocols, and Standards" of this Mobbing FIRST. Never mind that she knows that I was Framed in America, and knows why. So She never got "What she Wanted". Instead she got a Book.

I began writing about my Past to clear my name, and to get to the bottom of everything. Since I knew of this Techno, I could go back in my mind and find where it was used, and since I knew of this Corrupt Clown, I could go back in my mind, and "Look for Other Things" as well. Which is what I did, much to The Clown's chagrin. Now I already knew some things, as I was there for a lot of "Things said about Me" that were not True, but I had thought that for the "Most Part" the Feds had been nasty imbeciles and mistaken. Instead I began to find all sorts of Lies, Deceits, Manipulations, and all due to The Clown, and her own Mistakes and my own Observations of The Clown and Agent-Tactics. I began putting it all together like a large puzzle, and everything began to make sense. The Clown in the meantime was happy that I was "Looking at my Past" because she wanted to see if the Feds (in America) assessments of me, what they had written in my Files previous to 1997, were accurate. The Clown wanted to assess how good these Feds were. She thinks She is "Competitive", yet has to "Cheat like Crazy to Win". If you were to play a game of chess with her, she would simply reach over and take your king on her first move and say "I win". Clown Logic: "Winning is Everything" and "Cheating is necessary for Winning".

However her "Happiness" very quickly started to go downhill, as I put this all together, she began to get "Sour and Slip". As I put it many times, "She stunk of Desperation". Like a Detective, I was figuring all sorts of things out, that I was not supposed to know, doing it on "The Clown's

Watch", and would not have figured out any of this, never mind this book in particular, if she had not "Held Me In". Needless to say The Clown was no longer happy. And again, she had already created her own Cheap-Excuse for "Raising the Level of Techno", and was "Sour". The Clown began adding to her Extreme Behavior.

Heavy things, Many Things, and quite a few Permanent Things, began being done to me with this Techno. The Clown began taking out her Frustrations and Aggressions out on me, while hiding behind this Mobbing and her Job-Title: another Typical Clown Thing. Her Job-Title and this Mobbing, act as "Camouflage" for her Personal Motivations, and for taking out her Personal Frustrations and Anger. As She did so, I noted both more "Capabilities of this Techno", and more Government Tactics. I had already known quite a bit about "Tactics and Techniques" in America, but now everything had changed. I now knew how Corrupt-Agents do things, I knew the Feds in America had been absolutely Filthy Beyond Belief, I was "Learning more about their Dirty Tricks and Tactics", I now knew of this Techno, I knew it had been used on me in America, and I was now learning about the "Potential of this Techno", none of it "positive". Eventually in regards to this Techno, I began to "Figure a few things Out" as well—a basic understanding of the science behind it.

So the first thing that started after I found out about this Technology, was constant ringing in my ears, which I hear every time I go to sleep, EVERY TIME, EVERY DAY, for years now. Last time I checked that is Torture. The next thing to happen was my sleep began to be interfered with, and heavily—I was being "Kept Awake". At one point I was averaging FOUR HOURS of sleep for EVERY DAY, for three months in a row. Last time I checked that is Excessive and Torture. Whatever it is, it is not about "The Normal Rules" of this Mobbing. None of the other people I have dealt with, have been this way. This whole "Sleep Problem" has also continued to go on into the present, as well as been ongoing for years. I have not had normal sleep for at least five years now.

The next thing was constant interference with my DREAMS. Which these days are absolutely disgusting and TRAUMATIC. They do this no doubt with their ability to Stimulate certain parts of your Brain, their ability to Send-Images, their ability to Send-Thoughts, and no doubt other things as well, as Dreaming involves your Consciousness and they are able to "Alter This" via EME (Electromagnetic Energy). This has also been going on for years, daily and nightly, as I also take naps. There are other things as well that have been added on, each time I reach a point of being able to pass a "Test", and therefore reach a point of being free. It is a very long list of "Permanent Things" that are done daily, and have been "Left on Me".

Here are some more examples. Chronic Fatigue, they have damaged my eyesight/vision, given me Arthritis, put warts on me, damaged my sense of smell by doing something "Weird" to my nose, softened and damaged my teeth, Chronic Backache—in fact I believe this Idiot has "Curved my Spine", "Techno inspired Eating" with a twenty five pound weight gain, and a whole host of other things as well. In regards to my Kid, The Clown has "Put a Wart on my Kids Foot" so that it is uncomfortable for my kid to walk, and to detract from the quality of our time spent outside. And these are only the Permanent Things/Damage done, and not even all of them. There is a host of things, suffering, that I and my KID, go through DAILY, and all due to this Techno, and this Clown.

So eventually this Exotic-Technology started being put on my KID, and my girlfriend as well, although it was only on my girlfriend heavily, at the end of a particular "Round of Tests". Our "Personalities began being Re-Designed" with this Techno. A slow build up of things "Changed Permanently" with this Techno, so that we all eventually changed more and more, as more things with this Techno began being added on. Layer upon Layer of Techno Inspired Behavior, added on over time so that we would not notice these changes. After a while you "Look Back" and you see that "Very Much has Changed".

And in regards to my girlfriend, at "End Tests", my girlfriend began being "Targeted for Extreme Behavior". In other words it was put on her to create such Unusual, Unpleasant, and Extreme-Behavior, that I "Fail a Test" as a "Result"—she was "Affected to such an Extreme" that "Passing" was impossible, never mind that The Clown was "Creating Excuses for Keeping Me In" in conjunction with this "Remoting of my Girlfriend" (Remote Control Folks). At times she was near violence, and there could be no "Passing" when this was what I had to deal with. To top it off, like an Idiot, my girlfriend could not "Recognize the Affects" of this Techno, wanted to believe these Spooks were "Good People", and so ended up being used repeatedly to Traumatize our Family, in this manner, for Years. The Clown, seeing that she was "Less Observant", "Less Intelligent" (sorry but I had to go there), more Gullible, and Weaker Willed, Targeted her repeatedly and Traumatized all of us, in this manner. Eventually I was able to "Get Through" to my girlfriend, but only after she nearly Destroyed our Family. And I might point out that my girlfriend was doing everything this Clown asked of her, yet this was my "Girlfriends Reward" for her Trust. Just like the Feds in America were, with several people I knew who were doing everything the Feds wanted, yet they got Used and Manipulated Heavily—these Feds are Corrupt-Lowlifes.

My kid is also Targeted Heavily with this Technology, as opposed to my girlfriend who is Targeted Less these days. But then again my girlfriend is in league with these people because she feels she has "No Choice"—which is "Standard for this Mobbing", although how this Whorbat has "Treated Her" is not. The Clown has REPEATEDLY Stabbed My Girlfriend in the back for "Her Help". Only the very Corrupt-Agents use the Citizens, People around the Target, to such an Extreme and in this Heavy Manner. The Corrupt Agents will Misuse and Manipulate anyone to "Get to the Target", and in whatever Sick Way they can think of. The Feds in America did this with one of my girlfriends who also "Blindly Trusted" these Agents, Heavily Misused her to get to me. This "Type of Person", these Corrupt Agents, are Sick Lowlifes. In regards to my girlfriend here in Europe, it is a matter of help these people out, or sit in my shoes as well. I don't blame her, but it does make the relationship complicated and over-burdened. At times she has severely screwed up by going along with these people, who could care less about her. In the meantime, I and my Kid suffer, and very much due to this Technology.

"Things" still escalate. The price I and my Family pay for a Government Agency giving this Technology to a Psychopath and a bunch of White-Collared-Snobs. Someone totally not grounded in Reality to begin with, a total Manipulator, a total Power-Monger "Questing for More", totally Corrupt, capable of acting with Impunity, "Bubbled and Insulated" by Job-Title, and Out of Touch with Reality yet again because of this "Heady Mix"—which for many "Normal People" would be "Tempting", but for a Psychopath is an absolute "License to Insanity" and a "License to Violate". Again The Clown having "Power Issues", and "Abuse of Power Issues". This woman is "Blind" as a result of Her Mental Disorder, and TOTALLY OUT OF CONTROL as well. Thus many blunders and mistakes, very many due to "Showing Off", due to an Uncontrollable Ego and Self-Aggrandizement, and this is why I call Her The Clown.

Having said that due to Her Lack of Balance and Lack of "Good Judgment", never mind that She is actually a Psychopath, She is Dangerous. She already is Dangerous, but even "when she is careful", she is as well. I have had close calls, and my KID on several occasions as well. We are talking about a Total Inability to "Practice Good Judgment" regardless of "Consequence", and a Total Lack of Empathy. This is who I have at the end of this Electromagnetic-Technology. And Her Buffoons are obviously no different on a few points themselves. Her Buffoons are complete Suck-Ups, taking a paycheck from this Agency, while running around doing what they please, and not even capable of good decision making. These Idiots, are like people from the days of Nazi Germany, who stood smiling for the camera, yet "Did their Job" during the day of killing Jews. Brainless, yet Dangerous Suck-Ups, and Ass-Kissers, who are "Along for the Ride". In the meantime, this Exotic-Misuse of Techno, has Escalated even further, and did so in 2009, but that is for another Book I believe.

As I have neared completion of this Book, many more things have been done with this Techno, and I have learned even more. I can "Spot this Techno" being used on others as an example, and because it has been Misused on me and my Family so much. In fact I have seen them use this on a wide assortment of people, in a variety of ways—many from around our neighborhood. It has been this way for years now anyhow, but at this point I consider "My Education Complete" in regards to this Techno and my ability to "Spot It".

A Small List of the "Techniques and Tactics" with this Exotic-Techno with a Few Points and Observations as Well

Now that I have reached this joyful stage of discovery in regards to Exotic-Techno, allow me to share some observations and experiences. I think the easiest way to go about this is to number them and list them. I do this for people in my Exotic-Situation, and remember there are many more things not listed, and I have yet to discover them all. The most important thing to remember is that these "Capabilities" can be "Mixed and Matched", things like "Intensity Modified" as well. Think of a "Sound Mixer" in a Music Studio, or an expensive Stereo Graphic Equalizer, to get an idea of the possibilities. And keep in mind they can do "The Reverse" of any of these possibilities. They also do not need "Your Thoughts" to have an Identity-Marker, I believe that the "Markers", the Identity-Prints for many things are the same for all people. Keep in mind they have a Data-Base for many "Identity-Markers".

The most important thing here to remember is this: They can not see into your mind, what is not CONSCIOUSLY THERE. They can not, "Delve and Drive Around" inside of your Brain, pulling things out from twenty years ago for example. Things have to be CONSCIOUSLY IN YOUR MIND for them to "Access Memories" what have you. Even then they can not simply start "Seeing Things" instantly, they have to "WORK" at getting information from your mind. THEY CAN NOT SEE WHAT YOU DO NOT THINK ABOUT. THEY CAN NOT KNOW, WHAT THEY HAVE NOT SEEN—and even then, their "Perspective of what they See", can be very far off. Thus your biggest defense is not to dwell on things—having said that, once you do think of something, they have Tactics to try and get more, but if a Memory is really Buried, and you have to work at it, to remember it, they have no chance of getting a damn thing. So. As it is, they can "Dissect a Memory" and "Pick it Apart" based on "Different Identity Markers", and what "One Thing" means to one person, may have a "Different Significance" for another. Thus they have to get a "Feel for your Flavor" to "Dissect Things" and to make "Inroads into your Mind". Lets begin.

Number 1: The first thing is that your mind is a Radio for these people. They can "Tune In" to your Thoughts, and Transmit Thoughts to you—both Unspoken and Spoken. You are both a Radio and Receiver.

Number 2: They can Record Memories and Emotions. Like any radio with a cassette player, they can record and Play Back These Memories. In addition to this they can Play Back Emotional Responses and even Mental Images. The Rule of Thumb is they can Record ANYTHING in your Mind, and PLAY IT BACK. They can also take what they have Recorded and "MIX IT UP". As an example, the "joy you Felt when you got Married" which they have recorded from a Memory, can be "Played Back" while you remember an "Experience of something Disgusting", or while you "Eat a Pizza" as an example. They can "Mix" all things they can Record, and "Send it Back" all "Mixed Up", which leads to a "What the Hell is That?" moment. They can do this in terms of Strength of Radio Waves, Volume, INTENSITY, and "Mix It" like an "Equalizer" as well. When dealing with this kind of thing think of Common Technology to try and figure a comparison, in this case knowledge is power.

Number 3: When they Send Thoughts, there is a "Two Second Gap", between "Receiving a Thought" and "Being able to realize It". There is a "Two Second Gap" between an "Unspoken-Thought" and "Consciously Formulating a Thought"—or "Consciously Detecting a Thought". The Spooks are aware of what you are thinking before you "Speak Internally" to them, again by Two Seconds. This is not fool proof—and it is possible to "Throw Off" this Technology and The People observing it.

Number 4: They can "Pre-Send" a Memory as you react to something. For example: I wanted to turn the T.V. down because it was too loud for me and late at night. As I went to do so, they sent a "Memory" and "Thought" of me not wanting to be too loud for the NEIGHBORS. In other words they tried to substitute My Original Intentions, with Their Own, via a Recorded-Memory (my own) for the sake of Manipulation. They tried to SUBSTITUTE my wanting the music turned down, for a "FEAR" of the music being to loud. This was "Very-Quick" and almost "SUBLIMINAL", thus they are able to "Subliminally and Sublimely" affect your "Behavior or Final-Judgments". In this way, many times, they are able to get you to change your Behavior at the last moment in this manner.

NUMBER 5: Brain-Squeezing. This is a weird experience and can be totally inaccurate, as the Spooks around here have found out the hard way. For lack of better words they try to "Squeeze" an answer from your mind, about a particular subject you are thinking about, what have you. The thing is many times when you are "Squeezed" you can give very inaccurate answers, if not flat out cock-eyed ones. This has lead to serious Mis-Perceptions, and blunders, on the Spooks end. Basically they try to "Pull-Up" an answer which is in the "Background" of your mind, and near the surface of it.

NUMBER 6: This is also a big one. The Spooks can look at your conscious mind, and to a point your background mind or thoughts—but can not go deeper. They can not really go beyond the CONSCIOUS PART OF YOUR MIND, they can hear back ground thoughts though. For instance if they say the word "Elephant", your background thoughts show them a variety of words, thoughts, and feelings. They can not decide what it is that is "Significant or Accurate", on your end. They "Receive a whole bunch of Identity-Prints" which they then have to "Sort Through" and "Weed Out". They can decide based on a Psychological-Profile of you, but this Technology itself does not reveal all, or give all answers. It is NOT FOOL-PROOF. This leaves them blind. As a point you can look at a memory, but they can not access it, or understand it all. They NEED YOU TO DO THAT to analyze it for them. Don't do it. Keep in mind that with this Techno, they can "Understand Unspoken Thoughts", and can have someone "Sit Down and Experience" what you "Experience" to help them figure things out. The "More you Give" the quicker and easier it goes.

Many times when they are "Uncertain about Something", they will have someone you know "Ask about It" so they can get "More Information" and understand what "They are Seeing" better. By doing this they are able to "Label and Relabel" Identity-Prints based on your own character.

For instance if they see the "Image of a Woman" in your mind, they have to figure out what is "Significant" about this "Woman" to you. Based on a

Psychological-Profile they may say, well okay he likes "Blondes", and she is blond. Then they ask, did he know this person? For this there will actually be a "Signature/Identity Print" for a "Recognizable Face", a "Specific-Print" which automatically tells them "He knows this Person". But then who is it? This name they do not have as it is buried in your mind, etc. So say her name is "Liz"—then this name and image gets an Identity-Print and can be recognized next time. It is a complicated process but as time goes on, and they "Map your Various Brainwaves", and "Collect all of the Signature/Identity-Prints" for all of "Your Various Thoughts", it gets easier and easier for them. They build a Data-Base, a Collection of "Words and Tells" of what each Thought in your Mind Means—thus with time it is easier for them, and more complicated for you. As you are on the "Receiving End" it is more complicated anyhow...

Having said that these Spooks can actually have someone "Plug-Into" your recorded Memories and Feelings, and EXPERIENCE IT. Thus they can pick up on certain small details, "Certain Tastes" and "Flavors" from your Mind, certain significant points and details, but without YOUR HELP, without further explanations and CONNECTED MEMORIES, they can go no further. Again do not do it.

NUMBER 7: As an added point, as this involves Artificial-Intelligence and a Computer, all of these "New-Identity-Markers", Signature-Prints, get "Slotted" into a "File", and Agents update these within 24 hours, almost immediately as I have observed. If it is a "Big-Deal" and Relevant-to-the-Moment or a Particular-Unpredicted-Event, then they "Update" the "Identity-Marker" immediately and "Update Their Computer Program" immediately. Usually within an hour, small pieces sooner.

NUMBER 8: They can understand your "Unspoken Thoughts", in addition to understanding your "Worded Thoughts"—"Worded Thoughts" as an example being what we plan to say to someone, right before we do many times.

NUMBER 9: With or without audible voices in your head, they can and do bombard you with SUBLIMINALS. This is to affect your consciousness, and to try and bend your mind and will in whatever way they want. A lot of this is Automated-Response: the computer, responding to particular words, or brainwaves, and "Tweaking you slightly and Quickly" almost "Non-Detectable" and many times it is so, until you do a "Rewind" and take a good look. As time goes on, you get good at Identifying these Different

Things, Different-Tactics with Exotic-Techno, and so they mix it up, more and more.

As an example you may decide to do something, and just as you are about to, you get a "Quick impulse of Fear", almost subliminal, so that you decide not to do it, and right before you are about to engage in that activity. Thus you are "Subliminally Affected".

NUMBER 10: These Agents/Spooks have the capability to "Communicate with you Live", as in "Speak to you Directly". It is as if your Brain is a "Telephone" for these people. When dealing with a "Target" they can use "Pre-Recorded-Messages", but given the amount of Arrogance of some Spooks like the ones I deal with now, and their "Inability to Follow Rules and Protocols", I would not doubt if this has happened and been done at times for "Cheap-Psychological-Thrills" on their end. As I have noted and experienced, I have had "Phrases and Words" "Pop into my Mind"—no doubt on occasion, "Live". Again "PHRASES MAY POP INTO YOUR MIND" which would be these Idiots with Techno. They can, and usually do, "Flavor This" with "Your own Voice", an "Unspoken Thought" with an "Internal Spoken Thought" done in "Your Voice" to be "Convincing". As in you "Thought out Loud" to yourself.

NUMBER 11: Because of "Number 10", I know they have "Voice Changing Technology", "Voice Mimicking Technology". Remember the "Terminator Movie" where the Terminator is on the phone, and "Mimics a Voice" of someone else? They can do this. Again I have "Thought out Loud" in my mind, in "My Voice" but have recognized it is NOT MINE, many times.

NUMBER 12: They like to constantly harp on certain Themes, to keep you focused on these Themes. Sex and women is one very popular one with The Clown, who in Typical-Psycho-Clown-Fashion, put on a Sex-Show for me with her girlfriend (A Total Professional Obviously—I wonder if her Agency Trained Her in This). I declined. The Clown being a Corrupt-Psychopath loves "Demented-Themes" and this Techno is "Used to keep you Focused" on certain Themes, as they "Send Unspoken Thoughts", "Stimulate your Harmones", "Send Imagery", what have you—many times at "Inappropriate Moments" to "Try and get you to Modify your Behavior", but this is also a "Tell" it is not your own. The Corrupt-Agents are pigs, and love the sex stuff, and yes they are "There with You" when you masturbate or have sex. Yes it is disgusting and creepy, we are not talking about ordinary people here, not the ones I deal with today—The

Clown and The Buffoons that is. You can take my example of "Sex Stuff" and "Look at other Themes" as well.

They may also harp on themes like Work, Sleep, Illness, Fear, whatever is "Relevant in your Life" or what they want to try and "Make Relevant in Your Life", to try and Distract You, Disturb You, Scare You, Intimidate You, Stress You, or Bother You…Many times in these situations they push in an "Accompanying Feeling" to go with it, to convince you it is yours. As an example the woman theme: They may constantly bring up a "Particular Woman" and then make you very horny (Yes they can do this with this Techno) to convince you that you are attracted to that person. Obviously "Knowing Yourself" is an important thing for dealing with this Techno.

Number 13: The "Inside Out Tactic". While you are busy or distracted, they may try to send you an "Unspoken Thought" to convince you it is your own thought. I call these "Unspoken Thoughts" "Inside-Out" these days, and the "Phrases or Words" which pop into my head, "Outside-In". This is what initially tipped me off, years before I knew of this Techno, that something unusual was going on, that something "External" was influencing me. The best defense for this is to Know Yourself. Know your Core-Values, know who you are and you are less likely to get buried in an avalanche of their B.S. This, your Core-Values, your Deeper-Consciousness they can not see, they perceive your "Surface Mind". Profiling is used in conjunction with this to see your "Deeper Values" and Core-Beliefs.

Number 14: They can program their Computer, to respond to a Memory. I call these "Memory Bombs"—they are waiting to happen. At any given point, as you experience a Memory brought up by a question or answer (by a friend as one example) they can "Brain-Squeeze You" to try and get more information out of you, "To Pump You".

Basically they can mix and match every single Tactic I have given above, to make things more tricky to figure out. I have had days where I have had to deal with all, all day. As an added note, the example I give is "Specific to Memories". Actually they can do this with "Memories, Mental-Images, Specific-Thoughts, even Specific-Activities and these days I call this "Thought-Mining"—like in a war a "Harbor gets 'Mined'".

Number 15: A BIGGY. And not just because they used this on me last week (it is Nov.2 '09) so I would forget the "End-Date" and "Final-Test", as if this is even possible for someone in my shoes. THEY CAN MAKE YOU FORGET THINGS, THEY CAN BLOCK MEMORIES AND LITERALLY SUCK MEMORIES

FROM YOUR MIND. They once did this when I came up with a particularly good insult for The Clown. They literally began sucking the memory of it away, I could feel this, somehow I knew, so I kept repeating it in my mind. When I stopped repeating it in my mind, my Surface-Mind, the memory of the insult totally vanished. To "Re-Remember Something" you have to get there again all-naturally, and the fact of the matter is you won't know if you have come up with the same thing or not. So I learned an important thing here, that is that your "Surface-Mind" and Deeper-Memory, or access to these are two "Separate-Components". Thus the memory of the insult (and well earned) stayed in my mind as long as I repeated it, but once I stopped it vanished—thus the "Deeper-Background Memory" got erased and once I let go of the Surface-Memory, it ALL VANISHED. Thus a PERSONS MIND AND BRAIN CAN BE "WIPED CLEAN".

NUMBER 16: They can "Rewind your Thinking", to distract you and make you forget. For instance they can take a realization you have, and push your mind back to where it was a few moments before. Then if they "Don't let You" (in a variety of ways), you will not be able to access that particular memory, that realization from moments before. Thus you are incapable of "Independent Thought". Your ability "To Think" and "Function" becomes Debilitated.

NUMBER 17: They can replay Memories and Conversations. "Music" as well. Thus "Music can Replay" through your mind as an example. This happens to everyone but is a lot more "Common and Relevant" to one's life with this Techno around. Many times that "Music going through your Head" has "Themes" relevant to what they are "Doing in your Life".

NUMBER 18: "SCRIPTED DIALOGUE". They can "Over-Ride" your "Internal-Spoken-Thought" with their own words or vocabulary. If you are in my shoes you know "They can Hear You" and so you "Speak Aloud" in your mind to them. THEY CAN REMOTE-CONTROL YOUR WORDS. Internally or Externally. Keep in mind that a lot of "Dialogue is Scripted"—as in Pre-Written by Spooks. This includes "Dialogue, Responses On Your End" which you think are yours, but ARE NOT. This is actually difficult to explain.

In many a "Dialogue" the Spooks can "Predict your Response" to certain comments or questions. Now in order for this Computer to Respond, it has to RECOGNIZE YOUR RESPONSE. This may be Artificial Intelligence but it is dumb as nails—It does not actually understand anything. IT RECOGNIZES BRAINWAVES BUT CAN NOT "CONCEPTUALIZE" CAN NOT COGITATE, CAN NOT THINK ON IT'S OWN, and so merely gives Prepared-Responses, and Multi-Tasks other things as well. So if Spooks are going to "Dialogue With You" through someone, a "Pre-Recorded-Script" , this Machine has to Recognize your Brainwaves AND ANSWER/RESPOND.

If you do not have a "Predictable-Response" or better yet a "Predictable-Brainwave" then these Spooks can not carry out an "Interrogation or Profile You" in regards to certain things. So in order to do this, many times "WORDS IN YOUR VOICE", and based on "Guessing You, Profiling You" are "Put into your Mind" and "Mental-Mouth", or "Physical-Mouth", and you are made to feel these are your "Own Words and Thoughts". This is difficult, hang on.

IT IS AS IF THEY SEND YOU "YOUR OWN RECORDED WORDS AND FEELINGS WITH THESE WORDS" from before, as a "Response on Your End" so that it is "Perfectly Matched" to you, and FEELS AS IF IT IS YOU. You feel that you are indeed many times "Speaking Back" but in fact it is a "Perfectly Machined Response" on your end—in fact this Techno/Machine is "Speaking For You", has "Dialogue for Your End Already Picked Out" and so when a "Brainwave is Detected" is able to provide an entire sentence, as an example, which feels like yours. You "Feel like You are Speaking-Internally OR Externally" when you are not—or as a point, MANY TIMES THIS IS WORDING SIMPLY SELECTED BY YOU AND THIS MACHINE TOGETHER.

For instance I might Respond with "Piss-Off" in my mind to these Spooks in regards to a "Particular Event", which this Machine also has selected in regards to this "Particular Brainwave". However if these "Spooks Predicted my saying this Aloud in my Mind to Them", to a Particular-Event which they have "Staged", and decide that they want to "Screw-around-With-My-Head", then this Machine/Techno, as soon as it "Senses this Thought" might "Push In", in my Words and with my Feeling, "Sip-Off". And of course when I experienced this, I would have a "What the Hell?" moment. And then when "Looking" would find that "Piss-Off" had been substituted with a "Reverse-Spelling" just for the "Head Game" of it. And there are several variations of this.

So during a "Dialogue" in which you "Speak-Out Loud" in your mind with this Voice (Doing so takes TWO SECONDS, to Formulate Words—thus your "Reaction and Words are Predictable" by a TWO SECOND MARGINE) "Your Internal Words" are actually "Pre-Owned" by this Machine, and by a matter of TWO SECONDS. Thus it senses your "Thoughts" and "Provides YOUR words with YOUR Feelings" thus convincing you, that YOU are speaking "Aloud to these Spooks" in your mind. Or to another person you are having a conversation with, for that matter.

During this "Dialogue" while "YOU ARE RESPONDING", they can switch YOUR WORDS, because it has already been "Pre-Recorded", and "Predicted" because this "Event Itself" in the real world was already Planned. This was a very effective "Mind-Bender" for some time—"What the hell was that?", was my response many a time. So in your Mind during an "Internal Conversation" they can "Switch Your Wording" and screw around with your head based on "Predictable Responses" to "Stimuli". As another example, I TRY TO SAY to them in my mind, "You should be so Lucky", but since they have a Two-Second-Edge they can switch it quickly to, "I should be so lucky" (and this example is a real one). THIS LITERALLY APPLIES TO "EXTERNAL CONVERSATIONS" AS WELL, where you actually use your mouth in the real world for speaking to Family, friends, what have you. Your entire conversation with someone can be "Scripted" as long as they know what the other person will say. Around my flat, to do this, they "Script the Dialogue" for other Family Members as well.

In other words they CAN LITERALLY PICK THE WORDS YOU WILL SPEAK, AND EVEN HOW YOU WILL SPEAK THEM. In this case this Involves areas of the Brain, involving Motor-Control as well, the "Moving of your Mouth" in addition to the selection of words you will use, and the Feeling in which you convey with them. They "Turn down your Will-Power" to speak what you want to say, and then "Turn-Up their Will-Power" and "Insert" what they want you to say.

It is weird to experience, and weird to see this done with others. YOUR MOUTH AND WHAT YOU SAY CAN LITERALLY BE REMOTE-CONTROLLED. Around my flat I "Witness This" all of the time with my Family, Daily in Fact, and these days in addition to this, many times I experience some "Internal-Voice-Commentary" just to let me know that what came out of someone else's mouth was indeed "Scripted and Remote-Controlled". For instance my girlfriend might say in regards to characters in a movie, "Look at what they did.". Now unknown to her she got "Impulsed" to say this, it was a "Scripted Event", her saying this, the "Words and her saying Them" were chosen—she thinks SHE chose them. In my own mind, immediately after she says this, I receive, "Look at what WE just Did." just to let me know that she was "Machined" into saying this, "Scripted" into saying this, REMOTE-CONTROLLED INTO SAYING THIS. This "Scripted Dialogue" was already prepared, as was the "Selection of the Movie" by these Spooks. Thus my "Girlfriend is Scripted" into choosing the movie, and then with that in mind, these Spooks have this all prepared. My Girlfriend is already "Scripted" to say this, and these Spooks also have a "Recorded Phrase" prepared in which it responds to the "Stimulus of what my Girlfriend Says", or they simply choose that we are "Scripted in a Time Sequence" to do such. There are always many possibilities and choices for these Spooks.

So Dialogue, "INTERNAL OR EXTERNAL DIALOGUE", CAN BE REMOTE-CONTROLLED.

NUMBER 19: They send "Different Messages" (this can be "Images, Memories", other things as well) to then see which ones you react to, to Measure you Psychologically, TO PROFILE YOU. Many times to see what kind of Mood you are in, to test the "Psychological-Waters" of your Mind. Or for other purposes like to find something to scare you with, or things which are funny for you (which they will Misuse later) as an example. The THEMES VARY though and do not necessarily have to involve fear, although whatever the "THEME IS", this is a "Tell" as to something they may have Planned on down the road in your life. It is a "Tell" that they are "Trying to find something Out", or "Broadcasting Future Moves".

NUMBER 20: The Spooks have "Routines and Scripts" prepared for certain subjects. It is all used for Profiling—jokes as well, as they analyze your Humor. They will see "What Scripted Wording and Dialogue" you choose. The subjects, what you have to say (about someone as an example), and to get an idea of your Mood, Feelings, and even Information about any particular subject. For instance I may "Joke about a particular Person", well they will "Look for Tells" in my jokes, to find out what I feel about that person. Harsh Jokes obviously being an "Indicator of not liking Someone" as an example.

NUMBER 21: They can Auto-Pilot-Behavior, such as get you to look at a particular window every time you go by somewhere, look at a clock the same time every day, or what have you. They can also Auto-Pilot your Vocabulary so that you use certain words they have picked, over and over. In essence your "Behavior and Vocabulary" becomes SPIKED WITH "SPOOK CHOICES". This goes on DAILY.

NUMBER 22: THIS IS ANOTHER BIGGY. If you have ever meditated and tried to still your mind, you eventually hit a point of a stream of "Mind-Garbage". It is a "Stream of Audio and Visual Input", a "Subconscious-Stream" of Audio and Visual information and thought. This is what exists underneath, just underneath our Surface-Mind, our Conscious-Mind. People who experience Dementia are experiencing THIS SUBCONSCIOUS-STREAM BEING "PULLED UP OR RAISED INTO THEIR CONSCIOUS MIND". This IS the "Explanation for Dementia". I have had to wear this Bullshit, thanks to my Demented-Spooks, and it is a pain in the ass.

I call this the "Subconscious-Sub-Stream", or "Subconscious-Stream", it is where our Subconscious-Mind, and Conscious-Mind MEET. When you "Quiet your Mind", and go deeper, this is what you find. This is what the Spooks are able to see as well to a point, and is as far as the Spooks can see. It also makes what they see confusing for them, is a blind spot. What is running through your "Sub-Stream" is a list of possibilities, and not even all of them, what you choose is another matter. For this your own thinking and analyzing is required to get to the bottom of it. As I used the example above, in number 6, with the word elephant.

A WARNING HERE: They can also take this "Sub-Stream" and bring it to your Conscious-Mind, while at the same time "Spiking It" with their own thoughts and words, I had a "Taste" of this, and it is absolutely scary and crazy. This Techno can be used to make a person absolutely crazy, and I have been pushed to the absolute edge of insanity, more times than I care to remember. One night in 2009, I thought I was going to have to "Check myself in at a Hospital", and this is after I laid down to sleep.

NUMBER 23: WITH THIS TECHNOLOGY THEY CAN MIMIC MENTAL PROBLEMS, AND PHYSICAL AILMENTS. All Physical and Mental-Ailments can be duplicated with this Techno as they all have an Identity-Print—never mind that these can be "Recorded from someone Else" and PLAYED TO YOU, be "Inserted into You".

NUMBER 24: This is not a "Beam Weapon", like a laser, it is not a directional pointed thing like a gun—this is done with Astral-Markers, or "Tuning In" to an Astral-Thumb-Print and then ENERGY TRAVELS TO IT LIKE A TELEPHONE, once you dial a number, your call goes to that number. Remember that the space between the Earth's Upper Atmosphere, and the ground itself, is one big Electromagnetic-Medium. Thus you can be "Reached Anywhere" by them "Dialing your Energy-Print".

NUMBER 25: They can affect all Emotional-States. An extremely Common Tactic is to arrange a "Situation", an "Event", and then "Machine You" an over-the-top Emotional Response. Sometimes they will precede this with a "Sent Thought", to convince you the "Feeling" is your own. In reality, neither the "Thought" or "Feeling" is yours. In essence you have to know yourself. The "Sent Thought" is actually a "Cover-Story" to make THE FOLLOWING FEELING "CONVINCING". This is how I caught The Clown with her Technology.

If it is an Emotion they can "Send It", or "Stimulate that part of your Brain" responsible for feeling it, to create it—and again "Intensity is a

Variable" as well. They can "Intensify Feelings" way beyond any "Normal Human Level" thus, you can be "Manipulated into Extreme Behavior". The example I gave earlier of this young man being "Remoted into Stabbing his Girlfriend" repeatedly with a knife is a very good example of this. As an added note for my readers, luckily she survived. They were BOTH VICTIMS, and without these Spooks the situation would not have happened. In this way Spooks, Government are able to KILL PEOPLE without being caught. Someone else wears the blame, it "LOOKS NATURAL". Obviously it is not. Having said all of that, the real Criminals are these Spooks.

NUMBER 26: This one is for people who are "Spiritually Inclined", and was given to me by the Spooks and in part due to my observations. It applies to one of the topics above. When you do something called Japa, the repetition of one of Gods names, or a Mantra, it slows down your "Sub-Stream" putting it, bringing it, to a SINGLE POINT. It stops the "Sub-Stream" and all of it's "Various Content" and puts the focus on God, in whatever form you believe. You can also pick a Positive-Word like "LOVE" which I recommend if you are not "Religious or Spiritual". By doing this the "Sub-Stream" becomes quiet, you become clearer, operating from a Deeper-Level, and you pull in the "Energy from Above" as certain words do indeed "Hold an Energy-Charge". A good example of words holding an Energy-Charge is a scientist who found that by holding water, and then focusing on a particular word, we "Affect the Shape of the Water-Molecules" what it looks like under a microscope. Keep in mind that we are, as Humans, made up of mostly water...At any rate this is the Science behind Japa and Mantras, and as a point, as I said before this Technology enables the Government to understand many things that are "Esoteric", "Spiritual", "Religious", what have you. They can see the "Science" behind a lot of beliefs and practices, the Physics, how these things operate on a mental, and even meta-physical level, in terms of Energy.

NUMBER 27: While "Communicating with You" they can send "Background Thoughts", Unspoken Thoughts and Feelings, to convince you that they are your own; they like to do this especially if you are distracted.

NUMBER 28: I add this as this is going on now while I type, with my two parakeets (they killed one in December of 2009, right before Christmas with this Techno). They can and do affect the Behavior of Animals as well.

NUMBER 29: The RANGE and DISTANCE of this Techno. It has the ability to affect people over a very large distance. I noted these Spooks deliberately

"Shaping Clouds" for "my benefit" to show me what they can do, and to "Show the Range". With this Techno they are obviously capable of affecting Physical Reality. In these instances they made it quite clear it was they who were responsible for these "Odd Clouds". Another example, perhaps more believable for my readers since people "Tend to see what they want in Clouds", is that these Spooks affected someone many miles away at a radio station—while they were on the air live. Due to their "Manipulating Clouds", because I was contemplating "The Distance and Range", I can safely say that this Techno has a range of many, many, kilometers and miles. The person at the Radio-Station was more than twenty miles away. They might have "Asked this Woman" to do this for them, but I doubt it. The last things these Spooks do is "Practice Good Judgment", and The Clown and her Buffoons both lack any degree of Self-Control, especially with this Techno. Perhaps they are simply enjoying one last time, before they all get in Trouble, and from this Book. As it is they have "Affected many people around Here" with this Technology, so it is a safe guess. There is no doubt that the range of this can be measured in miles upon miles.

Number 30: They can "Script" MANY PEOPLE at one time. This means that "Dialogue, Emotions, Reactions, Behavior, and how everyone INTERACTS can all be ACCORDING TO A "SCRIPT" and with this Techno. My girlfriend, my kid, and I, we have all three been "Scripted" like actors in a play, to interact and go about our day, in a way that has been "Chosen For Us". There are days when we follow someone else's idea of what and how we do things, what we feel, how we behave, what we say, for the ENTIRE DAY. It is possible to go "Off Script" but they usually have a "Fence", a "Boundary", a "Barrier Up", which makes this difficult.

Number 31: DREAMS AND DREAM MANIPULATION as I have already covered earlier. A clear sign of this, is Dreams that do not make sense, Traumatic Themes, Dreams "Which come out of Nowhere" in terms of "Theme and Content", Sleep-Paralysis, and as usual anything which seems "Off, Unusual, Weird". With this Techno, this is always a "Tell".

Number 32: THEY CAN BLOCK ALL FORMS OF FEELING. This can be very weird and disturbing in regards to both. For instance you should be "Angry or Happy" in response to something, but you are not—you are ROBOTIC. You become capable of observing "HOW YOU SHOULD FEEL", WITHOUT ACTUALLY BEING ABLE TO DO SO. This is the same as, the Reverse-Of, "Amplified Feelings". In all cases as far as I know, THE REVERSE OF ANYTHING CAN BE DONE AS WELL.

NUMBER 33: THEY CAN "BLOCK PARTICULAR THOUGHTS", DO THOUGHT-BLOCKING. They have an Identity-Print for "Each Thought", thus they can "Block Thoughts" simply based on this. As a result you do not remember such very simple basic things, and your "Personality can be Reshaped and Remolded" based on this alone. The thought which would normally be "Re-Occurring" and staying within your Focus, simply DOES NOT COME UP AT ALL. As a perfect example of "Thought Blocking" there have been quite a few times when I knew, had figured out the "End-Date", yet on the Last Day, forgot entirely about this "Slight Detail" which left my head "reeling" the next day.

NUMBER 34: This Technology involves an Intensity-Setting, a Volume-Setting, a Speed-Setting. Maybe other points as well. Look to current modern techno to get an idea.

NUMBER 35: This Techno can Interfere and Manipulate, Emotional States, Motor-Control, Physical States, and Mental States. As an example of "Mental-States" when I go to sleep, right before I "Fall Asleep"—right at that point, they can "Block-Me", so that I can not go any further, and therefore can not go "Into Sleep". They can literally keep you at that "Mental State", Block-It, so that you can not go any further. This is currently being used on me again to deprive me of sleep and slow my book down, and has been used repeatedly on my kid as well.

NUMBER 36: EXTENDED-FEELINGS. This is another Big-One. They can take ANY FEELING, play it to you, and Leave It On: anger, tightness, sadness, you name it. As a very good example, know that "Feeling you have of it's Friday?". They can play this to you on Tuesday as an example, the whole day, and you have to "Keep remembering the real Day", or eventually you get "Tricked into Believing" that it is Friday on a Tuesday. Of course this depends on the "INTENSITY". If they do it a "Little" you will find yourself checking the Day in your mind on occasion, if the Intensity of this is "High", by the end of the day, at some point you will have been "Tricked into thinking it is Friday".

They can "Single out and Record" ANY TYPE OF FEELING, and "Play It", and "Extend It", leave this "Feeling On" for days, weeks, months, years. You stop being "You" and "Turn into someone Else".

NUMBER 37: They do "PERSONALITY-REDESIGNING". They can REDESIGN YOUR PERSONALITY. They can "Script" your entire day with, "Thoughts, Feelings, Words, Reactions", ALL COMBINED, ALL DAY, ALL WEEK, ALL MONTH, ALL YEAR.

This is what I have had to "Endure" for some time now. As if you are an "Actor in a Play": thus "Scripting".

By doing so they have "Permanently Changed Things" about me, for years on end, and in other cases, months and weeks on end. A thing that you may notice once in a while, is the "Removal" of "Something they have Changed" with this Technology. They are "Assessing" HOW CONDITIONED you have become to your "Altered Behavior and Habits", to see if you have "Been Changed" by their constant Misuse of Technology, and looking to see if you still engage in these once they remove the Techno. They are seeing how Conditioned and Altered your Personality has become. Thus you are left wondering what is going on, "What the Hell?", and then at some point they resume it.

NUMBER 38: It is the "Ultimate Spy-Machine" where the Target is the actual Audio and Video Component. You are both a MICROPHONE AND CAMERA for these Agents. When they "Wiretap your Brain" they can perceive ALL STIMULI through you, that YOU EXPERIENCE. Thus you become a Component of Spy-Techno.

NUMBER 39: They can Literally Interact with the Physical-World, with this Technology. They can interact with the Physical-Environment the way a "Ghost Might or Can". As an example a few years ago I was writing about this very subject, and to make a point, they knocked down a bottle in the kitchen while I was doing so. "Thought is Electromagnetic-Energy", "Power of the Mind", which is how "Yogis and Others" are able to "Manipulate Matter", able to do "Mind over Matter"—well this Technology deals with Electromagnetic-Energy, it IS "Mind over Matter", "Power of the Mind". Thus "The Physical World", like Neo in the movie "Matrix", can be "Manipulated and Interacted With". I have experienced this very many times.

NUMBER 40: I have also noted that this Machine allows these Jackasses to "Astrally Travel". As in somehow these Spooks INTEGRATE THEMSELVES with this Technology and are able to "Leave Their Bodies". Thus going where they like and able to enter people's homes undetected. I offer two examples, one of when my kid was two years old, and another of when visiting a friend who owns a dog.

When my kid was two years old my kid obviously could "See someone in our living Room" and someone who was none to nice. Obviously this "Person" was nasty and "Making some kind of nasty Weirdness" as my kid began crying—and obviously because of "Someone in the Room". I spoke out loud telling them to get the hell out, which judging from the response of my kid, they did. No doubt my "Response" shocked whichever Spook had entered our house. Keep in mind that at that time I had no idea of this Techno, it was just so blatantly obvious that "Someone was There". Now I know why.

Another example is from a year or two ago when visiting a woman who lived nearby and well after these Spooks revealed their Techno. Her dog noticed a "Presence" as it "Flew into the living Room", "Landed Nearby", and then two or three minutes later "Flew out the Window". In this case the dog was "Captivated and Intrigued" and her head "Followed and Trailed" this "Invisible Presence" which left, perhaps because I was "Noting Them". So again "Astral-Travel" and as you can see the Government is well aware of such "Esoteric/ New Age Beliefs". Keep in mind that these Spooks are "Seeing and Listening" to my mind constantly (and can see out of my eyes literally), so this no doubt was "Merely a Demonstration" of what they can do. As they were already in this woman's house in one sense—keep in mind that I am "The Surveillance Equipment". It is the "Ultimate Spy Machine".

NUMBER 41: Now to the Basics. One is that there are several Agencies that have used this on me. I started in the States with this Techno, then overseas to India where I did not detect this (perhaps they did not trust the C.B.I., India's Federal Police), then to Europe where I did not know of it the first two years until The Clown took charge. Obviously this is used by Multiple Government Agencies, and used on "Citizens". Seeing as how I am not a "Terrorist or Domestic Threat" or anything else for that matter (I am trying to raise a Family), we can surmise that this Techno is not simply "Reserved for Important Situations and Circumstances" involving "National Security", rather it is used more "Commonly" than that, and for more "Mundane Purposes and Situations".

It is more "Spread Out" in terms of its uses, and is used on "Common Citizens", and anyone like myself who 20 years ago knew of two Cover-Ups by the "Authorities".

Because of The Clown we can surmise that Corrupt and Dangerous Agents have this Techno. As if Corrupt-Government Agents with all of their "Power and Authority" were not bad enough, now Government is going to start handing them Exotic-Technology as well. Having said that, we can deduct that the Government is not CAREFUL in it's "Screening and Assessments" of who it gives this Technology to.

In order for this to be happening, it must be far more "Common Place" than is comfortable to even think about. These Agencies are Stupid to say the least, as This Clown would have NO PROBLEM, using this Technology to find out what Her Superiors are up to, and specifically in regards to Her. To top it off, I have noticed that The Clown is "Interacting with People" in this neighborhood, and elsewhere, with this Techno, so it seems to be used very commonly, and for "Monitoring many People". Never mind "Interacting and Influencing" them. Having said that, the Extent to which it has been used on people around here, no doubt is a "Choice of The Clown's". She has also Misused this Techno on some of the Field-Agents of this country, and on people at the American Embassy when I contacted them, on two occasions (Never Mind I was Affected and Scripted during my Call). The Clown Misused this Techno for Spying and finding out "How to better Lie" to these people among other things.

Obviously the Agents who use this Technology, the Corrupt ones at least, do not have to "File Reports" on how they use this Techno as the Feds in America used this for CREATING False-Pretenses, Framed Me with it, used this Techno for CREATING a reason for following me over seas, and a False One at that. If no one has to "File a Report" for when this Dangerous Techno is used then I find this irresponsible to say the least, as the last time I checked (admittedly many years ago) even a police officer has to file a report if he even uses his or her baton, never mind Exotic-Techno—talk about ridiculous. Or of course THE REAL REASON: these Corrupt-Officers are simply Lying in their Reports. I do not doubt this for a moment as it is BECAUSE of Corrupt-Agents, that this ALL STARTED, and because of Corrupt-Agents that this continues on, and that I have written this Book. You would think I released all the Space-Aliens from Area 51.

NUMBER 42: ARTIFICIAL-INTELLIGENCE. A COMPUTER is being used, certain things are obviously Programmed and Automated. These Spooks have constantly been referring to ARTIFICIAL INTELLIGENCE for the last four years now, and they are either referring to The Clown and Her "Ego and Her Need to Look Good for Others" or they are referring to this Techno. Having said that based on observation this Techno is obviously able to "Juggle things on it's Own". There are what can be only described as "Automated Responses", which are "REACTIVE" to particular Activities and Brainwaves. This Techno, "Makes Decisions" based on Computer Programs developed by Spooks, which are "Reactive to particular Thoughts, Feelings, Actions". This Artificial Intelligence clearly CAN NOT THINK ON IT'S OWN, it is "Reactive and Mechanical" in nature.

NUMBER 43: This Techno can Interfere with ANYTHING ELECTRICAL. Anything with an Electric-Field can be Manipulated with this Techno. They can Interact with your computer, a house fan,your flashlight, your Hand Phone, your Nintendo, the water heater, VCR, you name it. They can "Manipulate These" in a variety of ways. As an example they have "Drained the Batteries" of my flashlight before, and did so "Instantly". On other occasions they have "Kept Batteries Charged", at one point I used my Radio for an extremely long time, all of the time, and the batteries never died. This was not a matter of good batteries, this was beyond "possible". They have slowed our fan down before—it running at speed not found on the "Settings". They constantly interfere with our computer, no doubt this is done with this Techno as well—thus there is no way to "Trace This".

NUMBER 44: ONE LAST POINT. Keep in mind that if you are in a Situation such as my own, all of this can be going on without you knowing it. Your Sub-Conscious at this point, without "Voices" can be "Spiked" with phrases, words, messages, "Feelings not your Own", what have you. Unusual Behavior, Fantasies, Emotional-States, Word-Use, what have you, that do not fit your own character and which you have a hard time explaining, or understanding, are very clear indicators of this. A STATE OF CONFUSION AND WONDERING, SHOCK, ARE VERY CLEAR INDICATORS OF THIS. Things which are "Weird and Unusual" are usually "Tells and Indicators" of this Technology being Misused in one's life—as someone else is "Choosing Things" for you. And again, they can and do, usually mix it all up, using a wide range of Techniques and Tactics. For this Techno, there are Tactics.

THE END OF THIS CHAPTER
THE REASON THIS EXOTIC-TECHNOLOGY IS KEPT "HIDDEN AND SECRET": IT IS OBVIOUS ISN'T IT?

This Techno is hidden from the Public and the Public-Sector, so that there is a vast difference in what the Public knows and what really exists. This is so that stories of this Techno, information that is released, remains unbelievable. So that when people who this Techno is Misused on, Speak-Out about it, it is not "Believed". It is a Tactic for creating Disbelief, and is a form of "Insulation and Protection" to keep this Techno Hidden and Secret. It is a way of placing a Buffer-Zone, between the Reality of this Techno, and the Public Eyes and Ears, so that it can continue to be Misused on the Public and Citizens. So that Governments can continue to get away with Using this. It is ALL ABOUT POWER, and Power Over the Masses.

The other reason is that what the Government has, this Techno and knowledge of it, would create Fear and a Public-Backlash because of the "Possible Misuses". Go figure. Speaking from experience for me and my

Family, I would say for good reason, they already are Misusing this, and Lying about how. Total-Criminals are using this Techno, their "Having a Badge", having nothing to do with Reality—you would think, given the amount of Power this Techno gives someone, that Agencies would do a better job of watching their people using it. Having said that this is not the case. This does not change anything about the Reality I, or my Family, deal with though.

Keep in mind that other Governments already know of this Technology, it is only the Public of many Lands, the Masses, who do not know. I have lived in Two Countries where this has been used on me (India is Uncertain to me), it is no secret around here, therefore Multiple Governments do know of this Technology, and have agreed to "Keep Quiet". Therefore this is only Hidden from the Masses, and the only purpose for this, is for "Misusing this on the Masses". Generally speaking, things like this are "Kept Quiet" to give an Advantage, and in this case, an advantage to Governments.

This ends my chapter on Exotic-Technology, I hope it does a lot of people being affected with this, a lot of good, and I hope this INCREASES PUBLIC AWARENESS. As an added note, people within Governments NEED TO START COMING FORWARD, and Speaking-Out about this Electromagnetic-Technology, as it is a threat to everyone. This Technology has the Power to Change the World, and it is being Used and Misused this way, but in the Wrong Direction and Manner. This should be shared with all, used for all, and used for GOOD, NOT EVIL.

CHAPTER 21

Normal Climate Change and "Elite Climate Change": Exotic-Technologies for Manipulating Weather, Prophecies, and the Masses: "Guaranteed Bad Weather" Down the Road Time-Wise

The situation involving Climate-Change is a mixed subject. On the one hand there are those who say there is no Climate-Change, on the other hand there are those who say there is Climate-Change. There are also those who say any "Climate-Change" is normal, a cycle, a natural pattern, and there are those who say it is due to Mankind and pollution. I think it is a mix, and beneath it all is one more layer, the Elite's Layer, involving a Plan to use, Misuse, Exotic-Weather-Control-Technologies and Exotic-Earthquake-Technologies. This is why such a loud and prevalent "Sales-Pitch" for Climate-Change is being made in the Media, it is something the Elite want. Something they are burning into the Masses minds. It is so when they do this, the Masses, think it is "Normal", not Elite-Made.

The Elite have PLANS for "Destructive Weather", and a "Change in Natural-Weather-Cycles", that has nothing to do with the "Normal-Weather", "Normal-Weather-Cycles", and "Normal-Climate-Change". Thus they know, no matter what you believe, on either side of the "Climate-Change" fence, there will be much shocking weather, volcanic activity (which will affect Weather and Climate), and an outright breakdown in the World's Weather and Climate. All of it will be, as usual when dealing with the Elite, for multiple reasons. In large part, if not the whole reason, this is why these Exotic-Weather-Technologies have been hidden, Exotic-Earthquake-Technology as well. These things have been hidden for Manipulation of the Masses, and for the gains of the Elite.

Now if we are going to speak of Climate-Change then we need to know a bit about it. I in no way go into detail here, I feel Ann Walker who has written "The Messenger The Messiah" has done it the best, however I do cover a few issues that I believe need to be more known and commonly accepted. I also think it is important that people really understand the dynamics of "Weather Change", of Climate-Change so that people, the Masses will be able to recognize the differences between what is Real, and what is not Real: ie. what is "Earth-Made" and what is "Elite-Made".

If you don't like an Elite reference then think of it as the Military, as no doubt they are the ones who have developed these Technologies and who are using these Technologies on behest of others, primarily the Elite. This is the price we pay for Secret-Projects, disappearing money for funding (black budget projects), Civilian Companies working privately for the Military, and Corrupt-Politicians. Governments across the World have developed Technologies that are far more advanced than most people realize.

I should know, I and my Family deal with some of this hard to believe Exotic-Technology every day. From a Spook's point of view there is nothing hard to believe about this stuff at all, they use it every day—it is simply Technology. It is "Hard to Believe" simply because Knowledge of it, and the Science and Physics behind it, are Hidden. It is no wonder they "Hide It" as they do not want anyone else developing this, and certainly not Privately or in the Public-Sector. They do not want anyone Developing their own Exotic-Technology Privately based on the Science and Physics involved, and they do not want anyone else having that amount of Control or Power. They want it Hidden so that THEY have this Control—and so it is Hidden. Which makes the every day person, the average citizen, very far behind in terms of being Informed and Knowledgeable. It is because of this that these Exotic-Technologies are Hidden for a reason and kept quiet for a reason: for the purpose of getting away with Misusing these Technologies. YOU CAN NOT FIGHT, NOR SPEAK OUT ABOUT WHAT YOU DO NOT SEE, AND DO NOT KNOW: and this Exotic-Techno operates "INVISIBLY". It is because of this that this

Exotic-Technology is so DANGEROUS. It is in the hands of people who Misuse it. But I digress, back to Climate-Change.

A good way to look at this is in terms of what we know, things from everyday life. My examples are perhaps not the best, and I have to bend Reality a bit here, but I think with that in mind the reader can get the idea. At any rate I wish to give a few basics and bring up the Earth's-Core, which is not yet a common theme of those looking at Climate-Change, but needs to be. If people are going to talk about Climate-Change then they need to start at the basics, and have an elementary grasp of the intricacies involved.

If you want an ACCURATE PICTURE of what is really going on with the weather, and weather cycles, then I suggest you read Ann Walkers book "The Messenger The Messiah". In it, at the back of the book, she covers all of the dynamics involved in Climate-Change. It is the best thing I have come across. If you are not a believer of her story it is irrelevant, it is the Knowledge that counts. Just go to the back of the book, the Science she presents speaks for itself, and if I remember correctly she even presented this information to the United Nations. In my opinion it is what the World Leaders are using as a hand-book and know to be true. Scientists working for Government, no doubt just like me, will look at the Science of it—the "Source Itself" is irrelevant. Logic, Deduction, Reasoning, Observation are important for looking at this here—to determine if it is "Accurate". She discusses in detail, and in terms of Science, what is going on such as Heat-Transfer, the hole in the ozone layer which allows more intense heat in, the effects of trees and forests on Wind-Patterns, the Earth's Circulation-Patterns for distributing heat, the causes of the breaking down of these Heat-Distributing-Patterns, what will happen as a result, and the heating up of the Earth on the inside—the Earth's Core, as a result of it all. So with that in mind lets get to Basics.

Think of the Earth as a bathtub of cold water. Now in this bathtub, at one end we have a blender, a mixer, swirling the water around. This swirling of water causes a pattern of motion, currents, which causes the water to travel all around the tub. Now lets put in extremely hot water at one end of the tub. Usually cold water and hot water do not mix very quickly, they remain divided, very slowly assimilating one another until one temperature is reached-until "balance" between the two is reached. However without mixing the water, creating currents, this takes time; and until that happens the tub of water remains with two extreme ends of temperature, with just a tiny dividing line of "balanced temperature". With the mixer in place, creating currents the whole time, the hot and cold mix quickly, and the water temperature is less extreme in terms of hot and cold, balance between the water temperatures is achieved quickly. In this case, for example, the mixer is the Earths Heat-Transfer-System: it is the air-currents which are constantly and reliably blowing about the Earth. Without it on and running, the water remains as two very separate and distinct, temperature zones.

These Constant and Reliable Air-Currents, Wind-Patterns, are very important as they Distribute-Heat more equally about the Earth. Without them Temperatures would be more extreme.

Of course we are dealing with the Earth here which is far more sensitive, dynamic, intricate, and complicated. Whoever invented the blender did so in a relatively short period of time, the Earth has developed her sense of balance as we now enjoy it, over the course of millions of years. Thus to affect even one system of the Earth, such as the Earth's Heat-Transfer-System, is to throw a monkey-wrench into it all. Just like a watch, it is delicate and intricate, and to affect one small cog is to affect the function of the entire thing.

Now aside from man, there are two key elements affecting the heating of the Earth. There is of course the sun, and there is the Earth's Core which is the one I wish to discuss as most people are oblivious as to the role it plays. Molten rock, magma a "Ball of It" at the center of the Earth. Now, in my flat we have floor heating, the heat comes up from the floor, so I will use this as an example.

Imagine I have the heat on, turned up a bit too much: the floor is warm and the room is warm, both getting warmer. Now on occasion, by accident, this has happened in our flat. The solution is simple enough, turn off the heat. However, the ceramic tiles under our feet, and the concrete underneath have absorbed a significant amount of heat, so regardless of the heat being off, the floor keeps giving off heat for some time. Simply making the heat off, and opening a door, does not restore the room temperature to a more "Comfortable Zone". It keeps rising from the floor.

So I make a fan on, which creates air circulation, and mixes the warmer air rising from the floor, and cooler pockets of air in the room. Now the room is more comfortable and bearable. However if the fan is turned off, quite obviously the room heats up again and becomes unbearable. In this case the fan is the mixer.

Now imagine this floor heat being on all of the time (Like the Earth's Core), a fan would HAVE TO BE used to keep things bearable, to keep the air mixed. Without the fan balancing the air temperature the air temperature becomes unbearable, a Hot-Zone is created, from the constant rise of heat from the floor. Without a Coolant-Mechanism the floor gets seemingly hotter and hotter. Admittedly my example is not perfect but I think we can take a few liberties here to get the point across.

As long as the fan is blowing across the floor, the heat removed quickly, and dissipated quickly, the floor remains at a relative temperature without becoming totally hot. The air currents are removing heat from the floor quick enough that it can not overheat, the concrete and tiles do not end up absorbing too much heat to the point of being uncomfortable. The heat is removed as quickly as it gets to the ceramic tiles, all things being relative. Air is mixed and circulated so that the room does not become totally unbearable. However if you make the fan off, it no longer blowing across the floor, the ceramic tiles get hotter and hotter as there is no "Heat-Transfer-System" in place, the floor absorbs more heat, to even lie down on it can become uncomfortable. So by turning off the fan, the mixer, the Earth's Heat-Transfer-System, we create a "Hot-Zone" from underneath which becomes steadily and increasingly warm. For whatever reasons as the Earth's Air-Currents, Heat-Transfer-System shifts (actually cutting down large amounts of trees causes this all-altering Wind-Currents) , or dissipates, due to Internal-Heat the ground underneath our feet begins to absorb more heat from the Earth herself, from the Earth's Core. Especially in areas where the Earth's Magma is closer to the mantle which is what we live on. Thus "Hot-Zones" can be created from underneath as well. A lot of people do not know this.

Now in this case we are letting heat build up underneath, without Air-Currents removing the heat, the Earth becomes "Blocked", pent-up, an outlet, a release for heat is lost—or rather a "Removal of Heat", so the Underneath-Temperature builds up. Much like if you have too much clothes on, if you do not unzip your jacket to release the heat, your body will ABSORB the heat, you too will become too hot. Your Core-Temperature will BUILD, and you will OVERHEAT. If the heat is not released, does not go out, it stays in, and begins to overheat everything underneath—in this case everything "underneath" being your body. And so it is with the Earth—as everything underneath is left to absorb the build up of heat. Thus the body becomes the new Heat-Transfer-System. Eventually this heat building up underneath begins to radiate out, through "the clothes", or in the case of the Earth, through the Mantle, the outer layer of the Earth's crust, what we live on. So what happens is that with the Earth, "Hot-Zones" are created from Underneath, by the Earth's Core, and depending on the "geology". Areas are going to vary on this point based on things like how close Magma is to the surface, etc.

Of course your body has ways of trying to cool off, like sweating. No doubt the Earth will also respond with back up cooling mechanisms like eruptions, but again this is a dynamic thing with many factors involved. I believe you get the basic idea. I am basic and general here.

As a result of Earth's Heat-Transfer-Systems shutting down or being affected indirectly (Man-Made-trees cut down) or directly (Elite-Made) thus becoming less reliable, the Earth's Core will also begin to rise as well. Heat will become "Trapped Within", and the Earth will ABSORB this heat. Thus a New-Dynamic is introduced: the heating up of the Earth from Underneath, and it's affect on the surface. New "Hot-Zones" will be created, and created from Underneath. This is all common sense.

It begins with large amounts of trees being cut down, the removal of forests, which affects "Wind-Currents", the Earth's Heat-Transfer-System. The way in which the Earth herself is "Cooled" from her own "Inner Heat" and the Distribution and Balancing of this Heat.

With each monkey-wrench tossed in, with each System affected, another is affected, things quickly get out of control and more and more complicated. So as you see Climate-Change is both simple yet dynamic, intricate, and complicated. And I might point out here that I have not even addressed the issue of the Elite here, and their Weather-Control-Technology, and Earthquake-Technology. The Elite will be totally disrupting the Earth's-Systems, and the Earth at some point will begin to seek Equilibrium again regardless of Technologies: which will lead to no doubt, very unpredictable weather even for the Elite. But back to Ann Walker who's understanding is far greater and detailed than my own.

Ann Walker's overall theme and cure for all of this is simple, PLANT MORE TREES. It may sound far fetched, but in the end something as simple as removing too many trees, can no doubt have a large effect on simple, but Huge and Delicate-Systems, such as Earth's Heat-Transfer-Systems. Such as AIR-CURRENTS; in short it makes sense. To do any of it justice you have to read the information in her book. Now that is the issue of "Normal Climate-Change."

Elite Climate Change

I believe there is Climate-Change, and I most certainly do believe that man has played a part in this, and has yet a part to play in dealing with the issue. However I do not believe it is all as bad as it is made out in the Media, in fact the Media (as is typical) is being used to SELL US Climate-Change. To top it off the weather is being Manipulated by World Governments, by the Elite, and no one knows what affect all of this tinkering with the weather will ultimately lead to. We are not talking about Humanitarian motives here, or "Save the Earth" motives here, we are talking about Self-Centered Motives and Goals of the Elite. Behind all of this "Climate-Change" and "Save the Earth" are Ulterior Motives and Agendas. And I am not yet even speaking of "Weather-Manipulations" here. The Governments of Earth have yet to take care of it's own peoples and species, never mind the planet, so what is going on?

I believe there are several factors involved here. First of all there is some Climate-Change going on, changes in weather around the World. There is a "Natural Version" and an Elite-Version, I believe. The "Natural Version" I noted as a teenager, my grandfather (a farmer) noted it then as well, when I was in India for ten months (2000) I spoke with villagers who spoke of the difference in monsoon season (and notably so) over the years, and I have even noted it here in Europe, within the last ten years. This is by my own observation, obviously I have not totally traveled the World, but seeing is believing. I believe man is affecting the environment, this is obvious, to what extent, and "What other factors are Involved" is another issue. My main focus is to give background info and info on Elite-Climate-Change.

Now of course there is much news of Climate related things going on around the World, and people in these areas are noting the changes in their regions as well: no doubt for them, seeing is believing also. However I think a "Panic-Button" has been prematurely pushed, things are being MAGNIFIED and SOLD beyond proportions, and Climate-Change is being shoved down our throats constantly. It is being constantly repeated like a Demented Mantra. It is being used, Misused to cause fear and panic, and to put everyone into a "Disaster-Mode", and "Catastrophe-Mode". It is to create EXPECTATIONS of "The Worst to Come." It is a "Sales-Pitch" for the worst to come: the Elite are "Advertising and Broadcasting" what they are going to do. It is constantly spoken of in the Corrupt-Media to burn it into our minds. This is by Design and for purposes of the Elite.

We are talking about "Man-Made Climate-Change" in a very different sense of meaning. Not a slow build up of exhaust fumes and cow farts (Folks if you are dumb enough to blame cows on the issue of Climate-Change then you need your head checked—in my opinion this is a joke, perpetrated by the Elite, and when I say "joke" I mean just that), but Climate-Change created by Hidden Exotic-Technologies, and the secondary affects of some of the changes made by some of these Exotic-Technologies: like volcanic eruptions which when triggered will affect Climate due to ash.

In short I believe the Elite are laying down the Groundwork, laying down a "Belief-System", a "Sales-Pitch", for Plans involving the Manipulation of Weather and other related things which will affect the weather, so that when it happens it will be believed and taken at face value. And more importantly, to affect people, to make the Masses more Fearful and Dependent on Government, to gain more Power and Control. This is the "Creating of a Convincing Story", a "Long Slow Manipulation", the Groundwork for Future Manipulations.

In short they are laying down the Groundwork for "Earth-Disasters", and "the End of the World". They are using "Fear of the Mayan Calender", Predictions, Prophecies, and Religious Texts, such as the Bible, as a "Smokescreen and Props" for Elite-Made, Technology-Made Weather-Events. This will be done using Exotic-Technologies, to create disasters of huge proportions, which will seem natural, and which will seem to fulfill many a Prophecy. Picture several hurricanes hitting America at once, as an example—and Elite-Made hurricanes at that. People would think this "Weird, Ominous, and Biblical". Thus playing on peoples Religions, and playing on Prophecy while at the same time using the Earth's forces to wreak havoc and destruction. All of it will seem natural, very much of it, most of it will NOT BE.

That there is Climate-Change (a natural form), there is no question for me (I consider it minimal), it is just that I believe that it is being overdone, a "Sales-Pitch" is being made for it, so that the Elite can get away with using Exotic-Technologies to make things FAR WORSE than they really are. A lot of this having to do with, as usual, more Power and Control. People will become far more Dependent on their Governments as a result. More "Survival Oriented", more "Rooted in Physical-Reality", and more Fearful—thus easier to Control. Never mind the World-Population becoming "Less" as a result of "Extreme Weather"—the less people to Control and Manipulate the easier it is to do...

People basing arguments on Science against Climate-Change as it is being now "Advertised", although no doubt technically correct, are doomed. Regardless, of "Incorrect Statistics" which have been discovered, which have been Manipulated to make things sound worse than they are, there will STILL BE Climate-Change. This is why these Statistics are Manipulated in the first place, they are being Manipulated to fit what the Elite are going to make. They are creating a "False Profile" of the Climate to sell to the Masses. These "Statistics" are merely a "Selling-Point", a Lie and Scam, to explain weather further down the road. They are laying down the Groundwork in various Layers and ways of Deceit. Thus anyone arguing against Climate-Change is doomed, regardless of Science and "Incorrect Statistics" being found, because the Elite have decided THERE WILL BE massive Climate-Change. The

Elite already know this, and the Media controlled by the Elite "Advertise This", and for reasons based on Insider Knowledge and Planning.

This is why the Media is confidently pushing on with this "Sales-Pitch", the Elite and certain others, know there will be "Catastrophes" of MAJOR PROPORTIONS because they PLAN on creating them. This will create a lot of death and chaos and push people, populations, into DEPENDENCE on Government, far more than today. People will be far more Dependent on Government and looking for answers, and willing to give up freedoms, for "Help" from their Governments. In Reality, much will be created by "Government" in the first place. An example of Weather-Control-Technology is a hurricane hitting Louisiana, three years later, near the anniversary date of Hurricane Katrina. The Elite, the Government, is no doubt fine-tuning and further testing it's Exotic-Weather-Control-Techno. This is one good example.

Another point to be looked at is the film "2012", although it is not the only film or T.V. medium to be looked at. This is again an "Advertisement" of things to come. The Elite commonly use films as a way of "Broadcasting" moves they are going to make, AND TO "SEED IDEAS" into the populace, by influencing the making of the film, directly or indirectly. They are "Seeding" the Collective-Unconscious of Mankind, with films.

As an example, the making of the "X-Files", the writing of the scripts, were often influenced by Spook input, that is to say that Government Agents were in contact with the script writers and gave ideas and themes for episodes. A lot of people do not know this. The Government's influence, the Elite's influence, on Hollywood is not only there, but larger than anyone realizes. In short it is a form of LOBBYING going on: Hollywood Lobbying. Movies and T.V. are a subtle way of placing ideas into peoples heads. It is no Coincidence that there are many films now being made about attacks on America by hostile, fanatical forces, Terrorists, what have you. The Elite are placing this idea out there, through their Control of the filming industry, to PROMOTE THIS.

Think this does not work, forget it. I'll give you another example which even if you do not agree with, may amuse you by the theory and example. It began sometime back with a person jumping into the bear-exhibit at the zoo. Obviously they were chewed on, whether they lived or died, what country it was, and what zoo it was, I can not remember. The story made a lot of news because of how bizarre it was. Now even if I had made a point of remembering the zoo, and whether or not the person lived or died, I no doubt would have forgotten by now. Why? Because since this first story, there have been plenty of other similar stories across the World. People have been "Copycatting" this experience and jumping into the bear exhibits.

This "Idea" becomes IMPRINTED (some of you might note the irony of me capitalizing letters here) in people's minds, and the minds of those who are not stable.

Heard of "Copycat Killings", well this is the same type of thing involved here, IMPRINTING. Thus films of Terrorist attacks on America will only "Promote the Idea" of doing so, and will become "Imprinted" amongst the more unstable—as yet another example. Thus films of Cataclysmic Climate-Change, doom and gloom, will become Imprinted in people's minds and be there as a "Book-Marker" in their mind, a doorway to be activated, a Doorway to Fear, when real events transpire. Which of course they will.

It will go like this. Sometime around 2012 I am guessing, things will really rev up in the World. I believe there will be multiple things going on, but one of the really big ones will be "Cataclysmic Events." Due to films, both theater and T.V. films, and due to heavy Advertising through the news, T.V., radio, and newspaper, the idea of Climate-Change and Global-Disasters is prevalent and well Imprinted on Mankind. By 2012 it will be even more so. At that point in time no doubt crazy things will BEGIN to happen, or shortly after—I am guessing a year or two perhaps. Things like massive hurricanes, heavy thunderstorm systems, tornadoes, earthquakes, and volcanic eruptions, which are already being slowly built up so as to look natural. I expect there will be many volcanic eruptions because it is a good way to pollute the atmosphere and thus affect Climate without much effort on their end. The Elite need only trigger eruptions which they can do with microwaves, but I will get to that a bit later.

As this begins to happen, people will become "ACTIVATED", triggering memories in their mind of all of this News-Hype and Imprinting, and of films like "2012" which are heavy and dramatic. Thus they will refer to, come back to, these "Book-Markers" in their mind, which are Doorways to Fear which have been cleverly and subtly placed in their minds: to be Activated when real events transpire. It is a matter of Predisposing and Conditioning the Masses. Of course there will be a lot of fear anyhow when these events happen. The Masses are being "Programmed" in this manner, Collectively.

I imagine it will be a time of escalating chaos, and all of this IMPRINTING will only serve to create more Fear, and build up Negative-Expectations, and create more of a "Survival Attitude." Since this will have been Advertised for so long no one will even question the idea of it being "Man-Made." Of course the other Manipulation going on here is that a lot of this Cataclysmic Stuff will be fitting in with a lot of Prophecies of many Religions as well: the Religions being used as a Smoke-Screen and as Props. Since most of the World is Religious in one form or another, a lot of people will buy into all of this Elite-Made-Catastrophes without even wondering, just simply based on Religion alone. It is all a Sales-Pitch, and all a Manipulation.

Now another piece of Exotic-Technology to be used, which will seem natural, is Exotic-Earthquake-Technology. Perhaps it is the same Exotic-Technology being used. I believe that through Artificially-Created-Earthquakes this will also be used to cause and trigger, volcanic eruptions. Scientists who study such things, if they have not yet, will no doubt notice inconsistencies in earthquakes and volcanic eruptions. Breaks from the usual patterns and routines of earthquakes and seismic pressure, and eruptions. Eventually, if not already, it will become clear to Scientists that something unusual is going on here, and this will point to Man-Made, Government Made, Elite-Made Manipulations. Of course it will be written off as "Climate-Change" and the "End of the World Prophecies" will be constantly brought up in the Media. When all of this starts happening the Media will be used to SELL US a constant Barrage of Doom and Gloom. But back to the idea of Man-Made-Earthquakes, and this Techno being used to cause Volcanic-Eruptions.

How is such a thing possible? Well here in Europe, I saw a documentary on earthquakes, and "Super-Explosions", "Super-Eruptions" of volcanoes of the past. Basically it comes down to Magma under the Earth, hitting an underground Water-Source. Like a large bomb, or pressure cooker, once Magma hits a large enough Water-Source, the steam created by Magma hitting water, under the Earth and thus under vast amounts of pressure, causes an absolutely enormous explosion. The kind that can bring darkness to the earth for days, or even months, and blowing up a large section of land with it. Thus all of the dirt and ash darkens the sky.

Now there is Technology that is said to be microwave related, it is called HAARP. Microwaves are capable of going through the exterior of things, and thus in the case of HAARP—a giant Microwave Machine, capable of reaching underground Water-Sources. With this Technology based on simple Science, water can be heated up to cause, trigger, earthquakes, as the water heats and expands. It is the same principle as Magma hitting underground Water-Sources, only more localized and no doubt on a smaller scale, although perhaps not. I have to speculate based on Logic. So, again Exotic-Technology, an "Earthquake-Machine", Exotic-Earthquake-Technology. This does NOT mean that this is how it is done, can be done, or will be done (if not already), however based on this alone the CAPABILITY Logically exists already.

I am not saying this is how it is being done, I am saying that this ability to create earthquakes and eruptions, already exists. For what I know the Elite have a machine which causes Earthquakes and Volcanic-Eruptions in a very different way, although HAARP seemingly "Fits the Bill."

Going with my hypothesis, no doubt, in one way shape or form this can be used to create Volcanic-Eruptions as well. It is merely a matter of triggering a strong enough quake in the vicinity of a volcano, or perhaps whatever is being used can affect a volcano directly anyhow. The Earthquake and Volcano will seem to go together, in reality, it will be Man-Made, Elite-Made. The effects of, the Misuse of this Exotic-Technology will be devastating.

It will indeed seem "The End of Times" as Volcanoes-Erupt and Earthquakes occur around the World in growing numbers and magnitude (In places NEVER heard of), never mind Climate-Change. The devastation can be huge, affecting Climate globally and world-food-production as well. No doubt it will be Advertised in the Media, more than Paris Hilton marrying O.J. , and the FEAR and PANIC World-Wide will escalate. Especially after all of the Fear-Mongering over "2012" and "Climate-Change." All sorts of problems will arise as a result, and the World's Populations will be clamoring for help from its Governments, pushed into a State of Fear and Dependency.

This is the Plan, and this is why Climate-Change is being Advertised so heavily now. It is to create Negative-Expectations, of the worst to come, and since the Elite already know this will happen, they will turn around and say "I told you so", to all who doubted, based on a lot of Science. That there is Climate-Change to some degree (a natural form of it) there is no doubt in my mind, but that it will reach severe proportions, I think is ridiculous. So if you have read Anne Walkers books, I agree to a point, the Science of what she explains is sound; the Earth regardless of cause IS being affected and changing in a manner which she describes, I just do not think that it will be all natural, or that it even is now.

No, the Elite are building up Expectations of "Cataclysms to Come", are going to look good and reliable as a result of predicting it first, thus gaining a lot of "Faith and Supporters", when in Reality they will be behind it all. People will look to these Elite-Politicians and "Experts" working for the Elite all over the television, with respect and admiration, as the Masses see that their "Leaders" were able to predict such dire circumstances, and as they see that their leaders WANT to do something for the Masses. No doubt with all of this new found respect for their Leaders, the Masses will piss away a lot of their freedoms, and lose their heads, being fully Manipulated and

Tricked by the Elite. The Masses will become DEPENDENT ON their Leaders and Governments for help and relief. No doubt this is why World-Leaders, the Elite, have built a SEED-STORAGE-VAULT on an island off of Norway. They already know what is going to happen, they have Planned it, and like a MOVIE-SCRIPT, are now enacting it.

To top it off, the Elite will also be playing on "Religious Beliefs and Prophecies", hiding behind these as a Smoke-Screen. "Long Expected Prophecies" spanning thousands of years, will come "True", and be toted in the Media. Have you heard the expression, "The Media is the fourth branch of Government."? Well that is because of it's power to affect the Masses, and it IS the mouth-piece for Governments, and thus the Elite. It is no coincidence that when you go to any Country, the Media at heart, is always "FOR" the Government of the Country in which it resides—for Manipulation, not facts. At any rate, the Media already is promoting Climate-Change for the Elite, and soon, around the corner, "Talk of Prophecies" will flourish in the news. Thus Governments will be gaining the "Faith" of a lot of Religious People as well. In essence The People of the World will be duped by the Elite, times will be chaotic, increasingly unstable, and Fearful, and the Elite will ultimately gain a lot of "Power over The People".

No doubt when all of this begins to unfold in full Elite-Glory, a push will be made for One-World-Government to facilitate "Restoring Order and Maintaining Calm". Governments across the World will become centralized under one banner. Of course to make this all extra TRAUMATIC for the Masses and to really rivet up the fear, no doubt all of this will coincide with other STAGED-EVENTS as well

Conclusion

These Exotic-Technologies are being Hidden to perpetrate this charade on Humanity. The idea of Manipulating Weather and Climate to create chaos, is from a conniving and psychotic Viewpoint, a good one—strategically speaking. In this manner THERE IS NO ONE TO BLAME: unless of course people begin waking up to this PLAN of the Elite. People need to wake up to these Exotic-Technologies and their purposes, and Misuses. Knowing of it and speaking of it can change things. By speaking of the future we have the power to change it.

A CONSCIOUS PEOPLE IS A FREE PEOPLE.

The difference between what the Public knows and what exists is unbelievable, which is part of the problem, and a Tactic in and of itself. This creates Disbelief so that the Public does not know about such things and can not defend itself. What the Governments of the World have, in terms of

Exotic-Technology, would create fear and a backlash with the Publics across the World, with Humanity. Humanity would revolt over the possession of these Technologies because of the "Possible Misuses". I assure you that these Technologies are being widely used already, and MISUSED at that. The Secrecy is to keep The People from knowing, NOT other Governments which are clearly united against their own people—and therefore Humanity.

Climate-Change is not an orchestrated hoax, it is a Planned, Guaranteed-Deal, which will be arriving soon and in a manner which will negate all doubt. I believe that 2012, the Elite-Deadline will be the "Begin Date" for many things and Events, no doubt Climate-Change and Disastrous-Weather being one of these. As time progresses beyond 2012, no doubt "Weather and Climate" will become increasingly worse and due to Elite-Manipulations.

I would like to say that there is a lot of evidence to suggest that a lot of minor "Climate-Changes" are a part of a Normal Cycle, because there are many variables involved, and no doubt man is affecting Climate. However with these Exotic-Technologies and the insanity of the Elite, it is a moot point. Things are going to get far messier regardless of the "Affects of Mankind". The Elite plan on "Overriding" the Earths Natural-Systems, exacerbating the Earth's Cycles, with Exotic-Technology (Chemtrails as well), for their own gains—which as always is to the Detriment of the Masses. To top it off, ultimately no one can predict the affects of such tampering with the Earth's Systems, it is ridiculous to even think it so.

In short 2012 is no doubt going to be the beginning of this Plan for "Bad Weather and Disasters" which will over time, progress and become worse and worse. No doubt down the road here, "Weather Disasters" and "Climate-Change" are going to become common themes in the Media and People's minds. I believe that this is a Plan of the Elite and that this is what the World is in for.

CHAPTER 22

Theory of Corrupt and Powerful People—Why Doesn't the Average Person Believe?

This is one of those subjects that is going to reoccur in this book. People always say "But why would they do that? I just can't believe it.", and the thinking stops there. Like their question has answered all. Allow me to make a few points.

First of all, your understanding of people like this is not a prerequisite for seeing and hearing what is going on in this world. So if I say there are Corrupt and Powerful people trying to benefit themselves at the cost of Humanity, that it is Deliberate and Planned, what exactly does your question answer: The "Why would they do that?" or "I don't believe it." thing? You are talking about not understanding the VIEWPOINT of these sick people and not being able to relate to the scenarios and circumstances that they create, that is normal, but not looking and not hearing is NOT. Not even considering such possibilities is also irresponsible as a Citizen and as a Human-Being. As long as we continue to respond this way, these sick people will continue doing what they do. Why wouldn't they? When they see that they can make a lot of money, have a lot of power, and do what they want with impunity, why wouldn't they? It is not as if these people are being caught; the Masses make it easy for them by being Lazy-Minded and Fearful of the Truth.

"ANY AMERICAN WHO TURNS A BLIND EYE TO INJUSTICE AND CORRUPTION AROUND HIM SHOULD NOT BE ALLOWED TO CALL HIMSELF ONE, IT IS BY DEFINITION FIGHTING THESE THINGS THAT MAKES US AN AMERICAN IN THE FIRST PLACE. IN BEING AN AMERICAN WE CARRY THIS RESPONSIBILITY.—Mr Lee", this quote points out the obvious.

Corrupt Powerful Psychopaths do exist. And unfortunately within Government, which is ironic when I look at other people's Viewpoint. Psychopaths are attracted to Power, and the Misuse of it. Where else would you expect to find them? Government is a "Honey Pot", a light in the dark, attracting this very type of person because the Masses are not Vigilant and have made it easy for this type of person to "Thrive in Government". People are blinded by a "Job-Title and Stereotype". The Masses are not, and have not been doing their jobs as Citizens, making sure that their respective Governments stay clean.

The "People" have set up an environment within their Countries and Governments, conducive to this Type of Person existing within Government. People do not want to "Trouble Themselves", do not want to be Responsible, or are Afraid and "Look the other Way". And by doing this, Corrupt-People see that they can get away with being Corrupt, and within Government. It is stupid and ridiculous, who else is going to "Clean Up Corruption within Government", the Politicians? How hard is it to understand that there are very Corrupt and Powerful People within Government? And how hard is it to figure out that they are going to "Band Together", "Form a Click", "Unite Together", for their own Selfish-Self-Interests and Gains, when they can make more money this way?

I assure you such people do exist, I'm dealing with some in my own life NOW, and yes I understand how hard it is to grasp the Mentality of people like this: never mind how unpleasant it is. That is because they are not normal they are Corrupt beyond most people's understanding, Psychotic, the average person is not. It took me years to grasp and come to "Relative Terms" with people like this and their Mind-Set. Normal they are not. Who they are is simply "Outside of most People's View of Reality", however your not believing or understanding them, does not "Negate Their Existence". They "Exist Outside of Your View of Reality" and therefore remain "Unseen".

However as an example we don't have to understand the mind-set of the Zodiac Killer or Jeffrey Dahmer to see that there is something wrong with them. We have only to look at their ACTIONS and listen to their words to realize something is not right with these people. Again not understanding these people is not a prerequisite to Seeing, Hearing, and Looking.

As an example there was a shooting in the city I live in, a very unusual event for around here. Several people died including a mother. It sucked. But the point is if you had been there you would not have stood there in denial, the great Idiot-Philosopher, wondering why they would or could do such a thing, BEFORE reacting. It's ridiculous, and counter productive to living, so save the "Why would they do that?" argument for someone less Intelligent and less Conscious, that argument doesn't fly with me. That line of thinking is for people who do not know how to survive, Idiot-Philosophers, people like this are Darwin's Theory of Evolution waiting to happen. If you want to be Intelligent you are going to have to start by looking at Reality for what it is, and following it to its Logical Conclusions, no matter if it makes you Uncomfortable. Especially in regards to Survival. Fear should not be an issue in your ABILITY TO LOOK.

Now what happens when you take a Psychopathic Mindset, a lot of Power and Money, and Network it with others of the same Mind-Set? Think about this a moment. The "Level of Craziness and Insanity" goes up several notches. I imagine that Manipulating entire segments of Humanity and Countries becomes the ultimate "Kicker and High" for these people. Caring about people, Society, Cultures, Humanity, is Not an Agenda of these Types. More Power by way of Manipulation, no matter the cost to innocent people is the order of the day. It is part of their Warped Need for Fulfillment. It is EXTERNALIZED-FULFILLMENT and since this is so, involves people like You and Me. We are a part of their Warped Fulfillment, whether we like it or not. We are OBJECTS to them.

To further their cause they Corrupt others around them, the numbers grow, and things slowly progress and avalanche as Long-Term-Plans slowly come into fruition. As an example of this look at the Economy in America and the World at this point. None of it happened by "Accident or Coincidence", and it's not finished yet. When the American Economic-System totally Collapses, it will be due to years and years of chipping away at it, although there is certainly a rush being made to get there these days. That's because these people are very near the end of fulfilling their Plans of Total Control and Power, and they have a Deadline set of 2012 for reasons of their own.

CHAPTER 23

The Rich and Powerful: the "Corrupt Rich and Powerful": the "Corrupt Rich and Powerful United Together"—the Elite: the Elite and "Their Team": an Easier Way of "Seeing" these People

This one starts off easy enough but quickly turns into "unbelievable" for a lot of people. I understand, I used to be one of these people who "scoffed at" a lot of these ideas mentioned here. It was specifically my looking at the Tobacco-Industry, and Food-Chemical-Industry, that I realized there are people behind the scenes Manipulating things, their Power, for their own ends and gains, and regardless of the costs to others. I realized that these people are out of touch with Reality, and for all intents and purposes, Psychotic. What else would you call killing off large numbers of people by creating harmful, deadly, Industries such as the Tobacco Industry and Food Chemical Industry? Once I realized this I began looking, and putting facts together. This book is my attempt to make a lot of these "Hard-to-Get" subjects easier to consider and understand.

This next chapter is a look at the group I call the Elite, and their "Team", a very large group of people across the World who work for them. The

"Henchmen" of the Elite who get the dirty work done. For all intents and purposes, this "Team" is a Fascist-Cult, which I also refer to as a World-Cult, and the "Esoteric-Nonsense" that goes with it, I essentially ignore. The esoteric nonsense IS IRRELEVANT. A Fanatic by any other name is still a Fanatic. The "Cult-Aspects" of this "Team" is a "Sales-Pitch", window-dressing, for a particular Type of Person, the kind susceptible to Cults and Fascism in the first place. It is to "Attract a Certain Type of Weak Minded Person". The kind of people who blindly throw themselves into a "Movement and Cause". The Elite in fact plan on doing away with this "Team" when this "Team" has achieved all of the Elite's-Goals. I will do several pieces on the Elite and their "Team", which I also refer to as a World-Cult, covering several angles and taking several approaches due to the difficulty in making this subject "approachable and believable".

The Rich and Powerful

Like any movie, the Rich and Powerful need people to do the leg-work. This is simple enough to understand; the Rich and Powerful make decisions and pay others to carry out their Plans to Fulfillment. They can afford to, and their job, the "R and P"(the Rich and Powerful), involves making plans and schemes for more money, big money, for these people think big.

This is how Rich and Powerful people make money. The "small people" carry out the small steps, often unaware of the bigger picture, they are interested only in their paycheck, their money and job, and so get their job done. What work they are doing for their bosses is none of their business. In essence they are "Compartmentalized" only aware of their own role in the scheme of things. The "R and P" sit at the head of the spider-web and tie all of these "smaller people" with their individual goals and objectives, jobs, together: each one a singular piece of a puzzle. The "R and P" know what the puzzle looks like, know what the end result is supposed to be. If all goes well, they will have an end result, picture, that is what they want. The end result being more money, and if you are into power, more power as well.

Now keep in mind that having a lot of money translates to having power, which is what I am really referring to here. There are other types of Rich and Powerful people who are into the "Power-Aspect" and reap money along the way, as a result of their Pursuit of Power. In this case money is a side effect of the "Pursuit of Power". There are clear differences and both "Legal and Illegal Versions" of these "Types of People with Money and Power" exist. For the sake of convenience I choose what is most efficient here, not covering all-possibilities, to do so would be wasteful.

Now keep in mind here that I am speaking of normal Rich and Powerful people, not the Elite who are Corrupt, and twisted beyond belief. People like this might head businesses, deal with the Public in this context, but they would not do so at the cost of lives—most are this way. But as you can see, whether you are an honest "R and P" person, or a member of the Elite, people, "small people" are needed to get different things done, in order to accomplish a goal, a business deal, which profits the "R and P". We will work our way up to the Elite, the idea of an Elite and their "Team", their "Henchmen" who get the dirty work done.

So as I was saying these are only the Rich and Powerful we are talking about here, not the Corrupt version. As people rise up the ladder, options open up to them, they have room for "Playing Life" on another level. They have "options available" to them for experimenting and making money on a grander-scale. Thus a lot of "R and P" begin experimenting and branching out in terms of investments. Pop musicians, athletes, etc. are good examples of this: lines of clothing, perfume, shoes, sporting equipment, what have you. These people have "Power" due to money, they are not necessarily after it, or even if they are, I speak of the legal ones here. A lot of "Rich People" begin to step outside of their "Job Titles" and begin to experiment with the idea of making more money on a Grander-Scale. Making more money becomes the Goal, and in what ever way they think of, Scheme, and Co-Create. The formula is Consciousness, Planning, and Energy. These are not people who see "I can do it.", as an impossible dream. They believe in themselves and what they can do. They "Accomplish Success" because they know they can: these people are Self-Empowered which is what a lot of the Masses are actually "Looking Up To" when they put these people on a pedestal.

This would be the problem of the Masses who tend to worship those who "Can Do It", putting it outside of themselves, never realizing they can as well. The issue here is Self-Empowerment and Action. Belief in oneself, Planning, and Energy invested in oneself and ones goals without harm to others.

But back to the Rich and Powerful, and now the "Corrupt and Rich and Powerful".

The Corrupt Rich and Powerful

Now we have another group to deal with, one more piece to add, as we build our way up to an understanding of the Elite. We have been dealing with a legal-version of the "R and P", now lets momentarily deal with the "Corrupt R and P" people, and then quickly from there get to the Elite.

If you are Rich and Powerful, you may fall into the trap of temptation, deciding that you can do as you will, deciding to break laws what-have-you, as a result of the Money and Power you have that comes with it. People like this may feel "Above the Law" and begin to do as they like, scheming illegally for Profit. Or it can be as simple as being Corrupt in the first place, getting involved in something that makes you rich, and therefore powerful, and not getting caught. In other words perhaps from the beginning the individual was a Greed-Head, hooked on the idea of money, lots of it, and decided to make money in whatever way one could.

If you are Corrupt and "R and P", very good at what you do, with a lot of potential, then an Organized Group of such individuals—Corrupt and Criminal, may approach you. These individuals being the Elite. They sit at the top of the "Food-Chain" with an accumulation of "Knowledge and Contacts", "smaller people" across the World. Thus they see individual "Corrupt R and P People", rather quickly, like rising stars, and see their "Worth and Value" amongst the Elite. Another asset to the Demented-Collective—a way for the Elite to gain more Power as a Group and more "Illegal Contacts", which translates to more Power. Note that this approach is not about friendship, but about Business and Power. There is nothing warm and friendly about this group, they are cold-hearted Businessmen, engaged in Illegal Operations: LIKE ANY ORGANIZED-CRIME-FAMILY. These people are Organized-Crime. So far most should not consider this a stretch of the imagination.

The Elite approach these "Individual Corrupt R and P Players" and they become a part of the "Elite-Organization" or not. Those who refuse may be ganged up on, Mobbed on this "Higher-Level". The Elite wanting an individuals talents, skills, resources, contacts, and "Specialty of Field", working for them, and a part of them. A "Rising Star" being a potential threat to their Base-of-Power, and cutting into their own "action". They also see individuals amongst themselves, coalescing too much power, as a threat as well. It is cut-throat, but simple to understand. These ARE the type of people we are talking about here. Criminals and Psychopaths, with Money and Power. None of these people, Elite-Members, want any one person obtaining more Power, Money, and Control, than they have.

As an example of someone, a "Rising Star", who was approached by these people and in fact did dealings with them, until he found out what they are really like, is George Green. He was an investment banker, a broker and dealer, who was invited "On board with the Elite" because of his "Specialty in Field" and ability to do very well at this. He sat in board-rooms with Politicians, Business Leaders, and Corporate Leaders, for the sake of planning schemes to "Milk the Masses". He eventually quit the Elite when he found out to what extent their Corruption went. He found out how sick these people really are.

These are not merely people after money, but people who have been Corrupted all the way through—people twisted and perverse from Power and their pursuit of it. As a result of quitting, he was attacked collectively (Mobbed) by these Elite, and "legally" at that, losing a lot of his fortune in the process, as they whittled him down to size. Remember these Elite reside within the Political-Field, and therefore have the LEGAL-SYSTEM at their disposal. Which they used to their advantage to attack and destroy George Green. He is now a very outspoken person in regards to these Elite.

Whereas there are people who are "Powerful" as a result of having a lot of money, these Elite are after more and more Power (not just money), and did not take George Green's "quitting" peacefully: every new member counts as added strength, added contacts for Schemes, added Power. It is a good example of someone on the "Shady-End" of things, who got invited to "Jump on Board" with the Elite because of his skills. However, where we were talking about "Individual-Corrupt and Powerful-People" we will now get to a "Collective of Corrupt and Powerful-People", the Elite. This brings us to the Elite, and their "Version of Reality". We are not talking about "Normal People" here.

The Corrupt Rich and Powerful United Together: The Elite: Organized-Crime amongst the Corporate and Political-Fields Power-Mongers: the Constant Pursuit of More Power

The difference here, in regards to the Elite, is the "Playing Field", in which they exist and from where they Manipulate from. People, the Public, the Masses around the World seem to think that "Corrupt Politicians" and "Corrupt Heads of Industry and Companies" are an isolated event, as opposed to being an ORGANIZED GROUP, which is ridiculous to consider. It might make us more comfortable to think this, but the odds are against a group of "Organized-Criminals" not being formed within the Political-System, where so much MONEY, POWER, AND TEMPTATION, exists in the first place. Quite frankly this is what attracts a lot of these Warped-Individuals in the first place.

A lot of these Warped-Individuals in their climb for Power and Money choose the Political-Field because this is where the Ultimate-Form of Power is at; and is what no doubt Corrupts other Politicians who went in clean, but perhaps were not strong enough in character. It is logical to assume (indeed against the odds for it to be different), that as these Corrupt-Individuals rub shoulders in Government, that they are going to work together to obtain more Money and more Power. There is strength in numbers and thus of course a group of like-minded Corrupt-Politicians will band together. Think of a group of thieves banding together to start a "Car Theft Ring" for more money, or a bunch of drug-users forming their own click so there is always something to smoke.

Naturally Corrupt-Individuals within Government are going to band together and Pool Resources, it is ridiculous to think otherwise, and this means a group is formed, a COLLECTIVE, it is in fact Organized-Crime within the "Political Sector", Political-Playing-Field. We are talking about CRIMINALS who pose as Politicians. We are talking about a "Gang" of them within Politics, within Government, like a "Gang of Bullies" at the school. We are talking about people who use their "Job-Title", Jobs, within Government to their Own Advantages, and use their "High-Positions of Authority" to establish "Contacts and Connections" within other "Fields and Sectors of Life". They do this to make more money, and to have more Power, by "Banding Together", by "Uniting Together", by "Net-Working". They also "Network" with "Others" outside of the Political-Playing-Field.

This is what Lobbying does as an example; it allows these Political-Criminals to intermingle with other Crooks, other Players, from other walks of life. Like say Businessmen: Heads of Industry, Heads of Corporations, Heads of Business's etc., where a lot of the Money and Power is at. Lobbying is Institutionalized-Bribery for this very reason. Schemes, Plans, Deals, Money, are made and exchanged for "Promises". This would all BE ILLEGAL except these Politicians make deals to pass laws so that what is wanted by Big-Business-Leaders, becomes "Not Illegal". Later, down the road the Politicians "Helping Out" get rewards, perks, MONEY and POWER, and INFLUENCE as a way of RECIPROCATION. Like with the Food-Chemical Industry in which Politicians are "Members of these Companies", affiliated with them in some way. Through these Companies the "Politicians" "Receive Money". It is a matter of, a System of, Pay-Offs for "Favors and Money".

As an example these Elite in Politics, these Corrupt-Politicians, make deals with Heads-of-Industry and Companies, who are also Corrupt and who are "Hired By Government" and OVER-PAID BY GOVERNMENT for their "Services". Later the Corrupt-Politicians get some of this "Extra-Over-Paid Money" from their "Friends" heading these Over-Paid-Companies with Over-Paid-Government-Contracts, thus we have TRILLIONS OF DOLLARS missing yearly. IT IS THEFT FROM THE AMERICAN-PEOPLE AND MONEY-LAUNDERING. It is ALL ABOUT MONEY-LAUNDERING, BRIBERY, AND ALL ILLEGAL. Or at least it

WOULD BE, if people would really catch on to this. The People are not holding these people, this Elite, ACCOUNTABLE. It is all a "Buffet Table" for Corrupt-Individuals, within Government. It is also a way for them to meet and incorporate people from outside of Politics into their "Club, Group, Gang, ORGANIZATION"—thus the Elite unfortunately are also heads of Corporations, Industries, and Businesses. The Corruption spreads like a disease, it is the weakness and desire of these Individuals, their QUEST FOR POWER.

We are not talking about people who are "Simply Rich and Powerful" or simply Corrupt Rich and Powerful Individuals, we are talking about an Organization of this Type of Person. I hope this makes a point for the existence of a "Group of Corrupt-Powerful Individuals" banded and united together for more Power and Money. I hope this CLEARLY ILLUSTRATES THE CORRUPTION WITHIN GOVERNMENT AND HOW IT WORKS—none of this is "Far-Fetched". It is common sense, "Evolution" in one sense, given the circumstances which DO EXIST.

The "Government and Blindness of the Masses" makes it all a "HONEY-POT" for Criminals and Psychopaths. To think that there are a few random Criminals, within Government, a few Corrupt Business Leaders, here and there, "Individuals" doing it on their own, is NOT very realistic. If you look at the Political-System, where many of these Elite are, the actual Political-System is set up to favor them and their Corruption, and Corrupt Practices. Again LOBBYING is a perfect example of this. It is Institutionalized-Bribery, Institutionalized-Pay-Offs. Only Corrupt-Individuals Banded Together would set up such a Scheme within Government, and take advantage of such a System in the first place. How hard is this to see? You think it is Coincidence? Why, because it makes you uncomfortable? Your being afraid to look at the Truth is not going to change it, or Reality. Only Awareness, Consciousness, and Action, will change things for the better. Self-Empowerment is key for changing all of this. Who else is going to "Fix This", Corrupt-Politicians? Only "The People" can, and only by being Self-Empowered.

Now to get back to my earlier comparison of different Types of People with Money and Power. The Rich and Powerful, the Corrupt-Rich and Powerful, and now The Elite. These Elite are after Power which separates them from other Types of People who have Money and Power as a result of money. They are also separated from many in terms of being united together as an Organization. No doubt there are many Rich and Powerful who are "Net-Worked" together to make a healthy profit, legal-business-deals, but this is not the type of people we are talking about here. We are talking about the Elite who's Mentality is very different from most others. They are cold, ruthless, Corrupt, and constantly on a "Pursuit of More Power": life and people are like a big Monopoly Game for them.

These Elite operate differently from the "R and P" and former athletes, pop-musicians, what-have-you, "branching out" to make more money in a different way and scale. They have entirely different options open to them due to their position in life and Playing-Field; the Political and Corporate-Playing-Field. They are Corrupt beyond belief, their minds twisted and bent with their pursuit of Power, and level and degree of Power, thus they are all together another breed and level to understand. The creation of the Tobacco-Industry and Food-Chemical-Industry are good examples of this. These are also good examples of their Pursuit of Power and money at all costs, at the cost of the lives of others. To top it off, "Their Playing Field" affects Entire-Populations: a "Heady Mix".

Now as with the "R and P", the Elite need people to do the "Dirty-Work", "Henchmen", a "Team". Seeing as how the Elite is a group of one particular "Mental-Type", cold, heartless, ruthless, Corrupt, and Illegal, they need a team of a similar mindset. Hiring priests, nuns, good Samaritans, will not work: hiring Criminals, Degenerates, people from all "Walks of Illegal Life" will. Thus Corrupt people are going to do the "Leg-Work", the Dirty-Work of the Elite. By logic, "Evolution of Thought", we can see that these "Henchmen", this "Team", is Dirty, Corrupt. It is logical that the Elite have this "Team" and that this "Team" is this way. Criminals hire other Criminals to do their work.

The Elite like any "R and P" also do not inform these "smaller people" of the Larger Scheme of Plans, they only tell these people of their own particular part, what they need to know to get their job done. They are COMPARTMENTALIZED, to keep Secrecy, and the Larger Scheme of things is none of their business. Now having said that, this Team and the Elite, are a bit different on a few points, very different than most other people. To some extent this Elite-Team does know of the Elite's-Plans, the "End Goals" for the Elite and Humanity, it is just that this Team does not know that it will be eliminated by the Elite, for very clear, logical, cut-throat reasons. If you are a member of this World-Fascist-Cult then I suggest you read on.

For starters the Elite have molded this Team into an Organization, a MOVEMENT of it's own. Using the psychology involved in creating Fascism and Cults, they have created a group of Hard-Core-Fanatics, willing to do extreme things. This is the Psychology used in creating a "Collective" with a "Single Mind and Vision"—which is what Fascist-Movements and Cults are. Members, Individuals of Fascist-Movements and Cults being called "FANATICS". There are "Several Layers" to this Team as well. Suffice to say this Team is a group of Fanatics who are some-what privy to the Elite's-Plans, and believe it or not are Sold on a "World-Goal", "World-Agenda", to give this Team a sense of Common-Purpose, Cohesiveness, and a "Collective-Objective"—to give them "Unity" and a "Core-Belief-System". This is about giving this Team a Collective-Form of Self-Identity as well. These people, this Team, has been sold a bunch of Lies by the Elite, just like anyone else these Elite deal with, to get "The Job Done". In short this Team can be thought of as a "Fascist-Movement" which is the same as "Cult". A Fascist-Cult with a "World-Goal", "One-World" under "Their Rule" which is typical of any Fascist-Movement—and which is what the Elite want for themselves. Like any Cult or Fascist-Movement, just like with my Mobbing, these people are only being Used for what the Leaders want, without actually knowing what the "Real Intent" and "Real Plans" are. My "Page Count" permitting, I will get back to these points again later. At any rate there are other issues that need to be dealt with as well, first.

As I mentioned before, due to the "Playing Field" this Elite has options open to them for "Playing Monopoly" that most others do not. This also means that they have a wide range of "small people", Team Members, working for them, across the spectrum. Having said that not all people who work for the Elite are willing participants, not all even knowledgeable that they are being used.

Since Corrupt-Politicians are involved this means they have the Legal-System at their disposal, and all that goes with it, as an example. This means that Government Employees of all backgrounds, Good-Politicians, Agents of Various Agencies, and Police alike fall under the Elite's Sway. Compartmentalization is a part of the job for these people, part of their discipline, for those in the Legal-System, and they with their Resources (spying, technology, informants), either willingly or unwittingly work for the Elite, as needed. When it is already part of the job, a lot of questions will not be asked. In other words some are Corrupt working for this Team within their Job-Titles, but the rest are not, they are merely following orders, happy with their job and what they do—they do not know any better. This is one of the advantages the Elite have over others and is a good example of their Resources which they "Pool Together".

The Elite know what their Plans are in any given Scheme, these "small people", Government and Legal-Dogs do not. The difference between a sheep and a dog is power, and a little bit of knowledge, NOT blindness.

The end result for these Demented-Elite, in regards to Plans, is always more Power and Control, Control going hand in hand with Power. A

World-Economic-Crash, followed by New World Order, being pushed for by the Elite and their Team, is a good example of this. It is the Ultimate Con-Job, and ultimate Grab for Power. Again due to options available at their level, and their Corruption, they think, live and act differently from those who are simply Rich and Powerful, or anyone else for that matter. They are Corrupt, act as a Collective, and are totally hooked, addicted to Power. They are also Psychotic, Psychopaths, so distanced from Reality, that they are willing to cook up Schemes which are responsible for hurting and killing people, LARGE NUMBERS OF PEOPLE, for the sake of their own gain. From all of the possibilities they choose things repeatedly, Schemes, which are to the Detriment of the Masses. THEY ARE ATTACKING THE MASSES. Their ACTIONS show this, they are insane from any observant intelligent Viewpoint. Naturally they are going to Hide this. How people can not see it in the first place is beyond me.

Having said that they are sloppy due to arrogance, occasionally being obvious, and these days they are being noticed more as they try to push their Plans, with a 2012 Deadline, into place. The other thing is that the type of "New-Political-System" they have chosen is starting to break the surface, as they approach the time for when they will put it in place—2012 is when it will all start. It is becoming visible, this New-Political and Economic-System, as they make some basic preparations for it, for putting it into place, as they begin to slowly integrate it with the Current-Failing-Sabotaged-System, for reasons and benefits of their own. I speak specifically of Socialism which is constantly pushed in America these days. It will also be pushed for the World, as a "Cure" to the failed World-Economy, which will Collapse as a result of America Crashing. It will be easier to skim money off the top with Socialism in place. They will have all money and all Power at their disposal with this System in place. This is the only reason why these Governments around the World have not announced to THEIR OWN PEOPLE that this will happen. They will Deliberately Crash the World-Economic-System, and then "Offer a Cure", a "Solution", which Benefits Themselves.

All of these Corrupt-Political-Figures around the World are deliberately keeping their mouths shut for their own gain, and committing CONSPIRACY AGAINST THEIR OWN PEOPLE. This is an obvious Economic-Crisis, it will happen, it has to based on simple economics: ALL OF THESE GOVERNMENTS KNOW, just not The People. THEIR KEEPING QUIET MEANS THEY WANT IT TO HAPPEN, AND HAVE ALL AGREED TO LET IT HAPPEN. In my opinion it is for creating One-World-Government, a One-World-Economic-System, a New World Order. During this state of Shock across the World, once the World has been Traumatized by Economic-Crashes, there will be rioting and fighting. It is inconceivable to believe that these Corrupt-Leaders across the World are not prepared for this, thus something WILL HAPPEN YET AGAIN, something Big, Captivating, Traumatizing, to steer people where these Elite want them. These Traumas, World-Traumas, will be for Greater Control of the Masses—it will be to steer the Masses into a particular direction, one which these Elite want. The same way that ranchers might use guns and dogs to steer their herd of cattle into a corral. It is to get people Emotional and Non-Thinking.

The Masses have no idea, for some reason this is too difficult to see, yet it is easy enough to see and understand when you try, and follow LOGIC to it's natural conclusion. Again Logic, Deduction, Reasoning, and Observation are key for seeing the World around you as it really is. Again if something is brought up, about something that is Hidden, you do not have to "Reinvent the Wheel", simply be your own Detective: this works.

CHAPTER 24

What Separates the Elite from "Other Crime and Criminals" is their Knowledge of Science and "Meta-Physics": The "Team" of the Elite—The "Henchmen" that Get the Dirty Work Done: The "Team" Working for it is a Fascist-Cult: The Elite's "Sales-Pitch" to this Cult—a "Pyramid-Scheme"

Now lets discus another "point" of these Demented Nit-Wits, the Elite: lets discuss their "Team", and their "Sales-Pitch" to their Team. The Elite's "Team" being the ones getting all of the Dirty-Work done, the "Henchmen". As you may see these Corrupt-Politicians and Heads of Companies are not the ones personally running around getting all of the work done and committing all of the Crimes, this is what they have money for: they pay other people to do this. They have other people who's "Job" it is to get things done, "Henchmen", who do these things for reasons of their own. As an example President Nixon did not get personally caught invading the offices of others, he had others do it—people who's "Job" was to do such things. And like any "Job", in the end people such as these, Henchmen, have their own reasons for doing such things—many times it being Greed. And in this case of the Elite, these "Henchmen" come with a twist.

In order to get the type of "Dirty Work" done that these Elite are achieving for, aiming for, you need to do more than simply throw money at people, you need people who are FANATICAL, willing to "Sacrifice Themselves" for "The Cause", to ensure success. People who are willing to "Sacrifice Others" as well—you need cold, cut-throat people, willing to do anything to get a "Job Done"—FANATICS. You need a "Certain Type of Person" with a "Certain Type of Psychology". I like to refer to these people as "Loserfarians", they are "Typical Weak-Minded-People" found in Fascist-Movements and Cults.

First of all I will cover the Elite in a little more detail; although I have used Corrupt Politicians as the "Poster-People" of the Elite, there are others as well. These Elite are made up of many of the big Business People, they are the Rich and Powerful of the World. People who own Companies, Industries, Media, and Banks, in addition to those of them in Politics. If you look you will see there are many connections between Politicians and these "Others" outside of Politics, it is all about "Favors and Money", where both "parties benefit"—it is very much all a Scam where money is "siphoned off of the Masses". Thus these people are well connected across the World, in terms of Economics and Politics: they are the Economies and Policies and Politics across the World.

Companies and Industries are one of the subtle ways for introducing laws into the World, without being obvious, without bringing animosity towards Government. Acting as a "Buffer Zone", "Insulation" for Government. Thus these Elite can, and will, no doubt "Sneak Undesirable Laws" through Companies onto the Masses—as in down the road no doubt you will have to become "Micro-Chipped" to be able to work at your Job as one example. Playing on the "Fear of Survival" which is going to become more of a "Theme" in the near future, as the Elite "Plan to Play on This Theme".

Actually many Companies are owned by members of the Elite, and working for other members and in conjunction with other members of the Elite, within Government. Corrupt-Politicians within the "Honey-Pot" helping to protect their "Corrupt-Enterprises" with "Laws and Favorable-Controls". These "Laws" allowing what should be "Illegal Endeavors" to go on (like Lobbying) and therefore protecting "Friends", fellow Members of the Elite, as they go about making Money. Money which of course comes back to the Corrupt-Politicians for "Helping Out". For example Corrupt-Politicians awarding "Government Contracts" to "Friends who own Companies" in which the Company is EXTREMELY OVER-PAID. Later down the road these

"Politicians" get money back from these "Companies and Friends" in one way, shape, or form—really it is MONEY-LAUNDERING. Money is being Stolen from the Public, and then "Cleaned" through Over-Payment. When the American People wake up to such Basic-Obvious-Facts, no doubt America will become strong again as a Nation. Corruption being the CAUSE OF AMERICA'S DETERIORATION and the World's.

Thus we have Rich and Powerful People within Government and outside of it in Companies and Industry, playing a game of "Robbing The People" with Schemes, Scams, and Plans, sitting on both sides of the fence and MUTUALLY BENEFITING EACH OTHER at the cost to the Public, at cost to the Masses. These Elite in Government and Companies are continuously passing Money-Making-Schemes back and forth, and continuously passing Money back and forth. In all cases THEY BENEFIT, the Public and Masses do not—in fact the Masses are merely objects, a means for making money. The last thing in the heads of this Elite is "The Welfare of the Masses". This is an idea of "normal people", the Masses, not an idea of the Elite. If you look, very many of these Money-Making-Schemes kill people off, you would think the Public would be more vigilant in regards to their lives. The "Tobacco Industry and Food-Chemical-Industry" being good examples of Two Industries Created For Profit, which "Kill Off" large numbers of people. No, the Elite are about Their Own Money and Power, not the "Welfare and Benefits to the Masses". In fact the Elite view the Masses as Stupid, which if you think about what the Masses allow theses Demented-Jackasses to get away with, is an accurate assessment.

In short these people, Elite, are united as a Team, in terms of their Planning and what they are doing. They are also united with others of their kind, across the World. This is all about Criminals and Criminal Enterprises: what is different for most is that this is in Politics and Industry. But why is this hard for people to believe? This is where the Money and Power is at, which makes this ATTRACTIVE AND LUCRATIVE. This is the Ultimate-Honey-Pot for Criminals. Who cares about the "Job-Title" if you are a Criminal who stands to make a lot of Money and Power at this? Especially when "The People" are not watching, are not vigilant, and are doing nothing about any of the "Corruption and Crime". This makes it easy for these Criminals to do, Lucrative, and makes the "Ultimate-Honey-Pot" even more tempting. Since when did "Your Version" of how things are supposed to be, the "Masses Version", matter to people like this? Take a look at the World around you, IT OBVIOUSLY DOES NOT.

The "Job-Title" of Bankers, Business Owners, Industry Owners, Media Owners, Politicians, is all an Illusion, a Facade, and exists only on the outside for appearances. These are TITLES, NOT REALITY. By any other name, a Criminal and Psychopath, is a Criminal and Psychopath. ACTIONS should be looked at, what these seemingly "Different groups of People" benefit from, not Job-Title, Job-Labels, and APPEARANCES. This is SIMPLIFIED-THINKING, STEREOTYPES, for Idiots who do not actually want to think. This is LAZY-MINDED. ACTIONS tell us about the "Type and Quality" of people we are dealing with here. You have to see how these "Different Groups are Benefiting", and which Specific-Individuals, to see that they are United. For instance how a particular Industry and certain Politicians benefit one another. There are plenty of books documenting all of these nefarious connections, so I digress, this is not the point of my book.

The point of bringing up the above is that these Power-Mongers, although not all agreeing with one another, and not united on all points, have indeed agreed to a general Game-Plan and Agenda. That involving the Downward Spiral of Cultures, People, the World for the Final-Goal of New World Order, which will result in ALL MONEY AND POWER going into THEIR HANDS. As a result these Elite, this Criminal-Organization, are focused on "Goals to Achieve" and have an immense amount of Power, Money, Energy, and Resources to go about doing it. The agreed to Game-Plan being New World Order, which amounts to the Ultimate-Power-Grab, a Centralized System, a Dictatorship. They have gotten away with so many other Schemes, this is the "Next-Level". And when it comes, like every thing else, they will SELL IT as being "GOOD FOR THE MASSES", a "STEP FORWARD", as "BENEFITING THE PUBLIC, MANKIND". What a joke, is that how things look in the World today? It will be about BENEFITING THEMSELVES. This is what happens when Corrupt-Politicians and "Others", the Elite, are NOT HELD ACCOUNTABLE by The People.

Now having given a "Complete Summary" of these Dementoids, there is another important element involved here. This involves the actual "Date and Planning" of these "Events" for "Spiraling the World and Humanity Downward". These "Planned-Events", these World-Traumas and World-Shocks, to happen shortly down the road time-wise, are "Timed and Based" on Hidden-Knowledge, Meta-Physics, what the average person does not know. It will all begin in 2012. Based on Hidden-Knowledge, what many would refer to as "Esoteric Knowledge", Ancient-Astronomy, knowledge which has in very large part been Suppressed.

The other end of these Elite and thus separating them from Mob-Families and Mafia in terms of Mentality and Psychological Make-Up, is that these "people" are "Esoteric" in one sense, and have many Hidden

Esoteric-Beliefs and Knowledge: or rather a deeper understanding of Science and Meta-Physics than the average person. This is a result of them sitting at the "Top of the Food-Chain" and having Access To Knowledge that the average person does not. These Elite are the reason that MANY TYPES OF INFORMATION IS HIDDEN AND SUPPRESSED. The idea here is that "Knowledge is Power" and so it is. They know more than most people and have the ability to Manipulate others based on this: "Aliens" as one example of Knowledge amounting to Power.

They have a Belief-System in regards to the World around them, akin to Meta-Physics, and are much more Knowledgeable about what REALITY ACTUALLY IS. This muddies the waters for the average person trying to understand these people, and throws a lot of people off. "Corrupt-People" okay, but start throwing in "Esoterics, Ancient Knowledge, and Meta-Physics" and you can forget it, right? Well WHY? These people hold KEY POSITIONS IN GOVERNMENT AND "LIFE", they sit upon the Knowledge Accumulated by the Masses, Scientists, the Military, the Knowledge of millions upon millions of people, they ARE going to know things that most people do not. To top it off, as a result of their "JOBS" they are able to use their Power to further HIDE AND SUPPRESS this Information and Knowledge and keep it to themselves. These people have ACCESS TO THE GREATEST SCIENTISTS AND MINDS IN THE WORLD, you do not. They keep all sorts of things to themselves, they are Selfish, Self-Centered, in ways that the average person is not. The idea of "Aliens" is one simple example.

THEY HAVE A MORE "COMPLETE PICTURE OF REALITY" AS A RESULT OF HIDING KNOWLEDGE. As a result they have the ability and Power to Manipulate Reality more fully.

Now having said that, what I am specifically referring to in regards to the Elite is Science and Knowledge, of Body and Mind and the World around them, again Meta-Physics, many Esoteric Beliefs fall into this. For instance the Elite know that the subject of "Chakras", a commonly talked about Hindu Belief, is real, along with the existence of Aliens, Bigfoot, and other questionable things. Another good example and relevant to this book and our World, is their understanding of the Mayan Calender and Hindu Cosmology—which is why they have chosen 2012 as THEIR DEADLINE DATE. As a result of being at the "Top of the Food-Chain" they have access to Knowledge most do not, they have access to Knowledge that "Governments Hide". They are able to "Verify and Prove" a lot of this Ancient-Knowledge and know that it is real as a result—from this they can also make "Decisions". They have a much wider and deeper understanding of the World around them than the average person, and are Criminal and Psychotic to boot. They make decisions at times, based on this Knowledge, which is sometimes "Ancient Knowledge" but Knowledge nonetheless—their Deadline Date of 2012 is one very important example of this. This makes it very hard to understand what "Motivates and Propels" these Elite at times, as what "Motivates" them many times may be unseen and unknown as it is based on what most people do not know of.

So the Masses have a hard time "Understanding the Mentality of the Elite" and they have a hard time "Seeing the Elite" due to a "Lack of Knowledge". Having said that the Masses are not exactly "Trying to See", either.

Now underneath this group of Elite is their "Henchmen", and THESE "HENCHMEN" REALLY ARE CULT. They believe in Cult in a Religious Sense, practicing many ritualistic things, and believe in being sadistic, cruel, and evil for the sake of being evil; but there is a very real difference between the two groups. The Henchmen believe in the "Religious Nonsense Aspects", they are "Fanatics" resembling "Religious Fanatics", whereas the Elite can be thought of in terms of Psychotic-Businessmen, many of whom really are. In one sense the Elite are "More Grounded in Reality" than these Henchmen (Clearly I mean this as I mean This), and do not simply believe in things because they are "Attractive". We are talking about two very different Types of People, with two very different Mindsets. The Henchmen are Followers and the Elite are Leaders.

The Elite, needing people to get their nefarious work done, are Leaders and Creators of a Secret-Fascist-Movement, a group of Fanatics with a "Team Identity", a Cult, which is what any Cult is. These Elite are not followers of "Cult" itself. The Elite have created this Cult based on an understanding of the Psychology of Fascism, Movements, Cults, and Fanatics. They are in it for the Power and what they are able to obtain from this "Movement" they have created, they are not actually "Followers of this Movement" itself. They are not a part of this "World-Cult" or "In This" for the "Subservience of Cult"—again they are Creators of a "Cult and Movement" not "Believers" of it.

I "Hammer on this Point" because it is a "Point" which many do not seem to get. There is a lot of "Talk about Illuminati", and a lot of "Belief that this Elite is Cult", but in my opinion they ARE NOT. Their "Henchmen Are", these Elite NOT. Which goes against what most believe in regards to an "Illuminati". If you look at "The Psychology Involved" these Elite are obviously ABOUT POWER, and SELF-GRATIFICATION, not about "Cult". This is a "Myth". They are "POWER-ORIENTED", not "Cult-Oriented", and this is where many people go wrong. The "Henchmen", this Elite-Team are "Cult", but should not be confused with the Elite themselves. This will only "Throw

People Off" in their "Assessments" of the Elite. They are so absolutely "Power-Oriented", there is no "Other Version".

These Elite are people Manipulating an entire "Movement of People" which they have "Created and Support" to get what they want: this Fascist-Cult is a "Tool". Again these Elite have created an "Underground Movement" serving their cause, to get what they want which is MORE POWER AND CONTROL. They have an "Ultimate Goal" of One-World ruled under themselves in which all POWER, CONTROL, AND MONEY, is in their hands. This is called New World Order, and it is their Final-Goal. Just as when Hitler got started, he had a Group of Hard-Core-Fanatics working for him, doing Dirty-Work, getting the Filth done, to achieve his Goals. When this is your Goal, Total-Power and Control, then you are going to need FANATICS, and a GROUP OF THEM, to get the "Job Done". This is a matter of Necessity for such "Goals". And so the Elite have done this, by creating a "Fascist-Cult" which is founded on "Greed, Self-Fulfillment, Lust, and Corruption" which plays on "Human Weaknesses", and they have given this "Cult" a "Team-Identity" with a "Common-Goal". It is all common, Basic, Psychology, for creating such a Movement in the first place.

It is easy for me to see and understand, because of this Mobbing I have experienced for almost twenty years. I have "Experienced and Seen" the "Mentality and Tactics Used" for Creating such a Movement, in the first place. And I have also "Seen and Experienced" the "Difference in Knowledge" between those "Who are Controlling the Movement" and those "Who are In It" and a "Part of It". Those "At the Top" are in it for "The Power" (The Corrupt Ones—like The Clown) and Self-Fulfillment, unknown to those "Doing the Dirty-Work" who are Clueless.

This Team of Henchmen actually believe all that they have been "Sold" like any Fascist-Movement or Cult. They believe all of the "Window Dressing", all of the "Esoteric Nonsense", all of the "Team-Sales-Pitch". Just like these Idiot-Citizens taking part in this "Mobbing", believing it is all for a good cause, and that I am the "Bad Guy", when in fact this started because of a Pedophile-Cop and a Racist indirectly involved in a Murder. These Idiots are Misused by Corrupt People telling them what to do for purposes of their own: like these Idiot-Citizens who "Help The Clown", never knowing that She is engaged in a Cover-Up. This Team actually believes that it is working for this "Common Goal" of One-World under Their Rule. And the thing is they actually are working for the Elite towards this Goal: the Elite just don't plan on "Keeping this Team", once it has "Fulfilled it's Purpose". It is like a Pyramid Scheme: when it collapses, The People at the bottom of it who do all of the work get "Ripped-Off", while The People above get all of the money and "Go Free".

It is just that like any Cult, or Fascist-Movement, this Team is being used for what the "People at the top Want". This Team believes it is also "Working for Itself", for IT'S GOAL of One-World under Them, when in Reality this Elite Plan on Eliminating Their Team when they have what they want. Like any movie, the "Middle-Man" will be "Removed and Discarded" once he has accomplished his objective. And so it will be with this Team. Common Sense dictates that if you are going to have Total-Control, you can not have an "Army of People" who are trained in being subversive, underneath you. To top it off this is all stated in an Old-Text written by the Elite from more than a hundred years ago, that they will "Eliminate this Team".

The point here is that like any "Fascist-Cult" this "Cult/Movement" has been Sold on a common Purpose and Goal for giving it "Unity" and a "Sense of Identity". The "Goal" is actually what this Elite want, and like any "Fascist-Cult" this Team is being used. They are "Taking People from All Walks of Life" and "Uniting Them" under "One Cause". In the meantime this Team of Idiots is going about doing what the Elite want, and helping the Elite to set up this New World Order. This Fascist-Cult believes it will get to live in the "Aftermath" of such a creation, New World Order, "New Atlantis"—which it will not. You are either a "Member of the Elite" or you are not, everyone else is Compartmentalized and Expendable. There is no "In-Between". But these FANATICS, this Fascist-Cult, this Team, do not get this and so things in this World continue to Deteriorate as the Elite and this Team go about their "Business" and all across the World.

These Elite are Psychotic-Criminals, who may practice "Ritualistic Nonsense" with their "Henchmen" at times, but are Non-Believing in "Their Own Sales-Pitch". In other words this Elite practices "Cult-Nonsense" (Believe it or not Folks there is evidence of this—Look at various World-Leaders "giving a particular hand-sign", like some Gang) with its Henchmen at times, to SELL it to its Henchmen and to ENCOURAGE it, but does not actually believe in any of it. A psychotic person might kill someone, it does not make them a believer in "Cult-Nonsense" and so it is with these Elite. The Elite are merely selling a "Cause" for Indoctrination, the way the Nazis sold a "Superior-Race", "NATIONAL PRIDE" to it's people. A lot of people do not get this about the Elite.

This is necessary, to "Sell a Cause", to SELL A PURPOSE AND IDENTITY to a "Team", for "Team-Spirit", and "Cohesiveness". This is to create a Core-Belief, a Self-Identity for a Collective—this is about "Selling a Psychology" to a particular Weak-Minded Type of Person, people with an IDENTITY-CRISIS, people lacking any Individual Self-Worth and Esteem. This is a necessary element for Manipulating Groups-of-People, into a Single-Unit, a Single-Identity, for creating a SINGLE COLLECTIVE. In "Star Trek Terms" this is how sick people, the Elite, go about creating the BORG. A GROUP OF PEOPLE WITH A SINGLE MIND AND VISION—again what we speak of here is FANATICS. In this case these Henchmen really are "Cult" and really believe in it, as in the Stereotypical Definition, whereas the Elite are simply Dangerous Criminals, Psychotic, Manipulating a "Team of People", a Movement, into getting what they want. Like any "Fascist-Cult" these members are blind to how they are being used, and in what way. Of course "Followers" do not get to "Call the Shots", and are always used by The People over them in Cults and Fascist-Movements.

This Movement of Henchmen exists all around the World, doing the Bidding of the Elite, capable of all things sick and twisted. They are Fascist and Fanatics, who believe in THEIR VISION OF THE WORLD. A Goal of "Their World", a "World Under Their Rule" which this Elite also "Sells" to them. If you look, all Fascist Movements are CULT, and all all CULTS are FASCIST MOVEMENTS. And in either case each believes that how it "Sees the World" is "THE" way of "Seeing It". So it is with these people. They are the perfect people for the Elite, Fanatics are BLIND and therefore EASY TO MANIPULATE. Think of Robbin Williams character from the movie "The Survivors" to get an idea.

These Elite, are "Godless", "Non-Religious" in any form. They believe in what amounts to Science and Knowledge only, and ARE Psychotic and Evil. THEY BELIEVE IN POWER ONLY. No matter how you slice it, these Elite already are totally evil, by anyone's definition. They are "SELLING A RELIGION"—Cult, that they DO NOT BELIEVE IN, to a bunch of Weak-Minded-Beings, for Indoctrination, for Mind-Control, for Compliance, and to create Fanatics—those who will do anything for a "Cause" and "For the Team". In this case "The Cause", in the end is merely a matter of what the Elite really want, everyone else—including this Fascist-Cult, is Compartmentalized and in the dark as to what is really going on. What the Elite really want is for THEMSELVES ONLY, and everyone else, including this "World-Cult" and "Team", is simply in the way, people to be used until they have what they want. Everything else is a Smokescreen, an Illusion, Window-Dressing for the Blind, to gain "Subservience" and Blind-Compliance. These "Henchmen", this World-Cult, feel they are a part of the "Privileged Few", in Reality they are being used and "Putting their Faith" in the hands of Psychotic-People. How intelligent is this? The whole thing is crazy and these Elite and Their

Team are "Aiming For" a 2012 Deadline Date. This is when "Everything Starts".

These Elite are like SALESMEN, Sales Representatives, convincing people to work for them, to sell their product, when the whole thing is a PYRAMID-SCHEME. Only the ones at the top of this Pyramid ever really benefit long term, and in the end the whole thing always comes down. These Elite are Psychotic Business Men, rather than "Dark and Evil Cult-Types", these Henchmen just have not figured this out yet: along with a lot of people running around "Talking about Illuminati". To be sure they are evil, and in every sense of the word, just not in the sense that most who have knowledge of the Elite, believe. This Cult-Stuff is a SALES-PITCH for a Pyramid-Scheme. These people are the Ultimate-Con-Men, and are so, because they are Psychopaths, Total-Manipulators. They could care less about The People under them, thus there are two separate classes: the Elite and their "Team of Henchmen", who are Fascist-Cult. The Elite are only "Cult" in terms of Science and Knowledge, which really translates to "Esoterics", and Meta-Physics. In Reality they are "Psychotic Business Men", "Reality Merchants", with a lot of Hidden Knowledge and Power, Hiding their Intentions and Goals from all, because they are using all.

In 2012, based on their Knowledge of Ancient-Cosmology, Meta-Physics, Hidden-Science and Hidden-Astronomy, the Elite have picked a "Deadline-Date" for Implementing the Beginning of their Plans. Their Plans are NOT based on "Esoteric-Cult-Kook-Nonsense", rather on Science, Astronomy, Ancient-Knowledge, Physics, Divine-Law, which these Elite understand and therefore have picked a Deadline-Date, based on this. They have Access to Knowledge, that most people Do Not, Hidden Knowledge. This involves Real-Science, Real-Astronomy, Real-Processes that the Elite get and understand, and which they have Scientifically Observed, Studied, and Verified.

The average person, does not know of this "Material", never mind the "Scientific-Facts" which are out there that Verify This. Never mind the fact that the Elite have picked a Deadline-Date, based on this material, and the Science behind it. In fact what Motivates this Elite and what they do, what they have Planned, and WHY they have "Plans" is all based on this understanding of Hidden-Science, Ancient-Astronomy, and "Meta-Physics". Again this all gets down to these Elite, having a more Complete-Picture of the World and Reality we live in, and having access to Knowledge that the Masses do not. It all gets down to "Knowledge is Power" which many times it is: especially in this case. With that in mind we now move on to the next chapter "The Elite Deadline" in which I explain, or try to explain, what it is that is going on, the "Bigger-Picture" that most do not see, and why. I explain what it is that is Motivating the Elite these days, and why. If you "Get" this next chapter, a lot of what you see in the World, will make a lot of sense based on this—this is what is motivating the Elite these days, this is "At the Center of it All".

CHAPTER 25

The Elite Deadline: 2012 and Why: A Cosmic-Cycle, Yugas and a "Central-Sun": An "Electromagnetic Space Cloud", Called "Local-Fluff" by Astronomers—Which Contains "The Space Ribbon"—A Cloud of Light: An Opening to "Units Of Consciousness", the Path of Consciousness or the Path of Experience?

Of all of my chapters in this book, this is one of the most important. This explains why the Elite have a Deadline-Date of 2012. Since none of this Knowledge is "Common Knowledge" this makes the subject difficult and hard to believe for many. The material itself is difficult for some, never mind the "Documented and Factual" end of it. Having said that I do my best to explain. This involves Ancient Knowledge, what many commonly refer to as the Mayan Calender, or as I prefer because of the amount of knowledge, Hindu Cosmology. This involves Science, Physics, Energy, and Astronomy, and from a deeper perspective this involves a Cosmic-Cycle and a Divine-Plan. I am talking about a Conscious-Design of our Universe and it's "Rhythms", it's Cosmic-Cycles. A Conscious-Design by an "Energy" commonly referred to as "God, the Divine, Allah" what have you—in the end these are all names for the same thing, CONSCIOUS-ENERGY.

This chapter can be considered a brief summary of the subject. I have a far longer, more detailed chapter in my second book, which I will get printed later this year.

These Cosmic-Cycles are not known of or very well understood by many in our day and age, yet if we look at Past Cultures and Civilizations, we can find that these Cosmic-Cycles were indeed known of and understood. The Mayan Calender and the Hindu Concept of "Yugas", are good examples. I will focus on the Hindu Knowledge as there is much to be found. This is Old-Science, Old Knowledge, Ancient Astronomy, describing Earth's orbit through our Galaxy and Universe, the orbit of our Solar-System around what is called a "Central-Sun", a "Sister-Sun" in regards to our own. In terms of Astronomy this is not a new theme, this perhaps seems like a "Strange Idea", however it is not.

Since we are talking about "Space and Time" here, we are talking about large units of time. We are talking about time in "Astronomical Terms"—thousands and thousands of years. We also happen to be talking about "Types of Energy" in Space, which we as a people are not necessarily familiar with, as our planet and Solar-System does not often encounter this "Energy"—a certain type of "Space Anomaly". Having said that there are several "Space Anomalies" we encounter in one sense, as this Larger-Cycle is about the "Lining Up of Galaxies", and an even "Bigger Picture" than the one I go into detail here.

Having said that our Earth and Solar-System do follow a regular route through the Universe, a regular orbit, just as our Earth follows a regular orbit around our sun. Think of it this way, our Earth orbiting the sun is a smaller version of our Solar-System orbiting a "Central-Sun", a "Sister-Sun". There is a Larger-Pattern here, involving a larger amount of time—our Solar-System is not haphazardly traveling through space, it has it's own orbit, and rhythm. Again "Space, Planets, Cycles, Orbits, Galaxies, Suns" it is all an "Ordered Universe", there is a "Method to the Madness", planets, suns, Solar-Systems, what have you follow regular patterns—Cycles and Orbits. Occasionally these different "Space Bodies", different Planets, Solar-Systems, Galaxies interact with one another creating yet another type of "Rhythm or Cycle". Smaller Patterns and Rhythms interacting with one another to create Larger Patterns and Rhythms. It is like an Orchestra, a Symphony, a weaving of different patterns to create a Larger One—all the different instruments with their own particular part to play, blending to create a piece of "Classical Music". It is all like the inside of a watch or clock, intricate, each piece working in conjunction with another to create a regularly functioning "piece", watch, clock.

These infrequent, yet regularly occurring "Interactings" between different "Space Bodies" can be called "Cycles". Different "Orbits, Cycles, and Rhythms", different Space-Bodies, meeting infrequently to create yet another Regularly-Occurring-Pattern, a Larger-Pattern involving a Larger-Amount of Time. Smaller "Rhythms and Orbits" working together to create "Larger Rhythms and Orbits—what I call a Cycle, a Cosmic-Cycle.

Like our Earth and Solar-System completing a "Cycle" every 24,000 years which involves several Processes every 24,000 years, and involves "Certain-Space-Events" every 24,000 years. Every 24,000 years for example our Planet and Solar-System encounters certain "Space-Anomalies", certain "Bodies in Space" which interact with Earth and have an affect on Earth and our Solar-System. These "Space-Bodies" are currently Known Of, and documented through Science and Astronomy. Having said that these "Space-Bodies" are NOT being presented to the Public of the World in a way that is appropriate, or excusable. Well, one in particular, although there is "More than one Involved" when looking at the "Bigger Picture". But one in particular is the one that concerns us.

This has always been going on with the Earth and our Solar-System, and always will. The "Problem" for most people is that they do not know of such things. Seeing as how these Cosmic-Cycles take thousands and thousands of years it is no wonder: Knowledge becomes lost over such large periods of time. Mankind is constantly "Moving up and Down" in terms of "Civilization and Consciousness", thus Knowledge of such things, Knowledge of these Cycles—Earth's travels through the Universe and it's "Encounters with other Space-Bodies", becomes lost. Having said that such Knowledge of these Cosmic-Cycles DOES EXIST, has been passed down through the ages through various Cultures—the Mayan Calender and Hindu Cosmology being good examples of this. This fact is not, and has not, been lost on the Elite and Governments run by the Elite. In fact they have picked a "Deadline Date" for implementing a lot of their "Corrupt-Plans", for Manipulating the Masses, based on this.

Ironically our "Modern Age" does not get that this is REAL KNOWLEDGE of a real regularly occurring Astronomical-Cycle, does not get the "Earth's Encounters" with "Other Space Bodies" along the way as we travel an Orbit through the Universe. This is all very normal, it is just that we are talking about a large amount of time, and an "Occurring of Things" on a "Very Rare Occasion" from a Human-Lifetime perspective. Having said that there is Ancient Knowledge and History of such a Cosmic-Cycle, there is Ancient-History recording such a Cycle and the "Results of It" on Earth and Her Inhabitants. The average person does not know of this, the Elite and Governments DO. Science and Astronomy can verify this, and in fact HAVE. Unfortunately the Elite, and Governments who are Controlled by the Elite, are keeping this Information to themselves. For reasons of Their Own.

Anyone can find this "Hidden Information", and draw the appropriate conclusions. This Information is out there, it just has not been "Advertised for what it Is". There is something in space known as Galactic-Fluff, with a Strong-Magnetic-Field within it, and this has a Ring of Light within it as well, the Giant Space Ribbon, which seems to be "Reflected off of this Magnetic-Field". We the Earth, are approaching this. No doubt this Magnetic-Field is what is causing Earth's Magnetic-Field-Reversal, and no doubt this "Ring of Light" is what many refer to as the Photon Belt, and as "The Rapture". This is also affecting the Earth's Electromagnetic-Energy, the Schumann Resonance—"Thoughts being Electromagnetic-Energy", the Earth's Consciousness is rising, for a Scientific-Measurement of the Metaphor. This "Light Ring" in my opinion is a Higher-Energy, what I call "Divine Light or Energy", and is made up of "Units of Consciousness", the "Total Collective" of these "Units of Consciousness" being what many call God, Allah, the Divine, what have you. As we and the Earth near this "Higher Energy", like ALL MATTER, we ABSORB IT, and RESONATE WITH IT. It is Conscious Energy, and it is no "Coincidence" that we are approaching this at the "End of a Cycle"—what many refer to as "The End of the Mayan Calender". It is a "Sign" and "Heralding" of a New-Cycle. But for now, I digress.

The Elite and Governments have verified this Ancient Knowledge, have studied and measured it Scientifically, and are aware of this Cosmic-Cycle and aware of "What is to Come". This Cosmic Cycle being about an Energy-Process which the Hindu's call the Night and Day of Brahma, and this "Magnetic-Field and Light Cloud" being a separate part of this Process. "What is to Come" being our approach to this "Space Fluff" the Strong-Magnetic-Field within it, and this "Cloud of Light", and our ENTERING IT. We are in fact nearing this, and "Our Course and Path" is already predictable. So there are Two-Parts to what I speak of here. An "Energy Process" that occurs every 24,000 years which the Hindu's call the Night and Day of Brahma and which is "The Bigger Picture" for these Elite, and there is a separate part of this Astronomical-Cycle this Magnetic-Space-Cloud, which will be the "Astronomical Part" that everyone comes to know about. For all intents and purposes, one is more "Physical" than the other.

As this is all part of a Regularly-Occurring Pattern, it is Predictable and Knowable. It is also Scientifically Measurable and has been done so. These Elite have Knowledge, Scientific at that, and know the Affect these "Other Space-Bodies" will have on Earth and Humanity. In fact the "Affects" are occurring NOW and are even being "Observed and Measured" NOW—by Scientists, Governments, the Elite. So the Elite are aware of such things, are aware of this Cosmic-Cycle, are aware of this "Bigger-Picture" and knowing what is going on, have Plans for "Using this Cycle" to their own advantage: including this Light-Cloud.

As one example, shortly in the future there will be one Cosmic-Event that will seem very strange and no doubt "Terrifying" to quite a few people on Earth. The Elite and Governments in typical fashion are remaining QUIET about this event so that it comes as a "Shock and Surprise" to the Masses. Having said that they have "Publicly Released" information about this "Future Space-Event", twice in the Media—I cover one of these in my next chapter. This "Space Event", this "Cosmic-Event", is also mentioned in many "Prophecies". This Space-Event will be the Earth entering this Cloud of Light, and when we do, there will be light all around the Earth, "Night and Day". There will be NO DARKNESS at all. Now.

At the end of this "Bigger Picture" this Cosmic-Cycle, is a Divine-Plan, a Conscious-Plan, and this Cycle, although "Mechanical in Nature", is Deliberately-Designed. By what many call the Divine. Again it is all different names for the same thing—Conscious Energy. Thus our Solar-System, Earth, Humanity, are following a Divine-Plan, an Actual-Divine-Road-Map, through Space and one which involves Consciousness as well. It is all about "Creating and Experimenting" with "Consciousness and Creation" from a "Higher-Perspective". This is about the Divine, God, Allah, experimenting with it's Creations, and Experimenting with Consciousness.

In essence there is a Plan, a Road-Map for Mankind, Mankind's Growth in Consciousness, and no matter what, Mankind is traveling on this "Path and Road" which is now heading "Upward". This is all a part of a Regularly Occurring Pattern, a 24,000 year Elliptical-Orbit and Cycle around a Central-Sun, and depending on where we are in this "Orbit" determines the "Direction we are Going". This Central-Sun which is composed of Higher-Energy like this Light-Cloud, and whether we are going away from it, or towards it, determines whether our "Energy and Consciousness" is going to "Lower or Rise Higher". As it is Now, the Earth is traveling in a "Direction" which will eventually, over the course of thousands of years, Evolve Mankind's Consciousness "Higher and Higher" as Mankind and Earth approach this Central-Sun.

The Elite and Governments know of this, and in fact Hindu Lore is a good place for getting facts about this: this is in fact Recorded History and Science. The Elite with their Satellites, Scientists, and Super Computers, know that our Earth and Solar System orbit a Central-Sun, and they know that this is composed of "Higher Energy". They also know that it has an affect on "Consciousness" and I am sure have confirmed quite a bit that most people simply take for granted as being "Myth and Lore". They have "Verified Past History", Ancient-Scientific-Knowledge, and Verified Ancient-Astronomical-Knowledge. They understand why there have been "Advanced Cultures" in past history, before ours, and understand that our Earth is "Destined for a Higher Road" once again. This "Higher-Road" being about a "Higher-Consciousness" for Mankind—something the Elite do not want.

This Central-Sun is composed of a Higher-Energy, and all "Matter" all things physical, absorb Energy: this is simple Physics. So as Earth approaches this Energy Source, the Earth and Mankind, absorb this Energy. The Earth and Mankind absorb it, become Saturated with it, Resonate with it, Vibrate with it, thus Mankind's Energy and Quality of Thought, is "Raised": Higher-Consciousness is Experienced, and on a Collective-Level. Which is the reason for "Legends of Advanced-Past-Cultures" like "Atlantis and Lemuria", as they were closer to this Higher-Energy, this Central-Sun, and their Consciousness was "Higher". In fact "Yoga" is a Science of Body and Mind, which was developed by these Past Cultures for "Maintaining this State of Consciousness", for being able to "Reach It", even as the Earth began to get farther away from this Central-Sun, as it was on the "Other Side" of this Cycle, and getting further away. Yoga was developed so that The People could reach this State of Consciousness, for when this External-Source, which acts as Consciousness-Booster, was no longer present. This Light-Cloud is having a similar affect on Earth and Humanity.

This Journey "Upward in Consciousness" for us has already begun, we are already on this New-Road, as our Planet and Solar-system are on a "Return Journey" towards this Central-Sun. We as a Planet have "Passed the Half Way Mark", only recently, and we are now going in a direction towards this Central-Sun again. This will result in a long, slow, gradual, "Growth in Consciousness", but "Accelerated Growth" nonetheless. However due to "Cosmic-Cycles", the "Meeting and coming together of Smaller-Cycles", this Journey does not "Officially Start" until other Cosmic-Cycles have occurred. In short what is talked about by the Mayans and Hindus, is about to "Officially Occur". The "End of an Old-Cycle", and the "Beginning of a New-Cycle" which will be "Heralded" by this "Light Cloud". We are talking about Science, Astronomy, Energy, and Physics. There are Processes involved here that most people do not understand, however the Elite do.

This "Ending of an Old-Cycle" and the "Beginning of a New-Cycle", The Night and Day of Brahma, will be marked by the "Happening of a Space-Event", which happens once every 24,000 years. You can consider it a "Book-Marker of Time" of sorts, a "Sign along the Road", that lets Earth's Inhabitants know where it is along this Cosmic-Cycle, in one sense. Having said that this "Light Cloud" is more than that, this involves "Consciousness" and a "Realigning and Retuning of Mankind's Consciousness" so that it is "In Line" with this "Divine-Road-Map". So that Mankind's-Consciousness is also traveling in the "Right-Direction", in line with a Larger-Plan. This Light-Cloud is in one sense, like a Tuning-Fork for Mankind, so that it can "Re-Resonate Correctly", get "Re-Tuned" with the Direction and Energy it is going in. The direction it is going in, is towards this Central-Sun, and therefore "Towards Higher-Consciousness", and part of a "Bigger Conscious Plan". Again this is the Divine's Experimentation with it's Creations and Consciousness.

This Light-Cloud is meant to act as a Consciousness-Booster-Shot for Earth and Her Inhabitants so that "All are In-Line" with this Cosmic-Plan of Higher-Consciousness and the 12,000 year Plan of "Getting There" to this Central-Sun: which is when Mankind will be at "It's Peak" in terms of Consciousness. This is all "Planned" in one sense by a "Higher-Power", and you can consider it "Symbolic". If you like the "Scientific Approach" we are nearing an "Energy in Space" which is already affecting the Earth and Mankind's-Consciousness, and when we totally "Enter this Energy" the affects will be quite noticeable. This "Energy in Space" has already been noted by Scientists and Astronomers, and is a known documented fact. This Light Cloud is a "Smaller Version" and "Taste of This" Higher-Energy, and will interact differently with "People and Consciousness" than this Central-Sun, when we enter it. Having said that it will indeed have a very noticeable affect, and for some, already is.

As you can see this "Light Cloud" is known of Scientifically, however nothing is being said about our Earth entering this "Cloud of Light", or of our Earth nearing this Strong Magnetic-Field. In fact no one is saying we are going to enter it at all—well mostly. I have seen only two different sources, stating that we will. Needless to say this is already having an affect on Earth, a Scientifically Measurable one, and is already "Affecting Earth and Mankind" although most do not know it. Again the Earth's Magnetic-Field Reversal is a sign of this, and if you look, this is being Covered-Up and Lied about. Another example is of the Schumann Resonance, which again is being Lied about and Covered-Up. This is Proof that the Elite know of this all, know of the affects, and is Proof that they are Hiding Everything for the sake of Manipulating the World and Humanity. All of these "Anomalies" are explained by ONE SOURCE, this Magnetic-Space-Cloud and Ring of Light within it, which quite obviously no one is speaking of—the Affects and how this is all Interconnected and Related, that is. When there is so much Science to it, it is not simply a matter of "Kook Talk". It is Scientifically Measurable and you need not be a genius to see how it is all Interconnected and Related.

The Elite know of all of this, are aware of all of this, and so have Plans to "Hijack this Event" and to try and "Hijack this New Cosmic-Cycle". They are going to try and Manipulate several Processes here, and Processes which occur only once every 24,000 years. Two Different Ones actually. Not wanting to "Miss This", not wanting to wait another 24,000 years, they have a "Strict Deadline" in place for 2012 which they have Every Intention of adhering to. My previous chapter on the American and World-Economy-Collapse is all a part of this, it is in fact about Manipulating Mankind's Energy due to this Cosmic-Cycle and the Processes involved.

They know a "Course has been Plotted" for the "Raising of Mankind's Consciousness", a Divine-Road-Map, and the Elite WANT TO SUBSTITUTE THIS, with "Their Own Road-Map".

They want to substitute "Their Own Cycle" for the "Divine's", they want to put "Their Plan and Road-Map" in place by trying to Manipulate "Divine Law", the Meta-Physics involved, for their own ends and therefore "Hijacking the Divine's Plan and Cycle". Since this "Space Event" involves Cosmic-Cycles, Divine-Law, and Divine-Plans, involving the "Elevating of Consciousness for Mankind", the Elite have come to the conclusion and rightfully so, that they are eventually going to become "Extinct" for a lack of better words. The Elite will not be able to hide or exist with a Conscious People, a Conscious People en masse at that. In essence the Masses will be on a "Road to Higher Consciousness" and the Elite will no longer be able to live or function in such a World—thus they have to do something. Something for the "Sake of their own Survival". Instead of a "Conscious People", they prefer a people that is on the "Path of Experience" instead. Instead of letting the Consciousness of the Masses slowly rise, they Plan on "Lowering Mankind's Energy", replacing it with Negative-Energy,

they Plan on "Bringing Humanity Down" simply so that they can Survive and Thrive. At cost to the Masses of course. They want to "Install a Negative Timeline" and Negative-Road-Map.

Now as this "Old-Cycle" officially ends and a "New Cycle" begins, there will be an "Energy-Process" going on, the Night and Day of Brahma. If all goes "Normally", "Negative-Energy" will be "Collectively-Cleared" by this Night and Day of Brahma, and then "Reoriented in a Positive-Manner" with a bit of "Help and Tweaking" from this Light-Cloud, which will have a "Positive-Affect". That is the "Divine Plan" for "Re-Orienting Mankind" onto this Positive-Road-Map. Mankind will become "Realigned and Re-Tuned" by this Space-Cloud, and begin a Journey towards this Central-Sun on a much more "Positive Note". Negative-Energy will be "Cleared" and Humanity and Earth will begin this Road-Map with a Positive-Inclination and Orientation. You could say that the "Overall Energy" of Humanity and Earth will be "Positive Oriented", and "Set in Stone" that way. However the Elite do not want this.

Now it is much more complicated than all of this as there are several "Cosmic and Divine Processes" going on here. There will be several things happening, and the Elite understand all of this, they get the "Larger Processes" involved. They get that there is The Night and Day of Brahma, that it is a matter of Energy and Timing, a "Re-Balancing of Energy" on a Cosmic-Scale. This involves the actual "End of a Cycle" and the "Beginning of a New-Cycle", based on "Energy and Timing": it is like a Re-Setting of a Stop-Watch to zero, so that all can begin "Anew" again. This is in Fact what is about to happen. This "Event" will in and of itself occur in 2012—this Event is what most people are referring to when they speak of the Mayan Calender and/or the "End of Time". The idea of the "End of Time", the "Non-Existence of Time", being a reference to this Process. Having said that the Gregorian Calender is "Off" and so this actual "Resetting of the Clock" the "End and Beginning of a Cycle" will actually occur a few months before the expected date of "December 2012" or a few months after, which would actually be in 2013. It will be one or the other.

This "Resetting of the Clock" to zero, this "Clearing of Energy" in 2012, actually involves Bigger-Processes and Divine-Laws, and the Elite understanding this have chosen 2012 based on this. This involves "Laws of Creation", Divine-Laws involving Energy, Creation, and the Physical-Reality we live within. This involves Meta-Physics which the Elite also understand. There will be a "Clearing of Energy", a "Zero Point", or in our case for Earth and Humanity—there will be a "CLEARING OF NEGATIVE-ENERGY". This is to "Reset Mankind's Energy" to Zero, to Realign Mankind's Path onto a "Positive Energy Path". So in "Theory" if Humanity does not screw up, which it will, first "Negative Energy" will be cleared, and then a Positive-Energy and Path will be followed, in which Mankind "Grows in Consciousness" slowly and progressively over thousands of years. This "All Begins Soon".

This will be a Removal of "Collective-Negative-Energy", a "Removal of Negative-Karma" for Mankind, and a "Resetting of Energy to Zero", a "Resetting of Energy to Neutral". This is a very Important Process for Mankind and Mankind's Future, and if you understand it like these Elite, a very important Opportunity for trying to CHANGE THINGS and based on Divine-Law. This "Zeroing of Energy" will be for a Very Short Time, and once the Earth "Leaves This", what ever Energy is Present, will become "Cemented in Place": a choice made about "What Path We Are On". A Path of Consciousness or a Path of Experience. Humanity is being "Given a Chance" to "Realign Itself" and to "Choose a Higher Road", to choose the Path of Consciousness. It is in fact what is Intended. Having said that there are the Elite who want to "Change This", and so Plan to Manipulate this Energy-Process, to Mis-steer Mankind, so that a Negative-Road-Map is chosen.

If Mankind "Rejects this Natural Process" by "Staying with Negativity" and by "Remaining Fixated On It", then Mankind will HAVE TO CLEAR ALL ACCUMULATED NEGATIVE-ENERGY ON IT'S OWN. This involves the Path of Experience, and learning the Hard-Way. This is in fact what the Elite Plan on getting Humanity TO DO. By Manipulating Humanity with World-Traumas, they Plan on Hijacking this Divine-Process, and hope to Create a Negative-Road-Map for Humanity and Earth. Humanity will be "Creating Negative-Energy" with "Power of the Mind", will have Negative-Expectations, and so will be Manipulated with Negative-Energy, into choosing Negative-Energy, and a Negative-Road-Map: all based on the Elite's Understanding of Laws of Creation, Divine Laws. It is a Stupid Plan, in Denial of very Basic Principles, Physics, and Reality, and in the end, will NOT work. The Elite will only "Delay the Whole Process" and Screw It Up for a while but it will suck for everyone.

If you are going to try and change this Divine-Plan and Road-Map for Mankind and Mankind's Consciousness, then you are going to have to operate within the parameters of these "Divine Laws" and Manipulate them. This is what these Elite want to do, and PLAN TO DO. This is an Opportunity which happens once every 24,000 years. This means that the Elite will have to "Affect this Resetting of Energy to Zero", by "Inserting Negative-Energy", and they will do so by Creating Large Traumas for The Entire World—the Economy-Crash being the BEGINNING OF IT. Therefore when Mankind leaves this "Zero-Point" Mankind's Path and Energy will be "Set on Negative". Say what you like but these Elite do not think small.

They want to Manipulate this "Zero-Point of Energy", so that when this "Night and Day of Brahma" ends and a NEW 24,000 Year-Cycle begins, it will begin the WAY THE ELITE WANT, and this "Reality" they want, a Negative-Reality will become set in "Stone" from the beginning. Like "Quick drying Cement", or "Glue", as we emerge from this "Neutral-Point" they are hoping to Reset Mankind's-Collective-Energy back on "Negative", and to begin a NEGATIVE-TIMELINE. Everything from that point onward, in their minds, will be a part of Their-Negative-Timeline, a "Negative-Reality" run and Controlled by them. Ultimately it will not work, but these Elite could obviously care less about Reality. They have not looked at the Bigger-Picture fully enough. Quite frankly they are Desperate.

In essence they want to Create a World, in which Mankind's Consciousness is LOWERED and CONTROLLED by Themselves, the Elite. An Elite-Utopian-World in which everything is Run and Controlled by them, and in which the Masses have no clue as to what is going on, and have absolutely no power or future. In essence all will be "Worker Ants" while the Elite live in their idea of a "Dream-World". They are hoping to Seize-Control, a Dictatorship with no room for "Free-Thinkers", and to "Keep this in Place" permanently, so that any road to "Higher-Consciousness" is "Thwarted".

Science-Fiction movies in which the Future is run by Oppressive-Governments who Control All, are what we are talking about here. They want to Create a Dismal Future in which they are the Ultimate-Rulers and the Masses are Total-Servants. And this will be a "Future" in which they are totally obvious about this, or "Barely behind the Scenes" in terms of how they rule over the Masses. And they want to do this all by Manipulating Cosmic-Law, Divine-Law, the Laws of Creation and Laws of Reality, to get what they want. They Plan to "Hijack this Divine-Process" which is Positive, and Plan to substitute a Negative-Road-Map by doing so. So as you can see, there are very few who get what the Elite are doing, there are very few who get what Motivates the Elite and what Motivates their "Deadline-Date" of 2012. In fact there are very few who even know that they have a 2012-Deadline-Date in place.

We are talking about Manipulating Divine-Laws, Manipulating Reality on a Cosmic-Scale, by way of Divine-Laws so these Elite can continue to live in a "Less-Conscious-World" and continue to exist the way that they want. So that they can remain in Power and Manipulate and Control others as they please. We are talking about substituting Elite-Plans for Divine-Plans, altering Timelines for Humanity (the Future-Timeline for Mankind is in fact meant to be a Positive-One), and Creating a Miserable Future for all of Mankind, so that an extremely small number of people can remain, Fat, Filthy, and Corrupt. This is what all of this is about, this is why these Elite have a Deadline-Date of 2012 picked. They Plan to "Overthrow" this "Positive-Process" and to substitute a Negative-Road-Map. And while I only know that this Divine-Process, the "Night and Day of Brahma" will occur a few months before or after December 2012, I bet these Elite have a much more accurate date.

Now for a comparison lets look at the "Divine-Plan" here. The Plan is to Re-Orient Mankind, Mankind's Energy, with this Divine-Road-Map for Mankind and Earth. We are already on this Road, but it will officially start in a couple of years—after this "Night and Day of Brahma" has occurred (2012 or earlier 2013). The Plan is for Mankind's-Consciousness to Rise, the Raising of it, for many thousands of years—in fact this is a Positive-Timeline which Mankind is traveling upon. This will happen no matter what as we travel along this Road anyhow, but we are given a chance to START on this Road in Harmony with this Divine-Plan. Thus there are Processes going on to Re-Orient Mankind's-Energy and Consciousness in line with Mankind's Journey along this Positive-Timeline: that is what this Light-Cloud is for. This Light-Cloud is to "Stimulate Mankind" in a Positive-Fashion and to "Re-Orient Mankind" in a more Positive Manner. So that Mankind's-Consciousness is in "Tune, Balance, Harmony" with this "Divine-Road-Map" Mankind is traveling upon. It is this in fact that has led the Elite to "Conclude" that they are indeed, down the road, going to become extinct. There are other points as well, but this is the gist of it.

So to be clear here, this "Light Cloud" is a lot MORE than simply a "Cloud", a lot more than a "Physical Event". This Cloud is for helping us to begin the "Path of Consciousness" and thus in Harmony with this Larger-Divine-Road-Map. Having said that this Energy is "Optional", and has more of an affect if you are open to this Energy. This is not the same as being "Saturated in the Energy from this Central-Sun" non-stop, this is going to be a different process. When the Earth is near this Central-Sun, this will all be guaranteed, the Road already Paved and Chosen at that point. With this Light-Cloud, it is another matter—Mankind is being given an "Option and Choice" to start "Early and from the Beginning" on the "Right Road". It is "Optional" and is in fact a matter of "Choice". If we do not Collectively choose a Positive-Road, if we reject the Bigger-Plan, then this Light-Cloud is for those who STILL CHOOSE A POSITIVE-ROAD. It will be for those only, who stay on the Path of Consciousness. Thus in one sense this is a "Test" as well, and a "Rescue Package" from the Divine for those who stay on the Path of Consciousness: regardless of the Majority of the Masses choosing the Path of Experience.

This is what many in the New Age Community are speaking of, when they speak of "Their Energy Raising Them" into another "Vibratory Realm". This is what many have Prophecized when seeing and speaking of "The Rapture".

For a moment, for a comparison you can think of drugs, and whether you want a "tranquilizer" or not, decides how effective it can be. If you want the sedative effect you succumb quickly, if you do not, then you fight it off and may not succumb at all to it. So it is with the Energy of this Space-Cloud, which is a very Different Type of Energy than any ever seen or experienced by any on Earth at this time. It is in fact a Higher-Vibrating-Energy, what I call "Units-of-Consciousness" to borrow a term from Seth and Jane Roberts, an "Entry of God-Energy" onto and into our "Physical-Plane of Reality". For some you might call this the "Entry of another Dimension" onto our "Physical Plane" of existence. Scientists from afar are calling these "Particles and Energy" radiating from this Space-Cloud, ENAs, or Energetic Neutral Atoms, which I find interesting as the Earth is going to reach what I call a "Neutral Point" in terms of Her and Mankind's Energy. But I digress.

Already the Energy of this Cloud is having an affect on Earth and Her Inhabitants. In regards to Humanity, some are affected and some are not (not noticeably affected) based on THEIR OWN ENERGY. One's Energy, based on a leaning towards Positive or Negative having an affect on one's "Absorption" of this "Light-Energy". One's Consciousness being "Open-Minded or Closed-Minded", Conscious or Non-Conscious, having an affect on one's absorption of this "Light-Energy". Shortly down the Road we will all be immersed in this "Light-Energy" this "Light Cloud", quite literally, and this is commonly referred to as the "Rapture" in many Prophecies. The entire Planet will be IMMERSED in it, AND THE ELITE KNOW THIS.

Thus depending on how you are MENTALLY, determines what affect this LIGHT will have on you. If you are "Positive" you are affected more intensely than if you are not: again like my "Tranquilizer Example". If you are "Negative, Close-Minded, Non-Conscious" then you will be affected far, far, less to the point of "Nothing that is Noticeable". It is in one sense a "Choice", and a "Choice of Consciousness". In fact this Energy already is affecting many and the Consciousness of many is being "Raised", "Helped", being given a "Spiritual Booster-Shot". These are The People on the Path of Consciousness and this is a matter of "Thought and Energy". Keep in mind that one's thoughts are Electromagnetic Energy, and one's "Aura" is simply the "Overall Pattern and Flavor" of one's thoughts: the flavor of one's "Energy-Field" created by one's thoughts, which surrounds an individual.

When this time comes, those Truly-Conscious and of "Higher-Energy and Consciousness" will totally RESONATE with this Higher-Energy, and "Vibrate Away". Higher-Spectrums of light for example, can not be seen by the Human-Eye, because they "Vibrate too Quickly", so it will be with "Certain People" down "The Road". Thus Stories and Prophecies of the "Rapture" which is about "People Disappearing"—"One Disappears while the other Remains", because of "One's Energy" and depending on whether or nor they RESONATE with this "Divine-Energy". Thus the "Truly Good People", the Conscious People, those on the Path of Consciousness, will RESONATE with this energy, their "Vibrations Speeding Up" to the point that they "Disappear", just as various spectrums of light are invisible to us. But for now I digress on this point.

The original point was to compare the Divine's Plan with that of the Elite's. Having said that, now you have something to compare. While the Divine's Plan involves a Raising of Consciousness and a Re-Orienting of Mankind with a Divine-Road-Map, a Positive-Timeline, the Elite's-Plans are quite different. Seeing as how they understand the Meta-Physics and Divine-Laws, like everything else they will try to Reverse-Engineer this Process to get what they want. In fact that is exactly what they Plan on doing. They Plan on affecting the Masses in significant numbers to achieve Their-Goals. They Plan on Manipulating the Masses to create a Negative-Timeline, a Negative-Future based on an understanding of Meta-Physics, based on an understanding of Divine-Law.

The beginning of all of this will involve the Elite Manipulating Laws of Creation, Manipulating Mankind's Energy en masse (BY CREATING NEGATIVE-ENERGY THROUGH WORLD TRAUMAS) to Create a Negative-Reality. Very much like the movie "Matrix" in which everyone "Collectively Dreams" a certain Reality. The Elite Plan on Manipulating Mankind's Energy, the Power of Belief en masse, to Create a World which they want, because they understand and know, based on Physics and Astronomy, that due to this Cosmic-Cycle there will be a change in Consciousness for Mankind. And they know that this "Zero Point" is a very important moment for Creating and Inserting this Negative-Reality. Once we get through this Zero-Point, Mankind's Path will be "Set in a Negative or Positive Energy". As we come out of this Zero-Point, like quick drying cement, our Energy will "Set in Positive or Negative", and minus interference of the Elite, it would "Set in

Positive" as this is "The Bigger Divine Plan". Having said that, the Elite can only delay this process, not stop it—they just do not get this.

For most people this picture is too big for anyone to see or know of, and so the Elite—what Motivates Them and Their Deadline, remains unseen. I hope this explains a few things for those who have been wondering what is going on. If you look, many things being done "Fit In" with this idea of 2012, and THIS EXPLAINS WHY 2012 has been selected. It is a rare opportunity, a matter of Energy and Physics, and Astronomy and Time. This is why the Elite have selected 2012 and this is why many things now seem to be "Reaching a Final Point"—the Elite-Plans are all about to begin. And when they do, it is going to be Chaotic for the World and Humanity, as it is all about "Hijacking this Divine-Process".

CHAPTER 26

Prophecy: Negative and Positive Energy: Consciousness: The "Path of Experience" vs. the "Path Of Consciousness" How to Change the Future in a Positive Manner

I am good at making "projections" based on data and I am good in the realm of Prophecies. I have studied them for more than twenty years. Having said that we've Collectively avoided a few, and a Good Prophecy is one that doesn't come true. Ironically when this happens many people say, "Well obviously that person did not know what they were speaking of", when in fact it may be that the Prophecy itself and Consciousness and Awareness of it, prevented such a thing from happening and coming True in the first place. SPEAKING-OUT, a theme of my book, HAS THE POWER TO CHANGE THE FUTURE. Just as Governments realize that the "Reversed-Engineered-Form" of this, Secrecy, does as well.

Having said this the content of Prophecies can be avoided, a Prophecy can be avoided, if enough people heed the warning: Awareness and Consciousness are key. Prophecies are possible "Negative-Events" (as a general rule they are "Negative" as they serve the purpose of being a warning), Probable-Realities that have yet to occur: Possible-Events, Possible-Realities.

They also represent the following "Negative-Timeline", that which occurs after the "Negative-Event", all that results as a consequence of the event: a changed future. Since Prophecies represent a "Possibility", they are a warning, "Book-Markers" through time, spanning and straddling time, and they can be avoided.

I would say one of the biggest factors in changing a Prophecy, on a more subtle-refined level, and in a way that uses a "soft form" of Self-Empowerment (it is powerful though), is prayer—a form of POSITIVE ENERGY: especially done in groups and at the same time. Getting together in numbers and "focusing" with the same "Mental Intent" is a matter of Divine-Laws, and doing so "Amplifies this Energy". If you think of a person being a "Loud Radio" now imagine more being added—the "Sound Amplifies", so it is with "Mental Intent and Focus"—this is a matter of Divine-Law. Quite frankly I think this is why a lot of Negative-Scenarios have thus far been avoided, due to groups of people getting together across the World and praying with the same intent and objectives.

These World Prayer Groups are a good thing, the Collective-Consciousness of these groups, the Collective-Energy, is a very strong materializer for good, on our plane-of-existence, the Physical World. "Thought comes first, then Creation"—again this is a Law of Creation. This is why scientific studies conducted on people who are very ill and who pray or are prayed for, show that PRAYER WORKS, not all agree, but the principles are sound. This involves POSITIVE-ENERGY, "FREQUENCIES", the Power of the Mind, and is Scientifically Proven to work, at least some say so—it is PHYSICS on a "Human Level". Thought, Action, Energy with Intent, gathered together in numbers—groups of people with the same intent, is a very strong mobilizer and Creating-Force in our World and Level of Reality. This energy affects the World we live in, in a good way. What we put into our World is what we get. Thought comes first on our Level of Reality then it "Manifests", thought comes first then "Creation"—again Laws of Creation.

I'm sure Criss Angel, who performs "magic tricks", understands this, in fact I am certain he does. Those who know who he is and how he does a lot of his "Tricks", I am sure, get the point of this statement. If you don't know who he is, look up his "Coin Trick" on the computer, the one where he has the coin travel up his arm. You'll see what I mean. Power of the Mind is everything.

Another way of diverting Negative-Events, Prophecy, is through Self-Empowerment in a more "Visible Way" in the Physical-World. Consciousness, Will-Power, Intent, Positive-Energy, Focus, being Active, Positive-Action, go a long ways to change things for the better in the World around us: to avoid "Negative-Events" which we see coming around the corner. SELF-EMPOWERMENT, THE BELIEF THAT YOU CAN DO IT, is key here. So is taking ACTION.

Thought, Intent, Will-Power, Positive-Energy, AND ACTION, affects our level of Reality. How our level of Reality turns out depends on what we want, and how and where we direct our Energy. What we put into it, is what we get. Again "Action" is key here, an expression of "Positive-Energy" being put out into the Physical Plane-of-Existence, into our World and Reality. Positive-Action COUNTERS Negative-Action, Negative-Events. It is very simple to understand it is about Balance of Energy.

It is "Physics", the Law of Cause and Effect, Darwinism and Evolution: which in our case is "Evolution of Thought". The "Path of Consciousness" vs. "The Path of Experience". The "Conscious" survive, the intelligent survive, seeing things before they happen, using their brain to avoid disasters, having taken ACTION. Those who are LAZY-MINDED, choose the "Path of Experience", choosing to learn the stupid way: waiting for "Events" to happen before correcting their behavior, and thus learning in the end, to do things "Consciously". In the end people like this learn to do it "Consciously" the second time around anyhow: it is Stupid and Lazy. And if the circumstances are dangerous enough, it can be FATAL. There is not always a "second time": thus "Darwinism", "Evolution of Thought", ACTION, and survival.

So as I said before a Good Prophecy is one that doesn't come true, our Thoughts and Energy serving to create "The Reality Around Us", a Prophecy not happening, being the result of the manifestation of POSITIVE-ENERGY: the result of POSITIVE-ACTION. Whether the Masses are Conscious of this or not—it is simple to see, this is PHYSICS. So does this book affect things in a "negative way" then? Well that depends on YOU. If you sink into FEAR and obsess on the Negative-Possibilities this is not good, and does nothing for Self-Empowerment never mind improving the Collective of the Human-Race. This Fear, Negative-Energy, goes towards feeding Negative-Realities, as our Thoughts and Actions are shaped by this. However, in order to change Negative-Realities on the horizon, WE HAVE TO BE WILLING TO LOOK AT THEM. This is where LAZINESS OF THE MIND COMES IN, "COMFORT", AND FEAR. Many people are afraid to really look at the World around themselves, afraid to see what is True. Thus the secret is to look at these Negative-Realities, without EMOTIONALLY EMPOWERING them, without feeding them Energy. WITHOUT BEING ATTACHED. Be aware of these Negative-Possibilities not "Immersed in Them".

Put your EMOTIONAL INVESTMENT into feeding POSITIVE-REALITIES, with an unattached eye on the Negative-Realities. Be calm, cool, collect, Non-Attached, in regards to Negative-Realities: yet be ACTIVE towards creating a Positive-Future with Positive-Action. In short be SELF-EMPOWERED and put your "Emotional Investment" into POSITIVE-ACTION, into "feeding" POSITIVE-REALITIES. Positive-Action, counters Negative-Action. Again it is simple to understand, yet if the Masses were more Intelligent and Conscious, I would not have to address things like this.

The fact of the matter is, the Masses are not Conscious, and led by people who are NEGATIVE. Thus it is simple to see that many "Negative-Events" have occurred throughout Mankind's History, and WILL OCCUR yet again as a result of people NOT BECOMING CONSCIOUS, NOT BEING QUESTIONING, NOT BEING SELF-EMPOWERED, and being led by Negative-People, who do not care about the Masses, about Mankind. The Masses are LAZY-MINDED and want somebody else to do everything, thus they throw away their Self-Empowerment, and everything else with it. The World we live in is a result of the LAZY-MASSES, as much as it is a result of "Bad-Leadership": "Negative-People" with a lot of Power and Control over the World. The Negative-Events of Mankind's-History points to Negative-Leadership, and LAZY-MINDED-PEOPLE, who are all to willing to be led through the nose to their own Destruction. Thus SELF-EMPOWERMENT on an individual level is KEY TO CHANGING THE WORLD WE LIVE IN.

The purpose of this book is quite obviously not to create Fear, and thus feed a Negative-Reality. I hope that this book helps to awaken people to the "Negative-Realities" going on around us now, how they are done, how we can see these things (Techniques and Tactics), to show people how and where to look, and to look at the possibilities around the corner, "Things on the Horizon". I hope that this book Creates Awareness and gives others "Tools" for Creating a Positive-Reality. I hope that others eyes open to Predictions, based on what I see. Another purpose of this book, reason, is that by knowing about the "Negative-Possibilities" before they get here, we can avert a lot of them and no doubt negate and tone down the effects of others. Hopefully for some at least, their eyes are opening to things going on now, and they are now able to look around a little more clearly at the World around them, and the possibilities of things to come. This book is in part to inspire CONSCIOUSNESS, ACTION, AND SELF-EMPOWERMENT. By Speaking-Out we have the Power to change the Future and I certainly hope this book helps.

The more that people know and understand, the better the chances of altering some of these Negative-Events, Planned-Events, World-Traumas, World-Shocks, prepared by the Elite. In theory anyhow, that depends on people on an Individual Level, and as a Collective. I hope that there is something here for everyone. Again a purpose of this book is to get people Active and Self-Empowered, to help people realize the importance of this, and the importance of Unity: the Collective, the Masses, Mankind, working together in a Positive-Manner. POSITIVE-ACTION AND POSITIVE-ENERGY. Quite frankly these Elite would not be hiding in the shadows, and Lying and Manipulating from the shadows, if they did not think that there is "Power in the Masses": if the Elite did not believe that THERE IS POWER IN A UNIFIED PEOPLE. In one very real sense these Elite are worried about the "Strength of the Masses" and this is why they have to hide. If they have their way, in the Future, they will not have to worry about the Masses or about "Hiding".

Consciousness is Power: A Conscious People is a Free People: a Unified People, Unified Through Consciousness, is an Undefeatable People, a Less Gullible People, Less Manipulated People

But back to Negative-Events, and Timelines, and Prophecies. To be good at this, at understanding Prophecies and seeing the World around you, live a Truthful Life (at least be truthful to yourself and why you do things) and put Truth and Objectivity first when analyzing the facts, data. Be True to yourself. Look at all information, in all sorts of places, because you never know where you'll find the Truth. Your Truth Detector is "Looking Within". It is that inner recognition, it is literally a particular feeling, that you can learn to recognize: living a clean life—an honest one, can help you recognize this "Feeling". A lot of people call this "Intuition", or having a "Sixth-Sense". I call it normal and it is a part of your "Being", a part of us that most people used to be connected with, it is your "B.S. Detector", your compass for finding Truth. A "Diet of Truthful Living", reaps rewards in this area. The expression, "The Truth is out There" is FALSE, it is WITHIN YOU.

Prophecy. As a general rule, if a Prophecy has been made across hundreds or thousands of years, and spans different cultures, then it is something to be looked for. It doesn't mean it will happen, because we have FREE WILL, but it does mean there is a STRONG POSSIBILITY of it occurring. One very strong probability, based on Prophecy across the World, based on the formula I have offered, being the "Anti-Christ" (I pick this name because it is commonly used and recognized), or as Muslims refer to this individual, Dajjal. Other strong Prophecies are of World-War yet again. If we do not become a CONSCIOUS PEOPLE, choose to do nothing to change things for the better (which IS a choice—and lazy), then it is more than reasonable to assume that "EXPERIENCE" will dictate growth, and dictate a new way of looking at things, CONSCIOUSLY. I call this the Path of Experience. Again this is the dumb way, it lacks intelligence. But it is also what the Masses have been deciding for themselves for an extremely long time now. Again what we get in the World, is the fault of the Elite, but is the fault of the Masses as well. What we get in this World is based on the QUALITY OF ENERGY we put into it. How hard is this to understand? This is all a matter of what I call "Human-Physics".

This is why Prophecy exists, it is to alter Human-Consciousness so an event does not happen. To motivate people into Positive-Action, to create Positive-Realities, to give people a CHOICE in regards to what happens. To alert people to the possibilities, and to give them a choice. By not looking, not seeing, allowing something Negative to happen, we make a choice. This is about "Darwinism and Evolution of Thought", "Darwinism of Thought", the Path of Experience or the Path of Consciousness.

Thus as we near a Negative-Event, and the following "Negative-Timeline", in conjunction with a Prophecy (which are usually negative), people begin to note signs, clues of it's coming, it's appearance on the horizon. If it is something Prophecized we note that certain signs and clues fit with this Prophecy. The Prophecy is a picture of the "End Result", where we are headed. As we near the impending "Negative Event", we note the building up and gathering of clues, showing the direction we are going in. Thus Prophecies are to avoid Negative-Events, and the following "Negative-Timelines". Books such as mine exist for the very same reason, serve the same purpose. Speaking out about what we see, does make a difference. By Speaking-Out about the future, we have the power to change it. It is a matter of Physics, Human-Physics: Positive-Energy put out by People, counters Negative-Energy.

We can see through time, in the near future that such a thing will occur, by LOOKING. If not enough people with POSITIVE-INTENT and POSITIVE-ENERGY get involved then a "Negative-Event" will occur: an event that is BIG, NEGATIVE, and thus changing all subsequent thinkings, Actions, Realities after it. As a result a "Negative-Timeline" is put into place. OUR FUTURE BECOMES PERMANENTLY AFFECTED, everything after steered by the Negative-Event, "Flavored by It", affected and steered by this Event and traveling in a new direction.

Like tossing a large stone in the water, the "Negative-Event" is the stone, the subsequent ripples are the Negative-Timeline, as a result of the "Event". These are "Nexus Points", "Forks in the Road", "Crossroads", and what we decide at these points in time, decides which direction we will take—the Negative-Timeline which follows. The 9/11 situation is a good example. Everything has changed since then, affected by this World-Trauma, this

Negative-Event. Think of the World NOW, if this Event had not happened, think of how different the World would be. No doubt the World would be less messy, and people living in Western Nations, would have more freedom, less infringements of their rights, by Government. Which from my point of view was one of the main objectives in Orchestrating such an event in the first place.

Now it is possible that when a "Negative-Event" happens, that a "Good Timeline" can follow, but generally speaking, that is not the rule. We are talking about Consciousness, Energy, and Action. We are talking about the Law of Cause and Effect. As a general rule if people were not Conscious enough to alter the path towards a Negative-Event, then they are not Conscious enough to change things immediately after to avoid a Negative-Timeline. However that is a point in time, a chance, to still change things for the better. At that point, learning, Consciousness as a result of the Negative-Event, can alter any subsequent Negative-Timeline, thus something POSITIVE comes out of it. Thus we learn from mistakes. If we learn from the Negative-Event, then no further Negative-Experiences are needed and a Negative-Timeline avoided. As with my example of the man with a hole in his roof who chooses the Path of Experience. Except in this case he decides that "Once was Enough" and he gets the hole in his roof fixed, after experiencing "Rain and a wet House", only one time.

It is very much a matter of Darwin's Theory of Evolution, in regards to Consciousness: "Darwinism of Thought". We either choose the "Path of Experience", or the "Path of Consciousness". In the end the "Path of Experience", if it does not get us killed, will lead us to the "Path of Consciousness" anyhow. It is just that the Path of Experience is the long, slow, painful way. If we choose the Path of Consciousness, we look and learn immediately without "Experience", and take Action to remedy things (Positive-Energy) from the beginning therefore avoiding problems before they occur. Positive-Energy is needed to "Counter and Balance" Negative-Energy. This is intelligent, the intelligent way of dealing with things. The Path of Experience means that we end up choosing experiences for ourselves, which then teach us to think in the first place, to become CONSCIOUS. With the Path of Experience we learn the hard way, we learn the slow way, we learn the stupid way. We continue screwing up, because we are LAZY-MINDED. This is how the Masses are with "Their Leaders"—Lazy-Minded and thus we all "Suffer from Experience". This is "Power of the Mind", en Masse, in a Negative Manner. This is where the Masses are today and has a lot to do with why the World is so screwed up.

Prophecies are a gift from "The Universe", "The Source", and a gift from those who are "Plugged Into" a deeper side of themselves, who are able to "Step Outside of Time", "Time" being an Illusion. Prophecies are reminders and give us the chance to avoid unpleasant circumstances. As a friend of my girlfriends, Andreas, has said to her before, "We can learn by experience, or by consciousness.". In other words we can think something through, and decide we don't want it, or learn the hard way.

I've seen some Prophecies avoided, but I still see the "Worst to Come", and because the Masses are ignorant and choose to be, not because there are not better possibilities. The FUTURE in regards to Negative-Prophecies involves TWO PROBLEMS: the Elite, and the ignorant Lazy-Minded Masses, the "Dis-Empowered". I see there are PLANS in place by the Elite, that are totally crazy, and involve Total Control. If enough people become CONSCIOUS about "Plans to Come" these Plans can be avoided, at any rate SHOCK can be negated, to some degree. The less Shock, the better, obviously. These SHOCKS AND TRAUMAS are always for Manipulation as people are not thinking clearly under such conditions. Having said that the PLANS of the Elite will not last long regardless, but it is a matter of how much harm they can do in such a short bit of time. Which will be HUGE if people, the Masses, do not wake up.

The Elite have many PLANS IN PLACE, around the corner, for DUPING AND SHOCKING THE WORLD. The Crash of the American-Economy, and the World-Economy is the "Mild-End" of it. Unless I am mistaken, and I do not think I am, an ENTIRE SERIES OF WORLD-TRAUMAS, will begin with this Economy-Crash. This will be the "Beginning of their Deadline"—it will all start off with "Negative-Events", WORLD-TRAUMAS, for Manipulating the Masses by way of FEAR. MK-ULTRA ON A GLOBAL SCALE for putting a "Negative-Timeline" in place, based on "Meta-Physics" an understanding of the Universe and how it is "Ordered" on a "Deeper more Refined Level"; an understanding of the Divine's Laws, Laws of Reality, and how this Reality is "Ordered and Structured".

In these PLANS and their understanding of "Laws of Reality" they are arrogant and confident of their success, but they are also going against a "Bigger Plan", the Sources, the Divine's, and they know that their days are numbered. They know of many Esoteric-Things, Ancient-Knowledge, and because of this understand that there is indeed a "Divine Blueprint" for Mankind's Success, which they are trying to HIJACK, and INTEND TO HIJACK.

As I have already covered, this involves our Solar-Systems travel through the Universe, and it's approach to a particular type of, and source of, ENERGY—which will have an affect on Mankind's Consciousness. In essence the Earth and Humanity, are now headed on a "Positive-Path", for the next 12,000 years or so, and this is based on a Growing Consciousness of Mankind. In such a World, the Elite can not survive. So understanding the Mechanics of this Universal Plan, Blue-Print/Map for Mankind, they are hoping to Hijack this Plan, and SUBSTITUTE THEIR OWN. One which is Negative, and which allows them to continue to exist. If you are waiting for these "people" to start "Caring about Others" you can forget it. People have got to DO THEIR PART to change things for the better. ALL PLANS OF THE ELITE, CENTER AROUND THIS KNOWLEDGE. To top it off there are certain Cosmic-Events coming our way, that make it imperative that they keep with this 2012, early 2013, Deadline. Their Deadline is in fact based entirely on this Cosmic-Cycle and coming Cosmic-Events.

Thus if you look, many Prophecies speak of, hint of, many strange things to come: Man-Made and not so Man-Made "Events". The year 2012 is important because this involves a Cosmic-Cycle and because the Elite have picked this as a DEADLINE DATE, based on this Cosmic-Cycle and their understanding of the Meta-Physics behind it. Both because of the Elite and their "Plans" and because of the Natural-Events which will occur, there are a lot of Prophecies about 2012. Unfortunately the average person does not get the meaning or importance of this Cosmic-Cycle, and what 2012 represents. Many are in "Fear" over it, or oblivious to it. The year 2012, or early 2013, does indeed represent a "Significant Moment", however it is just not something that should be "Feared", RATHER it is the PLANS OF THE ELITE, based on this Cosmic-Cycle that should be "Feared". Having said that be Self-Empowered, be Active, put out Positive-Energy, and do NOT "Act from a point of Fear".

People need to to "Put Out" more Positive-Energy by being Positive-Minded, not more "Negative-Energy" by being "Fear-Oriented". Rather SPEAK-OUT about these coming Plans, SPREAD AWARENESS of these ELITE-PLANS, and do not be "Attached to Fear". Be Non-Attached to Fear in Your Actions, and in Your Speaking-Out. Speaking-Out is the easiest form of being Active, and has the POWER TO CHANGE THINGS FOR THE BETTER. Remember to be Self-Empowered, do not feed a Negative-Reality by being Fearful, be Non-Attached, and put out Positive-Energy with Positive-Action, as this can and will change the Future. Simply by SPEAKING OF THE FUTURE, we have the POWER to change it.

CHAPTER 27

The Elite's Hijacking of a "Cosmic-Event"—"Sold as a Sign" for the Appearance of a "World-Leader": An Appearance of Light which Astronomers Reffer to as the "Space Ribbon", what Many Call the "Rapture or Photon Belt": The Elite's Plans to Install a "Puppet", One who Many Call the "Anti-Christ or Dajjal"

As a note for my readers, this Chapter could have been in far better shape, but I am rushed into printing here, in order to expose this Clown, who only days ago killed one of our pets. I include it because the material is important.

So right before Christmas (2008) I found an article that put me in dread. One of those I had been "Waiting For", yet had waited so long for, I thought the "Timeline" had been missed. Unfortunately not. The "Savior" of our times and tribulations, the "World-Savior" and "Messiah", the "Anti-Christ".

Call him what you will, I am not Religious. I use this term because it has been coined and used for a very long time, thus most people from across the World and of different Cultures know of who I speak of—the Muslims call this individual Dajjal. Look at the source of my initial shock and information. The source alone is enough to put any ones head into a tale spin. This article comes from Dow Jones Company's "Market Watch" and is dated December 12th 2008. Dow Jones and Company owns Market Watch and also publishes the Wall Street Journal. From an article of theirs, obviously released publicly, I have info on the "Anti-Christ", or as they advertise him, "The World Teacher for Humanity". Got your attention yet? Be sure to check out the article because it is surreal beyond belief. This is a "Kook Article" in a well known Financial Publication.

This article is entitled, "Share International Reveals Christmas Miracle", and came out December 12th, 2008, and appeared in Market Watch. Market Watch is well known and provides Financial Advice to the World. What is an article like this doing here? I'll give you a hint, it has to do with the World-Economy Crashing, which is why this is released in a Financial Publication. First a quick summary of this article.

It states many things which are interesting. I will address each of these points, followed by a point of my own, to give an idea of my subsequent writing. The first thing this article states is that people are hungry for "Signs" in regards to miracles. This can be translated as "People are looking elsewhere for answers because this World is a mess, and our 'Leaders' seem incapable of fixing anything." Which is partly True, except World-Leaders are creating a "Mess" of the World INTENTIONALLY, for Greater Control—like the American-National-Debt as one example.

The next theme brought up, and seemingly out of nowhere, is that a large bright "STAR" will appear in the sky in the NEAR FUTURE, visible during DAY AND NIGHT for all the World to see. If you pay enough attention to the "Subtle Wording" of this article, this "Star" is being described as being seen on both sides of the Earth at the same time. This "STAR" is also strangely enough labeled a "MIRACLE". It may be miraculous in that it is seen both Day and Night (At the same time—Both sides of the World), but there are already many other red flags in regards to this "STAR TALK". What is miraculous about a "STAR", an Astronomical Event? And how do they know it will be seen Day and Night when the rest of the World does not even know of it's existence? Obviously this "Star" is Suppressed Information, it has already been studied and it's "Imminent Arrival" predicted, an interception with this "Star" is already known of. It is this Giant Space-Ribbon, this Light-Cloud, and it is known that Earth will indeed "Enter the Light of this Star".

We experience night as a result of our star, the sun, being only on one side of the planet at a time—there are basic physical realities which have to be looked at here. This is not a "Star", we are being LIED TO IN THIS ARTICLE, it is in fact this Light-Cloud which Scientists and the Elite already know about. But lets play along here for a bit, I will continue to call this a "Star" for now.

The article then goes on with how this is a statement of "Simple Fact". How do they know, and why would they, and how could they? This company does FINANCIAL INFORMATION, NOT ASTRONOMY. Yet they are convinced enough to let this be stated as a fact in their Publication. The next theme, "Ball of Weirdness" in the article is that it states a "World Teacher", "Maitreya", will step out onto the World-Stage shortly after this "STAR" appears for all to see. This "Star" will supposedly be a "Sign of His Appearance" which is a LIE AND FARCE—the whole thing is a SCAM.

This Cosmic-Event is NATURAL, occurring every 24,000 years or so, and he will simply be an "Opportunist" using it to his Own Advantage: actually the Elite will be, as he will work for the Elite. This will be the Elite's Idea, and it is a part of Their Plan for Hijacking this Light-Cloud, and using it for Their Own Demented Purposes. The only people this "Star" will be a "Surprise and Sign" for is the Masses, because they are being kept in the dark about this. Later Humanity will be "SOLD" this as a "Sign of this Idiot's Emergence". And it IS a "Sign", it is actually a "Sign" of a "New-Cosmic-Cycle" beginning, and a "Divine-Event" for people on the Path of Consciousness.

The following weird fact is that this article states that when this "World Teacher" appears on the World-Stage, he will not be using the name Maitreya at first. Obviously this freak, and "Television Preacher Scam Artist", will use the name "Maitreya" after he has convinced the World of his brilliance, of his "Genuineness", of his "Messiahness", after he has duped the World. Who is this freak, and what will he be doing? Obviously he is trying to use a Naturally-Occurring Cosmic-Event to claim it is "ABOUT HIM" (no ego there...) and obviously HE IS HOPING THE FEAR PRODUCED BY THIS "SPACE-CLOUD", the Fear of the Masses, will push them into "His Arms". Already this guy is a SICK-FAKE. Maybe Chevy Chase should make "Fletch 3" and put this guy in it.

Obviously this individual has information about this "SPACE CLOUD", this Space Ribbon and Our Entering It, and will be using it for his own "Dramatic Entrance". For such a "Wonderful World Leader" you would think he would forewarn the Masses of this "Light Cloud" to negate the SHOCK and TRAUMA to the Masses, but apparently he is "Saving It" to claim it is about "His Arrival"—FOR THE SAKE OF HIS OWN EGO. I hope I have made a point with this last sentence.

In this case both the Media Company and this Individual have knowledge of this "STAR", this Light Cloud, the Space Ribbon. The Media Company, one with a reliable reputation would not be printing this "Tabloid Junk" in the first place, otherwise. This "Reflects Poorly" on them. Odd. Now the following strange point given in the article, is this "World Teacher" will be interviewed on a MAJOR American Television Program shortly after this "STAR" appears (For the sake of "Discrediting" people like me the Elite may move it to another Country). This individual obviously has connections WITH MEDIA COMPANIES, Media Giants.

This is extremely weird yet again, and points to "Connections with Rich and Powerful People" in the Media World, to say the least. And THE ELITE own the MEDIA or else no doubt the MEDIA would be doing A BETTER JOB of reporting the fact that our Planet and Solar-System is about to enter a Highly-Magnetized Space-Cloud which contains a "Light Ribbon", never mind they would be "Covering the Economy-Crash". Oddly enough there is not much being reported about Earth's entry into an Electromagnetic Cloud with a Light-Ring, which will TOTALLY FLIP PEOPLE OUT when we enter it. ODD. Almost like this EVENT and the MEDIA, and some "PUPPET OF THE ELITE", their Poster-Boy, are all being USED for Manipulating the Masses. Gee, isn't that strange?

The next piece of weirdness, and almost the last in the article, is that this "World Teacher" is said to be "Awaited" by all people of all different Religions: Christians, Muslims, Hindus, Jews, and Buddhists. All of the Major Religions. THIS IS ODD, and SHOULD BE A MAJOR HEADS-UP for anyone who has heard of ANY PROPHECIES OF THE "ANTI-CHRIST", OF DAJJAL.

Odd indeed as this "Piece of Deceit" will come after a World-Economic-Crash, no doubt offering to "Cure the World's Economic-Problems", and then will "PREACH" a "One World-Philosophy" appealing to people of "ALL RELIGIONS"—to top it off he will be "Performing Miracles" which will be what really convinces the Masses. Having said that "THESE MIRACLES" will actually be the work of ELECTROMAGNETIC-TECHNOLOGY, NOT by "His Own Power". Whatever this guy is, he is not a Representative of God. He Is, and will be EVIL, Deceiving the Whole World by way of Exotic-Technology, and with the support of Corrupt-Rich-Powerful-People: the Elite. Thus he will be SOLD TO ALL PEOPLE OF ALL MAJOR FAITHS and the CORRUPT-MEDIA will be involved in this—if you are paying attention IT ALREADY IS. He obviously will have a "World-Religion" prepared, if he is to appeal to people of all faiths. Again ODD.

He will "Preach One World-Family", and "One World Religion", a "Philosophy", appealing to the Masses, in which "He Convinces the

Masses" that "They Can Do It" that "They Can Have Self-Empowerment", by FOLLOWING HIM. In Reality it will be the same Psychology Used for the Creation of a Fascist-Movement: Fascist-Movements being Composed of Dis-Empowered People, people with an "Identity Crisis". One Large Fascist Wave, following this guy, the Masses. People will NOT BE SELF-EMPOWERED, they will be "FOLLOWERS", and He will Trick the Masses into thinking they are "Doing It on Their Own". He will have "The Empowerment", and he "Will be Convincing Others" that they can as well, by Following Him. In Reality this is NOT SELF-EMPOWERMENT, this is being Weak-Minded: especially when You Look at the Facts.

If you are going to be Intelligent and Self-Empowered, then you have to be Self-Thinking, and Look at the Facts on Your Own FIRST: anyone doing this, will see there are "Many Things Wrong" with the Bigger-Picture, such as I am showing you now. This will be a Sales-Pitch for Attracting Weak-Minded-People, the Dis-Empowered-Masses, who feel they can Do Nothing, without people "Telling Them How", and without "Someone Leading Them": people with an "Identity Crisis" who are Incapable of being Strong on their own. This is the Mentality of People who are in Cults and Fascist Movements: the Psychology that will be practiced on the World-Masses, by this Pied-Piper, will be the same. He will seem to be a Caring-World-Figure, having "Solutions which Work", "Earning the Trust of the Masses", and he will BE SOLD to the Masses; in fact this Light-Cloud will be SOLD as a "Sign from God", and "Everything he Says of World-Unity" SOLD as a "Sign to Follow His Mouth".

That someone knows that this "Light Cloud" will appear is evident, THEY ARE STATING IT IN A MAJOR PUBLICATION, and this in and of itself is weird. Thus my conclusions are logical. This is a "Higher Energy" that the Earth is moving towards in it's travels through the Universe, a Mass of "Units of Consciousness" as I call it (Borrowed from "Seth" and Jane Roberts), a known anomaly for the Public-Sector of Astronomy—the Space Ribbon and Local-Fluff, a given KNOWN EVENT by World Governments, the Elite, military, Vatican, and others, but UNKNOWN TO HUMANITY. But back to dealing with the "Facts of this Article" and this Pied-Piper's website. They are full of DECEIT.

Between the "Healing of the World-Economy" through SOCIALISM, the "Uniting and Healing of the World's Religions" through PHILOSOPHY (Which will actually be "Godless"), the "SELLING OF HIM" through the CORRUPT-MEDIA, and his "ARTIFICIAL-MIRACLES" by way of EXOTIC-TECHNOLOGY, he will indeed seem like the "Messiah". Which if you note, the word "Messiah", was commonly mentioned in regards to Obama because the Elite were using THEIR

MEDIA, and Obama as a "Cover-Story" for using the word "Messiah"—this was and is a "Broadcasting" of the Next-Stage of Plans of the Elite to their "Henchmen", their "Team". This indicates that the Elite are going to the Next-Level of Plans, this "MESSIAH-SCAM". Think this is all Delusional on my end? I wish it was. What do you think it is like on my end here, knowing that there are Demented-Idiots, Powerful-Fruit-Cakes, who are actually Planning to do something like this? Obviously people have got to wake up to the existence of the Elite, which is not much of a stretch of the imagination, and have got to wage "A War Against Corruption" if we are ever going to live in a NORMAL WORLD.

This is what happens when we let very Corrupt-People do as they want, and show them we do not care enough "To Look". They become Confident in their Ability to Manipulate the Masses, and "Push the Boundaries Back" more and more—"Give an Inch, Take a Mile". They begin "Going Crazy", and Experimenting and Finessing with the Masses. Quite frankly they are Laughing at the Masses, this is "Proof of how Stupid the Masses Are", and these Elite could care less. From their Warped Power-Monger View-Point, anyone NOT CARING ENOUGH to "Look", is "Asking for It". THEY would "Look", and They would "Look" for the Sake of Themselves: by their "Reasoning" anyone who would Not do the same is an Idiot. Thus they "Resent" people like this, "Treat them like Idiots", and as "The Idiots" accept more and more, they "Push this Point" more and more. They in Fact are "Mocking the Masses". Anyone not "Willing to Care About Themself" is "Asking for It". If you can't "Care about Yourself", then don't expect "Others To"—this gets down to Darwinism and "Simple Survival Issues". If you are too Stupid to Survive, then you don't.

So this is what the Elite do with the Masses, and as they have a 2012 Deadline Date in place, they have some "Crazy Ideas" which if you are paying attention, Will Work, because few "Care enough to Look". It is these more "Far Out" ideas that WILL WORK, because the Masses can not even see the "Easier Ones". By "Pushing the Boundaries of Reality", they "Step Outside of the Box" with Ideas and Plans, and BECAUSE OF THIS these Ideas and Plans Will Work. They simply "STEP OUTSIDE" OF WHAT IS "BELIEVABLE": and it is a DELIBERATE TACTIC. So from the beginning, most do not see things going on as they "Really Are", most only "See the Surface". Most "See things as they are Told to see Them", as "They WANT to See Things", not as they Really Are.

Also stated in this short article, and also typically weird of this article, is that it states the message of this "World-Teacher" will be to the World that it is "One Family", and that this "One Family" should (to quote the article here) "Create world peace through SHARING, ECONOMIC JUSTICE AND GLOBAL

CO-OPERATION". Which is interesting as this is an OUTLINE, A FORMULA FOR SOCIALISM, and WORLD SOCIALISM at that. And this article comes out of a WELL KNOWN AMERICAN FINANCIAL PUBLICATION. All of this is stated AS FACT WITHIN THIS ARTICLE, AS SOMETHING WHICH WILL OCCUR. Everything from this "STAR", to this kook "World Teacher", to his "Revealing Himself to the World" via an interview on American Television. So talk about weird. Now that you have an idea of what is going on with this article, a summary, lets take a closer look. To top it off I have looked at this "Kooky World Teachers" website, who is named as "Maitreya", and have added what knowledge and observations I have to the following part of the chapter. You will note that his name is "Maitreya" because no doubt a "Candidate" was not picked out for this role of "Pied-Piper", and this was the Elite's Way of "Filling in the Blank" later. Folks, for such a small news article, a lot is revealed and explained.

Some "Revealing Information" from "Maitreya's Website" "The Day of Declaration": The Biggest "Television Preacher Scam" of all Time: a "World Television Preacher Scam": The Elite's Pied-Piper and Spokesman : Dajjal, the Anti-Christ and "Chosen One" for Cultists

Now I have looked at this guys website and there is a lot of the Elite's-Plans STATED THERE. It is like looking at a MAP OF THEIR PLANS in regards to the future. One of the most interesting things I found is titled "The Day of Declaration", and man is it a load of DECEIT. I have Government Spooks in my life right now, while I write, and they have EXOTIC-TECHNOLOGY which they have been using on me and my Family for years. I have a lot of knowledge about this Exotic-Techno, and therefore an insiders view in regards to many a thing. Including this "Maitreya" and his "Day of Declaration"—which is what he calls his "Day of Revealing Himself to the World". This is what his "First Televised Appearance" is being called—at least he has no Ego, or sense of "Dramatics" about it.

So what is the "Day of Declaration" aside from him "Showing himself to the World"? It is THREE THINGS. Three PHONY MIRACLES, all done with Exotic-Technology hooked up to Satellites. I would know, I and my Family live with this Exotic-Technology and all of its "Exotic-Misuses" every day. These "THREE SIGNS" will supposedly be the way the World knows this "World Teacher", this Scam-Artist, is real. All of it will be FAKE, a Television-Preacher-Scam using Exotic-Technology as opposed to conventional methods, conventional technology, used by many television preachers. If you do not know what I am speaking of here, for a quick and fun way to find out, see the second "Fletch" movie with Chevy Chase. So here are the three listed "Miracles to Come":

1. The World will hear him "Telepathically".
2. The World will "Feel the ENERGY of Love", "Feel his Love", "Emanating from Him". To the point of being obvious, as in people will notice the effect, that something strange is going on. Thus "Convincing Us" this is "Godly Love" and that he is "A Messenger From God".
3. People all around the World, hundreds of thousands of people, will experience "Spontaneous Healings", "Miracles" on the "Physical Plane". This means in all likelihood that these "Healings" will be Physical Healings, not "Mental Ones".

Now I have a few things to say here about this Elite-Jackass.

These "THREE MIRACLES" will be to convince the World that this guy is real, and will happen during the course of his International Network Broadcast. It then goes on to say that our response to this Scam-Artists "Message" will basically decide the "Fate of the World", the future of the World. Go figure. This guy is the biggest Television-Preacher-Scam-Artist ever, so it is no wonder that "Our Response" to this Jerk will decide the "Fate of the World". If enough Idiots buy into this SCAM, we will ALL BE SCREWED. It will all involve the MISUSE of EXOTIC-TECHNOLOGY which is HIDDEN FOR THIS VERY REASON OF MANIPULATING THE MASSES. You can not fight what you do not know exists. Fortunately for you, I KNOW ABOUT THIS EXOTIC-TECHNOLOGY AND I KNOW OF IT'S SICK MISUSES, so I can already explain ALL OF HIS PHONY "MIRACLES". They are anything but "Miracles", and I have already experienced these "Miracles" in my life. I will get to these shortly after a few more paragraphs and points.

This is simply a Television-Preacher Scam, where the Preacher uses Technology to convince people that he is "Plugged in with God". As a "Convincing Example" many times these Preachers "know things" about audience members, and it is simply a matter of an "Ear Microphone" and someone sitting somewhere else with a computer. It is all a matter of Technology. They simply find things out very quickly, background information and pass this information on by way of the "Ear Microphone". A lot of Preachers have been exposed for this and "James Randi" is a good place to look, on You Tube, if you are curious.

There is nothing special about any of this, it is just that in the case of this "Maitreya" the Technology will be many years ahead of the Public-Sector and is unknown of, thus this "Maitreya's Miracles" will SEEM REAL but actually will NOT BE. This is simply a matter of people being IGNORANT

ABOUT THE LEVEL AND DEGREE OF TECHNOLOGY THAT EXISTS. It is a case of "The Gods Must Be Crazy", where a glass bottle seems "Miraculous and From God" to those not exposed to such "Technology". It is a matter of Amazonian Pygmies seeing an airplane for the first time, and building an "Airplane" out of branches and leaves, for "Worshiping".

THIS IS TECHNOLOGY AT WORK, NOT GOD. And The People behind it are anything but "God Oriented", especially this "Maitreya Character". You have to be sick, warped, and twisted beyond belief, to Deliberately Mislead an Entire Planet full of people into believing that you are a "Representative of God".

This "Day of Declaration" is the beginning of the "BIGGEST TELEVISION PREACHER SCAM" in history. It will be the BIGGEST CON-GAME OF WORLD-HISTORY, even out doing the ROBBING of the World's People's by this ECONOMIC-CRASH-SCAM. All of it will involve Exotic-Technology, hooked up to SATELLITES, to fool Mankind, the World. And people wonder why this Exotic-Technology is Hidden. It is not Hidden from Other Governments, it is hidden FROM THE PEOPLE, FROM THE MASSES, FROM MANKIND, this is why. People will react to these "Miracles" like a bunch of renegade natives living in the middle of the Amazon, seeing an airplane for the first time. Or like the movie "The Gods Must Be Crazy", when natives find a glass bottle. IT IS ONLY TECHNOLOGY, NOT AN EXPRESSION OF SOMETHING DIVINE. In fact how it is to be used is anything BUT "Divine". I hope my readers are clear about this, and I hope you pass this information on, the more people that know of this, the less chance there is of this happening.

If people in Government, who know of this Exotic-Technology, were not such cowards, if they actually lived up to their Oath To Their Country, then it would not fall on people like me to get this information out. What the Elite Plan on doing here, the Level of Deceit, by Misusing this Technology, is scary.

The "Day of Declaration" is when this "World Teacher", this SCAM-ARTIST, will step out onto the World-Stage, through television and no doubt other Media Outlets. He will be invited, as it is stated on many a website, BY THE INTERNATIONAL MEDIA, to speak TO THE WORLD. To make it more interesting, he will NOT. How is that for confusing, interesting, and "Ear-Grabbing". But you see we are not dealing with an ordinary being here, we are talking about a "World Savior", "The Messiah", "His Holiness of Deceit", thus we will see his image on television yet "Hear him Telepathically". All around the World people will see his face on television but each land will "Hear" his words in their own language, in their heads—"Telepathically", along with "Feeling the Love", and "Spontaneous Healings". All of it through Exotic-Technology, Electromagnetic Technology.

How's that for weird? And this is going to be the World you live in, shortly down the road, if enough people do not become Conscious of this Plan, if this is what the Elite choose. Already I am getting tingly and excited by it. In general when this Exotic-Technology is used to make this "Television Preacher Scam Artist" seem real, from my point of view, it will be a case of "The Gods Must Be Crazy" as no doubt a lot of people will buy into this FARCE AND SCAM. People not using their brains, a bunch of "Renegade Natives in the Amazon", with no previous contact with the outside world, seeing an airplane for the first time. And THAT is the idea, and again it will only be Technology.

Exotic-Technology and the "Three Miracles": Personal Examples from my Life: it is Technology not "God's Approval and Endorsement"

People will see this "Con-Mans" face on T.V. but each Land will hear his words "Telepathically" in their own language. No doubt he will be "Giving a Compassionate Look" while he does this—a "COVER STORY", appearances, to fit in with his "Miracle of Love"—"EMOTION-PUSHING" as I call this Exotic-Techno TACTIC. With this Exotic-Technology comes a whole range of TACTICS, A HAND-BOOK, for Misusing this Techno, and getting away with it. For "Fooling People's Minds".

Folks, in regards to "This Telepathy" (Which we all will experience if the Elite choose this Scenario/Set of Plans), based on my own experiences, and based on how obviously Fake and Corrupt this "Elite Pied-Piper" is, this will be Exotic-Techno. Let me give you a very specific example from my own life, as I have had this very experience.

I one day watched "The 13th Warrior" on DVD in the language of the land I live in (Here in Europe) but experienced it in English. Due to my "Mobbing" with these Spooks, I have been slow to absorb the culture, and do not speak the language all that well. In fact I have quite a bit of difficulty in both understanding and speaking it. However when I watched this film I began to understand it in English—PERFECTLY. Until I noticed what was going on, and the shock of the experience "Woke Me Up" from it—or perhaps closer to the Truth, these Spooks stopped doing it. These Spooks were being ironic and deliberate in their choice of timing, let me explain.

The main character an Arab, played by Antonio Banderas, who is traveling with Vikings, does not understand their language but begins to understand them after many days and nights of journeying with them. He begins to understand them not by "Normal Means" but through "Higher Consciousness", by Telepathy, "by listening". It was minutes prior to this, in retrospect, that I began to understand the language "In English". In short, there is no other way to describe it, I understood this foreign language, by way of "TELEPATHY". I would say the experience lasted 10–15 minutes total, long enough for me to be certain of it, and to make some Mental-Notes.

Wow. That is a Miracle! Incredible! I must be the one to save Mankind! I can experience "Telepathy"! NO FOLKS, IT WAS ONLY A MATTER OF TECHNOLOGY, EXOTIC-TECHNO, BUT TECHNOLOGY NONETHELESS. The same thing will be done with this Messiah-Con-Artist. This Exotic-Technology, in order to reach the Masses of the World, has obviously been connected to SATELLITES from which it can affect ALL people. It may also be interfaced with television, and television transmission, as a guess. Maybe this is why there is such a big push to switch to digital television in America. As an added point here, there are many who are capable of such things, it does not make a person "The Messiah". Criss Angel, a master of "Illusion", no doubt is capable of such a thing, along with many an Indian Saint, Yogi, in India, and this is off the top of my head. This is a matter of "The Power of the Human Mind", it does not qualify someone for Messiah-Hood.

So when I had this "TELEPATHY" incident, with the Spooks and their Exotic-Techno, it was subtle, I didn't even realize it was going on for several minutes. When I did I was SHOCKED, as will be the World when this Messiah-Con-Man uses "Telepathy", Exotic-Techno to communicate with the World. But back to my own personal experience. This is the Spooks idea of being funny and ironic. Just as an extra note, I've had this experience several times since then, I believe twice more. At any rate when people across the World start "Understanding" this "Messiah-Con-Man" for the Elite, via "TELEPATHY" this is why. EXOTIC-TECHNO. This FAKE is the "Chosen One" of this Elite, not the "Chosen One of God". Of course DECEIT and MANIPULATION is involved here on many levels.

So I have described HOW this experience of "Telepathy" will take place. Now lets talk about the second miracle, how "A Feeling of Intense Love" will take place.

This is where an EMOTIONAL-STATE is pushed into you, one which is not your own—and in this case the Emotion, will be "Love". Our minds, our thoughts, our feelings are all a matter of "FREQUENCY", "VIBRATION", and this Exotic-Techno is able to MIMIC many things based on this principle. Thoughts and Feelings are a matter of ELECTROMAGNETIC ENERGY and this Exotic-Techno is based on this type of Energy, able to Manipulate and Control it. Thus people can be influenced, affected, "CONTROLLED", by this Exotic-Technology. Thus many will think this is the "Messiah", AS THEY FEEL "INTENSE LOVE" outpouring from this guy, in reality this is Exotic-Techno, being used to Manipulate the Masses in the ULTIMATE CON-GAME. This will NOT be "HIS LOVE" it will be the PRODUCT OF TECHNOLOGY, a matter of FREQUENCY, VIBRATION. It in fact will be ARTIFICIAL.

This is a TACTIC used with this Exotic-Techno, one which I call EMOTION-PUSHING. It is where an Emotional-State is pushed into your mind. Obviously if it is not yours, it is someone else's, and obviously if someone else is choosing it for you, then this is a VIOLATION, and use of FORCE. It is a sick Misuse of this Exotic-Techno and has been used DAILY IN MY LIFE FOR YEARS. It has been used to try and destroy my relationship with my girlfriend, and my Family. People with a Sick-Mentality use this Exotic-Techno in my life, they are Lowlifes. As I put it, "Some people are the salt of the Earth, these people are the s**t of the Earth.". This is all about Manipulation with Techno.

I have seen the aftermath of it being used on a young man with his girlfriend, who these SPOOKS CHOSE to have stab his girlfriend. These Spooks also TRIED SOMETHING SIMILAR WITH ME, many years ago, in regards to me and my girlfriend. IT IS ALSO WHEN I CAUGHT THESE SPOOKS FOR THE FIRST TIME WITH THIS EXOTIC-TECHNO. I caught a "Thought pushed into my Head", the Spooks pushed this thought in, it was the "Cover Story", followed by INTENSE ANGER, DANGEROUS LEVELS OF IT. The thought was the "Spooks Thought", it was prepared in advance. To serve as "Faulty Reasoning", an excuse for attacking my girlfriend, A COVER STORY, as then an INTENSE FEELING OF ANGER was quickly PUSHED IN afterwords. To convince me that one went with the other.

The Cover-Story, being the first "Layer", the "Emotion-Pushing" being the second "Layer". Again this is TACTICS used, in conjunction with this Exotic-Techno, this is HOW it works. These Spooks as a result of this Exotic-Techno, have a Handbook on how to Misuse it. Thus I have given you an example from my life. Again sick Misuses of this Techno, used by sick people, trying to "Get something they Want". With this Techno, they can get away with any crime. Now to get back to the original point of Emotion-Pushing, this can be done with ANY EMOTION.

In regards to this "Messiah-Con-Man" the EMOTIONAL-PUSH will be LOVE. This CAN BE MIMICKED by this Exotic-Techno, ANY EMOTIONAL STATE CAN BE. It is a matter of FREQUENCY, VIBRATIONAL STATES OF ENERGY. Love like anything is an ENERGY with a FREQUENCY, with a VIBRATIONAL RATE. It has a "VIBRATIONAL LEVEL" which is it's own, it has a "VIBRATIONAL FINGERPRINT". This Exotic-Techno, operating on principles of energy, is able to mimic this. If you need total "Proof", I offer you the following example.

These Spooks in my life are sick people led by a very sick, Psychopathic Woman who is in charge, and who I call The Clown. Due to her being an actual Psychopath, she lacks balance, clarity, loves to "Flaunt Herself", and is Extreme in her use of Force as a result of her lack of Empathy. In short she, The Clown, and The Buffoons, use FORCE to compensate for a lack of Intelligence, and this book is a direct result of Her and Her Team of Idiots. She is The Clown.

She is literally a PSYCHOPATH within this American Agency, who has and uses this "Job", for her own sick, warped, needs. She is for all intents and purposes a "RAPIST", having Power-Issues, and Abuse of Power-Issues, inflicting herself on others. It is why SHE HAS HER JOB, AND CHOSE HER JOB. To give you a hint, you CAN TORTURE PEOPLE WITH THIS EXOTIC-TECHNO, and not get caught. You have no idea. It is NON-TRACEABLE. Obviously I, and my Family, have been TORTURED repeatedly by this Clown. I can not stand this "woman" for all of the right reasons, at all, and this IS A MAJOR UNDERSTATEMENT. And as I understand it, based on how most of the other Spooks from my past (who were in charge), used this Techno, it is meant to be used for "Profiling" and "Wire-Tapping a Persons Brain", not for "Holding someone in this Mobbing" or for Torture—no doubt one more violation of her Agencies Protocols and Rules. So with this in mind, the next part.

So to give you an example of this Exotic-Techno in regards to Emotion-Pushing "Love", in this regard, I had the same experience, a months worth. This "Situation" I am in, generally lasts about one month, thus I had "One months Worth" of being "In love with this Woman", The Clown. I have stuff written from this time, notes, that are sickening, mushy, and in retrospect totally disgusting. Having said that I have "Recovered", obviously. This can be done to anyone with this Techno, in regards to a "Particular Person", and these Elite obviously Plan on Emotion-Pushing "Love" with this Techno, on the Masses across the World. People will "Feel this Love" quite apparently, it will be obvious as "Coming from Elsewhere", as the "Intensity of this Techno" can be "Turned Up", and in this case will be.

So in regards to this Messiah-Con-Man, and his "Miracle of Love" which all people across the World will feel to an extreme, being quite noticeable as NOT THEIR OWN: that is because IT WILL NOT BE. IT WILL BE EXOTIC-TECHNO HOOKED UP TO SATELLITES, ENERGY, Vibrations, Frequency. THIS WILL NOT BE A "MIRACLE", it will be TELEVISION-PREACHER-SCAM, with the latest state of the art TECHNOLOGY, to be convincing—to convince us that it is a "Miracle". You will see his image on television, "A Caring Look" as the "Cover Story", followed by an "Emotion Push" of "Love", which will fit this Phony-Jerks "Compassionate Look". No doubt they will "Leave this Love On", what I call "Extended-Feelings", for the length of the broadcast. So that is the

"Second Miracle" to be addressed by me. Now on to the "Third Miracle", and again used on me and my Family all of the time.

The "Third Miracle" to be experienced by the Masses. Again it is a pile of DECEIT. It is stated at this Fake-Messiah's website, that many (hundreds of thousands) people will experience "Miracle Healings", obviously "SPONTANEOUS HEALINGS" in conjunction with this International Broadcasting of this Con-Man, to the World. "The Day of Declaration" as it is called, this International Broadcasting of this Jerk, will be one large Cover Story for his "Miracles". Again I have experienced this repeatedly along with Family members, but in a negative manner. As a recent example, yesterday, my girlfriend came back from the hospital (September 28, 2010), and was there in the first place, due to these jerks, and their Exotic-Techno. Nothing beats the "Gift of Giving" unless you are in my, and my Families shoes. So what happened?

A couple of weeks ago her jaw, her lymph nodes were slightly swollen—by way of Exotic-Techno. Eventually it went away. This was the "First Cover Story", and a building up of layers, of more to come. Shortly after, perhaps a week, she began having "heart palpitations", "heart problems", and discomfort, which in turn she began to worry about. Now the "swollen lymph nodes" were Techno related, and the following "heart palpitations" as well. But from a MEDICAL VIEWPOINT it looked as if "the virus" had gone from her lymph nodes to her heart, and infected it. To top it off, no doubt when my girlfriend began wondering at all of this, they were sending her "scary thoughts" ("disastrophies" as I call this), and every time she "bought into one" they raised the level of fear (subtle Emotion-Pushing) she felt, by one notch. It turns out she was fine, but the point is it was all done with Exotic-Techno. They CAN AND DO MIMIC PHYSICAL AILMENTS, and use "Cover Stories", to sell this. The "Cover Story" is simply the "Logical Cause or Pretense", that which "sells you" the rest of the Exotic-Techno Con-Game.

Obviously in regards to this Messiah-Con-Man the Cover Story will be something that happens on the Televised Event, the International Broadcasting in which we see him. Perhaps he will "Put his hands up to the television screen in a healing gesture" which would be the "Cover Story" to sell the following Exotic-Techno related event of "Healing", I do not know for certain. Perhaps it will be his "Loving Look" which is the Cover Story, it does not matter. The following "Healings", no matter the Cover Story, will again be this Exotic-Techno, used very briefly for good, but in the end used for doing harm. There is no doubt that this Exotic-Techno would change the World, IT CAN BE USED TO HEAL ALL AILMENTS, however the "Changing of the

World" that this Exotic-Techno will be used for, is for anything BUT GOOD. As you can see this is all a matter of TECHNOLOGY, not a "Messiah". People will need to keep their heads and not act as if they are in a new version of "The Gods Must Be Crazy". When you look, the pieces fit, and it is better that people start looking NOW at this Techno, rather than later. People need to spread the knowledge of the existence of this Techno, before it is too late.

A Summary of Several Plans: a Grim Future

Now to add to this "MIRACLE MAKING", and this Messiah-Con-Artist jerk.

There are several PLANS, THINGS WHICH WILL BE GOING ON, if this "Messiah-Con-Game" is chosen by the Elite. One is that there will be a New-World-Economy which this jerk will put into place on behalf of the Elite. IT WILL BE WORLD-SOCIALISM, thus this jerk preaches "Sharing with your Brother".

Two is that this jerk will offer a WORLD-PHILOSOPHY, based on many practices and tenants of Various Faiths. This will be a combination of many Religions, mixed in with "Humanitarianism", to appeal to all. HE WILL ALSO FULFILL MANY RELIGIOUS PROPHECIES, of many Religions by using this Exotic-Techno TO "PERFORM MIRACLES" AND BY "LITERALLY FULFILLING RELIGIOUS PROPHECY" right down to the exact wording. To be convincing. HE WILL NOT ACTUALLY HAVE THE POWER TO PERFORM MIRACLES, it will all be done by someone else CONTROLLING TECHNOLOGY—think about what a SCAM this is. Thus many Religions warn of this "Pied-Piper", this FALSE-MESSIAH, THIS IS WHY. Because he will be "Convincing".

Three is that this jerk is to be made to look real, a "Messiah Figure", therefore seemingly surpassing Governments in leading the World, and CENTRALIZING ALL AUTHORITY TO HIM. The World-Governments will seem to "Bow Down" to him, it will all be a SHAM, A LIE, A DECEIT. This guy is the "Puppet of the Elite", and BACKED BY THESE DEMENTED-IDIOTS. From this, Religious Prophecies, other Prophecies, and books (like Dollores Cannons "Nostradamus Books"), we can safely assess that this guy is a FALSE-MESSIAH, that this guy IS THE AWAITED ANTI-CHRIST, DAJJAL, and or any other name you want to give him. I simply call him "Jackass" among other things. Regardless this has to be taken seriously, and keep in mind he will be trying to appeal TO RELIGIONS because this cuts across "Boundaries of Nationality" and will be easier for "Uniting People" under "His Banner".

This guy will be used to usher in New World Order, Fascism, and EVENTUALLY A GRUESOME TYRANNY. This will be what happens if people do not begin questioning things and take no Responsibility for Themselves, their Thoughts, their Actions. In short this can happen if people do not become Self-Empowered, Self-Thinking, and Active.

A Few Added Points

Now an important point here: "who I speak of here" is a "western version": these are the plans in place. However, I believe that a "Figure will arise from the Middle-East", and that the Elite, will eventually "Put him in the middle of these Plans"—that they will Use Him for what I write of here. Thus there are for now, two possible versions. Having said that, this "Messiah-Program of Deceit" is already in place, and is obviously on the table for Consideration. If a "Figure from the East" who has United the Middle East into "One Force" shows up, no doubt these Elite will try to use him. Perhaps the Elite already have a "Deal worked Out" with such a figure. As of January 2011, as I listen to the News, and if you are aware of the "Conversations with Nostradamus" books by Dolores Cannon, this "Figure Nostradamus Speaks Of" is now making his "Emergence". There are, and is, Multiple-Countries-Rioting against Governments of the Middle East and Africa: Egypt right now, being big in the News. This is a sign, and Grim-Times are approaching.

I also believe in regards to this fake-Messiah there will be some nasty information coming out about the Church before hand, to weaken it's status in the World's Eyes. To make people of this Large Religion more susceptible to his "Charms and Sway". If the Church is made to look bad, members of this Religion questioning it, many will be looking for answers and a "New Faith"—many will be led into this Conman's arms. Again it is all Multi-Layered Manipulations and Deceit.

He will preach about Humanity-Uniting, "One Family" without nationality, and curing the worlds problems. He will try to unite all religions under his banner with a world-philosophy, and will have a great appeal to the Masses, because he will seem to offer everyone something. Never mind his "Magical Powers" and his "Curing the World's Economic Woes with World-Socialism". His World-Philosophy will be a synthesis, an amalgamation of all religions, one philosophy invented to appeal to people of all faiths. It is designed to appeal to Christians, Muslims, Hindus, Jews, and Buddhists, and will actually be "Godless". He will "Fulfill Religious Prophecies" by way of Exotic-Techno. He will be both a World-Leader, and a "Figure of Faith", a "Religious-Figure". He will preach Socialism, and Humanitarianism, while perpetrating the biggest hoax of all time. He will be the biggest Fake on the Planet.

He will offer World-Socialism, for a Deliberately-Collapsed World-Economy. "Economic-Cures" to the "World's-Economic-Ills". This is why the World-Economy will collapse, so a lot more people will be "Poor" as a result, in addition to the already existing poor. Socialism will be offered as the "Fix". It will offer something for everyone, it will be a "CASH-INCENTIVE" for everyone to go along with SOCIALISM: "If you want any of your money back, go along with this cure we offer you-Socialism will fix it all.". Thus it will have a lot of appeal to the Masses of the World, and it will appeal INSTANTLY. The amount of "Poor People" will be the "Majority". Already in the News I have heard of very rich people giving all of their money away (2010), this is part of the set-up for this World-Socialism, and World-Philosophy of "Sharing and Giving". It is a pile of Deceit.

All of this will be easy for him to do, because he will really be a spokesperson for the Elite, "Uniting the World", the Masses of the ignorant underneath him, and thus underneath the Elite. Seeing as how the Elite ORCHESTRATED this Economic-Collapse, in America and the World, they will thus be ready to offer "CURES" through this man. The "CURES" will seem to be his, in reality he is in league with these Demented-Nit-Wits; as I've said many times in my life in regards to these Spooks—"First they break it, then they fix it". The SAME TECHNIQUES these Elite have been using all along, will continue to be used. Every "Solution" will be one more step down the path to hell. More and more rights will be lost. People will allow themselves to be rounded up like sheep for the slaughter, and if you look at the signs, that may indeed be what happens. Eventually people who are "Useless to the Elite" will be eliminated as they will have "No Value".

So. The Elite plan on using this Exotic-Techno to convince the World's-Masses that he is the "Earths Savior", come to fix all of the problems which are going to "Spring Up" in the next couple of years, and based on this Cosmic-Cycle and 2012. All of these "Coming Problems" being Manipulated into existence by this Elite. Specifically Economic Ones, although no doubt other strange things as well.

This televised event to come is called the "Day of Declaration", and is A SCAM, LIE, AND TOTAL DECEIT. This character can be found on the Internet by looking up MAITREYA, and you can see many of the PLANS of the Elite to come, by looking at some of these websites. The reason I believe this guy has been chosen by the Elite, is that it is a Plan long in the making, and has already the endorsement of Major-People and Companies. THIS GUY HAS THE OBVIOUS BACKING OF THE ELITE, he is not some random kook in the woods, and he sure as hell isn't genuine.

I've looked at several websites touting this guy, and every "Miracle" I've seen is "Exotic-Techno" related. A lot of his followers "Experience Maitreya" in the dream state, as one example. Folks me and my kid have been experiencing a Manipulated Dream-State, every night for years (If you have the ability to listen to someone's thoughts, and you follow them into sleep, what happens?—See what I mean?). Again it is Exotic-Techno. This is "TELEVISION PREACHER DECEIT" taken to the next level, taken to the World Level, it is the biggest Con of all time. This guy plans on playing on the WORLD'S FEARS (Manipulated and Created by the Elite), and then the WORLD'S HOPES—solutions offered by him but Created by the Elite. The Elite create the Fears, he turns around and offers Hope (Of course the "Solutions" and "Hopes" offered by "Him" are also being formulated by the Elite), SEEMINGLY APART FROM THESE PEOPLE WHO CREATED ALL OF THESE PROBLEMS IN THE FIRST PLACE, YET ACTUALLY IN LEAGUE WITH THEM. It is pure, Television-Textbook-Preaching, the ripping off of the Masses, by playing on peoples "Fears and Hopes", and using Technology to Manipulate and convince the "Followers".

In my life, this Exotic-Techno is considered anything but a "MIRACLE". But if you don't know this Exotic-Techno is out there, it is going to seem real enough when the time comes. This Exotic-Techno is going to be used on the Masses to create a "TOTAL DECEIT" and to sell to the World a TOTAL FAKE. This Con-Man and his Elite Backers want everyone to put their Faith and Trust in him. The Masses, if they choose this experience, will suffer tremendously. Again the "Path of Experience" is stupid.

An example of this Exotic-Techno being used by satellite.

This Exotic-Techno, USED BY SATELLITE, was also already FIELD-TESTED, if I remember correctly, in the west or mid-west of America. See Dan Akroyd on that one. He experienced it himself, along with a lot of other people if I remember correctly. My belief is that this was one of the first times it was used by SATELLITE. Obviously this Exotic-Techno now has the capability to be beamed by Satellite now, for specific use on the Masses. As I remember Dan Akroyd woke up in the middle of the night and had the "IMPULSE" to go outside and look at the stars, the sky. If I remember right there were UFOs in the sky. All of this I have to go with by memory, because these Government Jerks are sitting on this computer. Things that I need to find I can't, sights are blocked, and this computer does not even begin to act normally. Some people around here call this "making a living", and get paid a lot more than the average citizen for doing this kind of BS. I'll have to rely on memory because quite frankly these jerks are not letting much happen on this computer. Entire subjects seemingly don't exist.

Now this "IMPULSING" I've experienced so many times it is disgusting. And yes I have been "awoken" by this bastardized Techno to fulfill an urge, an impulse (usually eating, others as well). Again all Technology, no miracle involved. This Exotic-Techno enables the Governments to Manipulate Physical Reality at will, people at will—both their mental and physical states, and the Physical Reality around you. It is nuts, and scary to think of, however it is TRUE. Hiding under a blanket or with ones head in the sand isn't going to change this.

Being aware of it and speaking of it to others will. Increased awareness will. The hundredth monkey syndrome eventually takes over. PEOPLE NEED TO SPREAD THIS INFORMATION.

Another point to make here. If you go looking for pictures of this guy, you aren't going to find many, for a guy who has supposedly appeared all around the World and performed "Miracles"—you will be hard pressed to find any photos of him. I was expecting to find photos and videos of this guy, from Technology-Washed-Devotees, but strangely enough I can find only a few pictures. And in those "few pictures" it is obviously at least two DIFFERENT MEN—this is stupid. Strange huh? Is no one bringing a camera? If everyone "loves him so much", why are they not taking pictures and why is this guy, and how is this guy, two different men? Obviously the Elite had not picked out a "Person for this Role" yet, and were simply using "Substitutes" until they did.

I've also read that when "He" does appear, he appears in "Disguise". No doubt to explain why he always looks like obviously different people. If he plans on appearing before the world, what the hell is the disguise for? Anyone "good and genuine" does not need to resort to DECEIT. No, the Elite have been planning this whole ordeal for a long time now, but did not have the right "Candidate" picked out for a long time. THEY HAVE HAD THIS OPERATION IN PLACE FOR A VERY LONG TIME, JUST NOT THE RIGHT PERSON. They have been laying down the "Groundwork and Foundation" for this Scenario, but did not have the "Right Person" picked out for the job. Obviously, they now do.

No doubt when the time comes, some "logical-nitwit answer" will be provided as a "COVER STORY" for this obviously weak deception. In fact it may also be so that people on the look-out for the Anti-Christ, Dajjal, see this weak ploy, and think it can't be this guy. Actually take another look because IT IS THIS GUY, it is just that his real face has not been shown yet.

Having said that this seems to have changed, they seem to have someone picked out now, to fit all of the infrastructure they have put into place for this World Wide Deceit. This Plan and Deceit has been in the making for many decades now. THIS ARTICLE ANNOUNCES THAT THEY HAVE SOMEONE PICKED, A CANDIDATE FOR THE JOB OF ANTI-CHRIST, this announcement is a message that the FINAL PLANS are near. And if you are paying attention you will see that they plan on "Hijacking this Space Cloud of Light", which is in essence an "Act of God", by claiming it is for this phony Con-Artist, this Pied-Piper of the Elite. They plan on claiming that this "Light Cloud" is an "Act of God" to "Herald in their Phony". In Reality this "Cloud of Light" has to do with very different things entirely: a Cosmic-Cycle, a Positive-Path and Growing Consciousness for Mankind, and "People who are on the Path of Consciousness". These Elite are going to absolutely Bastardize It, and Misuse It.

So expect the near future to be totally nuts, totally scary and full of insecurity—all of it by DESIGN. UNLESS PEOPLE SPEAK OUT. Enough people having knowledge of a PLAN, can alter it and change it. That is how PLANS are executed in the first place, by people knowing about them, and being Active in regards to carrying out a PLAN. The same principle can nullify a PLAN. At any rate until enough people become CONSCIOUS of such plans, things will get all sorts of worse, and this Fruit-Cake is going to be advertised as "The Great Hope".

Now go to websites advertising this guy as "Maitreya". If you want to know what the Elite have planned this is the best place to go to. I've found out all sorts of things.

First of all CAPITALISM, due to the collapse of the American Dollar and America the Country, will be claimed to "Not Work", Dysfunctional. Just like with the Old-System in Communist Russia-Communism was declared "Defunct" and not "Realistic". Which is ironic, because this Elite will basically be offering it as a SOLUTION TO THE WORLD-ECONOMIC-CRASH. Thus a "New-System" will be declared, and ironically it will be SOCIALISM (Communism under a different name in this case)—thus all money and power and faith, will go to Governments, One World Government, actually. WHO WILL THEN PUT, BESTOW, ALL AUTHORITY ONTO THIS INDIVIDUAL, THE "FAKE-MESSIAH"—this will CENTRALIZE ALL AUTHORITY UNTO A SINGLE PERSON, AND A CROOK AND CON MAN AT THAT. He may be used to "Propose this Idea" of a One-World-Economy, and a One-World-Government. Then in the end, and part of this Elite-Con-Game and Deceit, the World-Government will "Bow Down" relinquish control to this "Con-Man" and "Poster-Boy" of the Elite. All "Power and Control" will seem to go to him, when in reality it will be the Elite who are in Control. This will be to Centralize-Authority, and to give the World a "Leader", who will be a "Figure-Head" in terms of "Government" and "Religion". He will be able to "Fix All" and seem to even "Unite all Religions" under his "Cause of Brotherly Love" which will be a FARCE. A "Loving Dictator". As you can see this guy really will be the "Pied-Piper" of the Masses.

As Capitalism disappears, individuals will have no chance of obtaining real money, and therefore power—whether it be for good or not. The idea of "Climbing up in the World", the American Dream, will vanish. Everyone with any hard earned savings and cash, will be REDUCED TO ZERO, and made "EQUAL". SOCIALISM will be offered as a CASH INCENTIVE to all of the "POOR", which will be damn near everyone once these Elite Crash the World-Economy into the ground. But hey, on the bright side the "Cash-System" will be eliminated, and therefore a lot of theft and robbery with it. Especially once people start getting implanted with "Micro-Chips" containing all personal information, including Financial Information. You will be your own credit card, and no doubt this will start with this "Messiah Figure".

Think about it, YOU WILL BE YOUR OWN CREDIT CARD. It sounds like an episode of Star Trek, with the Borg. Let me give you an interesting example from my life, or rather someone close to me, that POINTS TO THIS.

A friend's colleague had an interesting experience with one of her bosses. She and the colleague she shares an office with, thought the experience was strange, as well it was. Her colleague went to the boss, one of three bosses, asking a specific question about taxes. As an example he used this "Idea"—"If the Employees were asked by the Company to go to a hospital to get an IMPLANT FOR WORK, for the company, and then bla, bla, bla". She said "What is going on here? Is this some kind of New World Order thing?" she couldn't understand why he picked THAT of all things as an example, it was out of the blue (or was it?), his example of being Micro-Chipped for work. He wouldn't respond in one way, form, or shape, or another: in response to her N.W.O. question—he said absolutely nothing, which is also strange. He was clearly speaking of something he knew of, and putting Info out. He obviously would have gone into more Info and Detail, except he was obviously worried about Consequences. Having said that the obvious thing is that THIS BODY-MICRO-CHIP WILL OBVIOUSLY BE IMPLEMENTED AT A WORK-LEVEL FIRST. IF YOU WANT TO WORK YOU HAVE TO BE IMPLANTED WITH A MICRO-CHIP, "FOR THE JOB".

Obviously this boss, was speaking of something he knew about. Obviously the Governments have a PLAN to use WORK, as an excuse FOR MICRO-CHIPPING PEOPLE. You will "clock into work, and out" by way of being your own "CARD". This is a sinister, cunning way, of using FEAR, SURVIVAL, to get people to comply with this MICRO-CHIPPING. If you DO NOT GET MICRO-CHIPPED, then YOU DO NOT GET TO WORK. No work means no money, which is an issue of comfort AND SURVIVAL. How can you exist in Society without WORK, WITHOUT MONEY? Thus this is TO FORCE COMPLIANCE IN REGARDS TO MICRO-CHIPPING. No doubt this will be introduced as one of the "GREAT IDEAS"

AND "GREAT INVENTIONS" of this Messiah-Jerk. At "his time" TECHNOLOGIES will be pushed as the "GREAT MIRACLES", in addition to himself. He would know, he will be using Exotic-Technology to "SELL HIMSELF" as a "Miracle Maker". "God, the Divine, Allah", will all go right out the window, replaced by a World-Philosophy, "Technologies" will be the "New God" and "New Miracle Makers".

But back to the point. THIS IS A CLEVER DECEPTION. IF THE GOVERNMENT WERE TO DO THIS THERE WOULD BE A "MASSIVE RESISTANCE". BUT IF COMPANIES ARE "ALLOWED TO DO THIS" BY GOVERNMENT, THREATENING YOUR INCOME, IT IS OKAY. Of course a lot of people with no backbone and no brain will give in to this. Anyone doing so is an IDIOT.

And I am sure that THIS SCAM will happen when this Pied-Piper is put into Power.

IF EVERYONE SAYS "NO", THE PLAN OF THE ELITE DOES NOT WORK HERE. Again it is plain and simple. The Governments can not force this kind of thing without looking obvious and exposing this, which they don't want to do, they want to implement things SLOWLY. This way no one is aware of their own Governments DECEIT until it is to late. Saying "NO" TO MICRO-CHIPS on an Individual Basis, and Massive-Scale, in regards to any situation and "Government Excuse" is the best thing people can do. This will seriously hamper the Elites PLANS. Hopefully people will sense there is something wrong with the idea of being implanted and be resistant in the first place. Let's hope so.

This is the reward to Jackasses for being blind to the system, and not listening to people like me, and thousands and millions of others trying to open peoples eyes. You could have learned by opening your Eyes and Ears and being Conscious, but the Masses choose to learn by Experience instead. A lot of people are total jackasses, having chosen to stick their entire head into flames, instead of getting a clue and realizing that fire burns. So this will be your reward, unfortunately people like me pay as well. In fact people like me get Ball-Kicked by people like YOU (The Ignorant Masses), for trying to open up your eyes in the first place. Then I get to live in this mess of a world, because people like you—en masse, have determined, due to your ignorance and gullibility, what will transpire.

This Fake-Messiah will be to busy "FIXING THINGS", to be able to tell you that all of these "Solutions" to "Problems" will be A BIG SCAM TO WIN YOUR FAITH in him. BLIND TRUST, BLIND TRUST FROM THE BLIND. This guy is the Pied-Piper and he wants to rule all through the nose, so do the Elite supporting him, after all he is THEIR "POSTER-BOY". The "SOLUTIONS" being the "NEXT STEP" TOWARDS LESS FREEDOM. Like with this Scam to put "Microchips in People". One big step closer for Mankind, towards hell on Earth. People are going to totally Piss-Away their Rights and Freedoms, and this "Micro-Chipping for Work" points to this.

They'll pay for it later, because it is the nature of the Power-Hungry to take and do, more and more. That Negative Energy will keep growing and growing, until there is an equal and opposite reaction to balance it. The LAW OF CAUSE AND EFFECT ON A HUMAN LEVEL. People can do their part now, or do it all at once later, when the "Situation" reaches "Critical Mass". At that point the only Solution will be violence. Past-History speaks of this, it is called Revolution, and has happened time and time again for the very same reasons. People need to decide if they want to learn through Consciousness or Experience, it is that simple. If you choose Consciousness then you need to get ACTIVE and start disseminating knowledge, the more people that know about what is going on, the less likely things are to occur. If People do not COUNTER and BALANCE this Growing Corruption and Oppression now, they will have to do it later. Having said that People are waking up, I even saw a guy on CNN refer to the "Illuminati", another name for the Elite (a kookier name in my opinion). So get busy and start waking people up.

Remember by "Waking People Up" you are Counter-Acting, and RE-BALANCING ENERGY: in this case Negative-Energy. But back to the "Fake Messiah".

The "Fake Messiah" will appeal to all Lands and all Religions, preaching "Acceptance and Self-Realization", much of this copied from Real Enlightened People, and a lot of the ideas taken from various Religions. In Reality he will be the Biggest Scam of all time, and this will be the only thing he is great at. While he is leading others to believe they are a part of a "Great Change for Humanity" he will be soaking up more Power and more Control. Having said that your power is WITHIN, not without, and not placed in the hands of some Fake. Eventually everyone will pay for their blindness.

Those who can not find Truth within themselves will be misled and suffer as a result. In the end, this Con-Man-of-the-Great-Hope, will show his True-Colors. The Elite, through him will start "Eliminating Non-Useful Members of Society". Guess he'll be obvious enough when that time comes. If that time comes.

That depends on The People speaking up, looking and being vocal, and getting the Info out. It also depends on people praying—asking the Divine for help, and focusing on the Positive, while being Non-Attached to the Negative and getting the Info out. Keep a "Positive Attitude", don't sink into fear over the Possibilities, this energy feeds them. Simply by knowing the Plots of the Elite, a lot can be negated and avoided, these people like operating from the shadows, not in the light of day. There is Hope, REAL HOPE, and it doesn't come from some Elite-Pied-Piper, it comes from WITHIN, from The People, from each person becoming Self-Empowered on an Individual Basis. One person can change the World, it takes less than we might imagine to make a Positive-Affect, especially when a Small-Collective is making good with "Will-Power, Energy, Focus, and Intent". So get busy.

Some "Final Weirdness"

One last note here. When he, "Maitreya", has been asked about specific instances of UFO sightings, he usually (always) says they are "From Mars or Venus". Which I find ridiculous aside from the fact that we are talking about Government Spooks operating this whole SCAM and they are trying to steer and predispose people to "Aliens from Mars" (and Venus apparently). Just like on a computer, you don't see him, don't see the person you are dealing with, you can and in this case probably are, dealing with anyone. Including Joe Spook, complements of tax-payer money, and secret illegal money-making projects by these Government-Criminals. Of course they work directly for the Elite.

No doubt the idea of "Friendly Aliens" who will "Fix All" is being SOLD to us here. Someone, that Humanity can "Put all of it's Dis-Empowered Faith" onto. Of course any "Aliens Arriving" who the Governments "Relinquish Control To", will obviously be "Government Made", Elite-Made, in the first place: you may have noticed they are about "Having Total Control", not "Giving Up Control". They simply want someone who works for Them, who the Masses will Blindly Follow. Any "Aliens" willing to "Take on All of Humanities Problems", should NOT be Trusted. Again see my "Path of Experience and Consciousness" piece. I have a Chapter on an "Elite Made Alien Scenario" in my second book, which I will try to print this year, when I have the Resources.

In the last year (2008) there was even a photo of "Someone" or "Something" on Mars. Obviously any "Staged Alien Event" will involve Government-Jackasses "From Mars or Venus". Government-Jackasses, regardless of their "Appearance", as this is what is being "SOLD" on this "Maitreya" website. Keep in mind the American Government along with others has been engaging in Genetics Programs since World War Two. People from the stars have been visiting Earth since it's birth, them coming from Mars and Venus are "Human Ideas from the Fifties" not Reality. The evidence in regards to Mars, suggests life was wiped out there. And yes I get there are other possibilities, it is just that in this case there is obviously Manipulating and Deceit going on in regards to Mars these days. We are obviously being "SOLD" on Mars and the idea of "Alien Life". Which means this is a controlled situation, meaning it is the Elite's situation, and in this case as with so many others, they are not being honest. They obviously plan on Manipulating the Masses with this stuff. It is plain and obvious, simple even. "Hidden in Plain Sight". And also on this "Maitreya's Website".

Notice that Mars, and facts of it are coming out more and more? Just today I saw an article, saying that they have found water on Mars, I think it was salt water, but water, nonetheless. Notice, that Mars is slowly becoming a more LIVABLE PLACE than at first claimed. Slowly bit by bit, it is a controlled release of information, a slow building up to the idea of "Life on Mars". It is no coincidence. If you look at, and believe in the Billy Meier material (which I do), he has a picture of an American satellite orbiting Mars, from the 70's...Who lives there "Martians or Humans?"...Toss in the idea that America has a UFO/Space-Fleet, and it is obvious that the American Military is going to explore the Solar-System and try to set up house somewhere else, as in a different planet. The "Ultimate Restricted Zone", this would make "AREA 51" obsolete. If any of the evidence from Mars and a past Civilization there is accurate (You can go to Google Mars and find things—again a deliberate slow release of "Past Life on Mars") you can bet the American Military would make a point of going there. None of this is hard to understand.

As I have said this "Maitreya" website is like looking at a Guide-Book, a Map of the Elite's-Plans, and this whole "Mars and Venus" thing is among some of the more "Obvious and Weird" stuff there. A subject that is "Outside of the Box" for Humanity, and therefore "Useable for Manipulating the Masses".

People should keep their Eyes and Ears open, and Speak Out. Post fliers in your town or city with information with what is going on with the Elite and their Plans, you never know who is going to take an interest—this is a simple way of getting Info out. Be creative and legal, but get the Info out. With effort, Energy directed in a Positive Manner, we can hopefully avoid the "Situations" I have mentioned above, and open up new possibilities to ourselves as the Human-Race. One thing is for sure, we'll be finding out before too long.

CHAPTER 28

"My History": How I Got in this "Situation/ Mobbing": A Brief-Simplified Version

The following chapters are really "Book Two" and will help you Understand and See how Government Manipulates The Public. The following chapters are also not as well Edited or as well Written due to "Unusual Circumstances", and my desire to get this Book out NOW. Much of the content is about my personal life, "Mobbing" which is a Corrupt-Government-Program, and about Corrupt Government Agents, and Government Agent Tactics. Having said that the following material is of benefit to All People. This Illegal Government Program of Mobbing, and the Government Tactics I have learned of, have made up my "Government Agent Education". These Tactics, and these Corrupt-Agents, this "Type of Person", can be seen all around us in the Political-System, Media, and World of Today. In my opinion, anyone reading the following chapters, especially the Government Agent Tactics Chapter, will benefit greatly in terms of how they see "The World Around Them". With an understanding of these Tactics, and of this Mobbing, you will be able to see the Corruption of this World far more clearly, and be able to see the Elite of this World, and "What and How" they are getting away with things. As a last note, the following chapters do contain some "Harsh Language", which I normally would be against, but felt it best as it makes some points by it's "Tone". As it is, I deal with things at this point in my life, that are far worse than "Harsh Language", so if this is "The Greatest Burden of your Life", try to get over it. It is more important you look at the Knowledge, and not get "Fixated on a Few Words". With that in mind, lets begin.

For the "Brilliant-Citizens" who take part in Mobbing:

"You gotta be responsible for your actions. If you want to give me your life, I'll take it. And I'll put it in the deck and I'll deal it, you dig? And if you don't like the way I deal it, then don't give it to me. You dig what I'm saying? But when you give it to me then it's mine. I deal it anyway I want."—Charles Manson

"Tell me what what you don't Know"—
My Response to Ignorant Citizens

As you may have gathered from reading this far, I am in what I refer to as my "Situation". My "Situation" is commonly referred to as "Mobbing or Street-Theater" on the Internet—and it is part of a GOVERNMENT PROGRAM. This is NOT just an American-Program, it is GLOBAL. There is an International-Agreement to do this and Implement this, in spite of the fact that this is Illegal in most Countries except Police-States, and against International-Law as people know it. My "Situation" has gone on for eighteen years, and has spanned three Continents and three Countries—and no doubt cost Millions of dollars at this point. In my "Situation", what "Got me into this Mobbing", is that I was FRAMED by Corrupt-Government Agents in America, who turned around and "Said" I deliberately interfered with a Federal Investigation. They Lied and Framed Me, for the sake of getting me in this "Mobbing". I was actually Framed more than that, many times, but I deal for the most part with the one that got me into this "Mobbing", one before this "Situation Started" that went on behind my back, and one other which occurred during this "Mobbing".

All of this started, began in one sense, because I knew of an Undercover Police Officer who due to his Negligence got his 14 year old girlfriend beaten and raped (the relationship already was Statutory Rape I might point out), and because of a Murder which happened outside of my window, and which was "Racially Based" as I will eventually get to. In both cases there was a Cover-Up, and in both cases I knew the Truth of the matter. They were "Watching Me" and "Investigating" me because of what I knew about the first situation, the Pedophile Cop, and then the Murder happened. To top it off, I was speaking to certain people about this, which obviously created fear for the Police Department and others involved. My "Situation/Mobbing", is the result of this, unbeknownst to all—the Public, Citizens, and even myself at the time (to some extent in regards to myself), had no idea. I was not used to this level or degree of Corruption, this "Type of Thing" usually only exists in the movies—when it happens to you it is hard to believe and see. Having said that I could see plenty, but there were some points later that were lost on me.

The first situation that happened involved this Undercover Police Officer, and this is when "Things began to Happen", behind my back, and unknown to me: COINTELPRO. The situation with the Murder took place a couple of years later. Having said that it was this situation with the Pedophile Cop, that got the "ball rolling" with the Feds. From their point of view they were going to "Neutralize a potential Threat"—me. In reality they were looking out for a Pedophile Cop, and a Police Department that should have taken more Responsibility in regards to it's "Employee". You will note that this Pedophile-Cop was never "Mobbed"—odd. I doubt that a single Citizen who later "Mobbed" me, has any idea of any of this—I can be certain of it. Having said that the Police Department and Feds "Thank You" for your support of their Pedophile Employee, with special thanks to Eve, who allowed herself to be Manipulated, Duped, and Suckered, and who not only put her "Head through the Ringer" but my own as well.

From this point on, based on my knowledge of these Events, "Things began to Happen" and initially (for more than a year if I remember correctly) all of it behind my back, again COINTELPRO. You can not "Expose" what you can not "See or Know Of", and so the Agents involved, Manipulated People behind my back, to avoid being "Caught and Exposed" by me: which is what I have currently going on in my life now, with The Clown and her Buffoons. Except the difference between "My Past in America" and NOW with The Clown, is that I can see what The Clown is doing and getting away with: but back to America. I was known by many as being "Paranoid", and by the police as being good at "Spotting their People". I have no doubt I would have figured things out, and apparently they had little doubt as well which is why this all took place behind my back so discreetly—as it is, more than twelve years later I began sorting it out, and have sorted it out. Thanks to The Clown. My process of "Discovery" was unpleasant to say the least, never mind "Going through this Situation" while I did so.

In the meantime, just like back in America, my "Mobbing" with Ignorant-Citizens goes on here in Europe, and again just like in America, for reasons that the Citizens and Clueless Field-Agents actually have no idea about.

And just like in America, for Corrupt-Agents that they do not even know. As it is none of this makes sense any more, like it did in the first place, and unknown to the Citizens here in Europe and the Good-Non-Corrupt-Field-Agents, I am being "Held In" this "Mobbing" Deliberately, by The Clown who is extremely Corrupt, and who has "Kept Me In" for Personal Reasons. She is "Caught" at this point, unlike the Feds in America, and is "Using this Mobbing" as an "Excuse" for trying to contain damaging information about Herself—She is doing a Cover-Up. Not because I "Have not Passed", which I HAVE, but because of Her own Corrupt-Nature; She is a Liar to say the least, and in actuality, without using this as "cliche", She is a Psychopath and "All that comes with It". This is unknown to the Citizens, and I am sure Her Agency as well.

As it is She is not following the "Rules of this Mobbing" and any idiot, or anyone from Her Agency looking at this, would be able to tell with looking at the "Inside Information" from my files. Having said that I am sure these have been "Edited" as The Clown has had more than a year to prepare for this book and Her Exposure—in the meantime I am assured through various "Innuendos and Insinuations" that this is the case. Because I have "Passed"—I should have been free years ago and "According to the Rules of this Mobbing", and the Standards of Her Own Agency as I understand this "Mobbing to be"—which I do. Instead I have "Passed" many times and been "Held In" many times and because of "Personal Desires", and no doubt "Her being Angry and Disgruntled with Me" at being Caught and Exposed. In the meantime I and my Family are "Attacked" and suffer—much of it with this Electromagnetic-Technology.

To top it off The Clown is disgruntled with being called The Clown.

None of the Citizens around here, know any of what I share here with you, and the few Citizens I have "Shared some of this information With" have been clearly "Whitewashed", "Programmed" to see things differently than they are. This includes Good-Field-Agents as well, many who simply "Trust a higher Authority with a Lying Mouth", and who do not have all of the facts. Again it gets down to the "Motivations and Quality" of the Agents—and unless I begin "Keeping Company with Lawyers (familiar with Fed Tactics) and Psychologists", most of these people are going to be too stupid and blind to actually get the Truth of what I am saying, and the significance of it. Quite frankly most are Manipulated very simply, and by "Taking Events without Context", and "Building a Picture" for the Citizens—like a "Nasty Divorce". But since these people are "Honest Government Agents" and I am the "Bad Guy", it must be so. But back to America and these other Corrupt-Agents.

Having said that these Agents decided to "Discredit Me", COINTELPRO, and drag my name through the dirt, behind my back, Character Assassination for the sake of this Lowlife Cop, and later this Racist 911 Dispatcher with "Connections to the Police Department". By Misusing Information and Psychology, and by Manipulating others in my life. Initially this started by them Misusing Evidence—taking it OUT OF CONTEXT and then Using and Manipulating a girlfriend, Programming Her, and from there they kept "Adding On" to CREATE AN ARTIFICIAL-PROFILE. Unfortunately I made this easy for them, and it all started with "Evidence" that was not even True, never mind that it must have been acquired Illegally.

An Undercover Police-Officer who was a Pedophile, Negligence, and Rape

To begin with I knew of an undercover-cop, a year or two older than myself (I was around 22 at the time), who was dating and having sex with a 14 year old girl. He lived out on an old country road, where there were crack-dealers living, crack-houses, as I remember it. I do remember him saying a few days in advance of one particular night, that he was going to his cabin in the mountains to have sex with his girlfriends friend—also a minor. Given his nature I doubt that she knew of his intentions. And she was the same age, fourteen, and this is considered "Statutory Rape" in that state, never mind the normal issues of this. Keep in mind this Jerk was, and no doubt still is, a cop. I say "no doubt still is", because like any good Police Department, and thanks to Clean-Government-Agents, this was all covered up. By Police and Government Agents alike, COINTELPRO. Talk about good people, look at who and what they supported by doing this (The Clown assures me Her Agency will do the same thing for Her). That I am in this "Situation" because of this is beyond me. They supported this Lowlife piece of crap, the Feds involved who wouldn't have let him near their own children, yet Framed Me several times, and eventually put me in this "Situation" for merely speaking out about it to a handful of people: in regards to This and a Murder. And "Speaking Out" to a handful of cops at that—to get them to take "Responsibility" for this Jackass, without it being a Total Mess on their end or mine. As long as he was going to be "Straightened Out" I was fine. For this I pay, and my Family pay, to this day. So do American Tax-Payers for that matter. What a Pile of Deceit.

So a few days later, I drove out to his house looking for him. I knew at the time the Police were keeping "Tabs on Me" because of my smoking pot and "Big Mouth", and I knew this guy was a cop. Now this is merely a matter "Of Keeping an Eye on Me", I speak of here—this is Different from what the Feds Later Did. To top it off he knew, I knew, he was a cop. Every time we hung out it was cat and mouse, but we did hang out together, both of us trying to get information from the other about a particular subject, or person, or each other, what have you. I can count how many of these "Sessions" we had on one hand, and they were a mix of business and pleasure. They always involved alcohol. But on this particular night, I drove out to his house, only to find his girlfriend there, with two extremely shady characters who were older than I. These two guys wreaked of danger, and it is no wonder, as you will see.

She told me Meik (fictional name) was not there. She was nervous (no doubt about me speaking to Meik) as these two guys were not supposed to be there, and would not even begin to be if he was there. I felt these two people were dangerous and left the house wondering how to contact him, find him, I felt she was in danger. As it turns out, and as I remembered later, he was in the mountains with her girlfriend, "Trying to get laid". As he put it, prior to his going.

Now to top it off, just to make the situation stranger, I worked with this guy at a pizza place, with a girl, Eve, I met there who became my girlfriend, and her uncle owned the store. To top it off Meik had given me the tip to come work there to put me in a situation where he could profile me, where the police could watch me. In addition to this within a very short time I realized my girlfriend was studying to be a cop, and that her uncle also worked for the police. They did not know that Meik worked for the police.

Well, I could not find him and had to hope for the best. The next day later the front page of the newspaper was about his girlfriend being beaten and raped. I was disgusted, I had been unable to find him, he should have known better, and this was the result. I pointed out to my girlfriend and her uncle that he was an undercover cop, why he went to the mountains, and the result, along with my efforts to find him. Having sex with her was also statutory rape. I was pissed.

My speaking out made waves for many people and had many consequences as a result—including the ones I live with today, my "Situation", this Mobbing, as this situation with this Cop, started it all in one sense. I was talking to people who were involved with police, in my girlfriends case she was becoming a police officer, about one (a cop) who was involved in statutory rape, and because of his stupidity and negligence, had inadvertently gotten his girlfriend beaten and raped. Needless to say the newspaper did not even begin to mention the information I give you. They "Covered it Up".

But I knew about it and spoke out about it to certain people, all working for the police to get them to take Responsibility for their Lowlife Idiot. This was my version of making the guy think and start being responsible—what was done, was done. The cops were stupid, if I had really wanted to make problems I could have gone to the Newspapers, and been loud. If I had I would not be in this "Situation", this Mobbing. In retrospect I am Pissed Off about this point alone never mind others. On the other hand I had not experienced such Corrupt-People before, and not on a "Federal Level" either, these people "Used and Did" a lot of things I was not familiar with. These nasty Corrupt-Tactics, Lies, Manipulations, and Illegal-Psychology used behind my back being the "Product of COINTELPRO". Having said that I consider myself more "Educated" these days.

A Few Added Notes: the "End Result" of Me Knowing about this Cop: "Consequences" and "Responsibility" from a Corrupt View-Point

"No Good-Deed goes Unpunished"—The Corrupt

This was the beginning of police departments working together and no doubt the Feds involvement—COINTELPRO. My girlfriend, and her uncles pizza shop, were in my city, but they lived in another. There is no doubt that she and her uncle spoke to Meik about "his mess", but that was the plan; to get him to think about his circumstances, and he needed to hear it from someone else other than me, the "Bad Guy" (because I smoked pot—note the irony). Actually hearing it from someone else "Aligned with the Police" was the best idea anyhow. I recognized signs almost from the beginning that the three had spoken together, I had an idea of what was going on, or so I thought. Ironically the Feds would later Use my Girlfriend who was training to be a cop, for the sake of creating "Dirt", as they will do with anyone they have contact with, who is "Close to You", by Programming them, to Alter their Perceptions. In this "Mobbing" or any "Investigation" they will use The People "Close to You". I put "Investigation" in quotes, because it was not really one. Quite frankly there was nothing to Justify the Time, Energy, and Man-Power being spent for the "Crime"—smoking pot, marijuana. Thus "Investigation" goes in quotes.

This was NOT about an "Investigation" it was a DIRT CREATING MISSION.

Again this was COINTELPRO: and this is merely about Creating a "Smear Campaign", Character Assassination, "Creating Dirt", it is about Manipulating Facts and Evidence to Create a "False Picture" and False-Profile of the Target/Subject. In "Layman's Terms" this is very much like a "Nasty Divorce" in which the two parties take "Things, pictures, videos, evidence, shared times" OUT OF PERSPECTIVE to paint a NEGATIVE PICTURE of the other to "Get what they want in Court"—such as all of the money, or the house, or the kids, what have you. And to do this COINTELPRO will do anything, including MISUSING PSYCHOLOGY and MISUSING PEOPLE to get at the Target. This is why, and when, COINTELPRO was exposed by the Media many years ago, there was a "Big Sensation" about this, and is why many and much of the Files of COINTELPRO was "Covered Up". It is EXTREMELY ILLEGAL, EXTREMELY NON-AMERICAN, and EXTREMELY DANGEROUS.

The funny thing is if I had really intended to get people in trouble, I could have gone to the Newspaper. What I had intended was that people take Responsibility for their Screw-Up, especially given the "Seriousness of it All". This point seems to have been lost, or never found, in regards to my "Significant Others" at the time. As it is, this is when all of my problems really began, and in retrospect I should have gone to the newspapers. At this point in time, Government Agents got involved, began "Investigating Me", and Interacting with me through their "People".

They also Used a Girlfriend eventually, who they were also "Investigating" (the Feds investigate "all parties" involved), who I was dating at the time, by Lying to her, and by Manipulating Audio-Evidence, and "Playing on her past with her Brother" I believe, to "Program her into Negative-Expectations", and to "Create a Psychological-Snap" with her—through Programming. This Audio-Evidence being "A Recording of A LIE" I told someone, Biff, who I was trying to catch at "Selling Me Out" (I'll get to this). Which makes my ex-girlfriend an Idiot, never mind what she put us both through, and never mind that to this day she remains Clueless as to what Reality is. It took long enough, but now a lot of things make more sense to me, and "thanks for the three days of shock".

This "Programming" is something Government Agents use, something that is Used and Misused by COINTELPRO, not local police, and it took me a long time of "Playing Detective" to sort it all out, never mind having this experience again repeatedly, and especially here in Europe. Having said that, the "Programmed-Event" they created by Using my girlfriend, was Used for Character-Assassination, and had nothing to do with me, or my Character, or even what Really Happened. This all had to do with what she CHOSE TO BELIEVE, by being Lied to by Corrupt-Agents, it did Not have to do with Reality. These Agents were Sick and Corrupt, "Misusing Evidence" and providing NO CONTEXT, to Manipulate People in the worst of ways, for "Dirt", including my girlfriend, because of this Pedophile Cop. These Agents who engage in COINTELPRO are Sick and Corrupt. They are CRIMINAL, and it is the ONLY KIND AND TYPE OF PERSON who would ever get involved in such a Disgusting Government Program.

At the time, and AFTER the situation with this Pedophile Cop, and his girlfriend being beaten and raped, I began to "Notice Things". I began to suspect that I was being "Investigated" and there was not a lot of "Reason to Be", minus this "Situation with this Cop" and his girlfriend being beaten and Raped. I kept my "Law Breaking" down to a Deliberate-Minimum, merely smoking pot and out of preference, so that no one had "Real Reason" for "Investigating Me". There were "Times" when I did use "Other Drugs" but this was "Minimal" and "Experimental" and they would have had to stretch the Laws to "Investigate Me" on these grounds—again because of my "Mouth" I kept things to a "Low Profile" legally speaking. But after I "Spoke-Out" about this Idiot-Pedophile-Cop I began to "Notice Things". Again this is why I say it all began here, as no doubt these were the same Corrupt-Agents who later Framed Me on down the road in Regards to a Murder. They were Filthy, Dangerous, and Corrupt and no doubt still are.

Around the time that these Agents started interfering with my life, behind the scenes, I began to notice that things were "Amiss". I already knew that I was being watched closely by the "Police" because of what I knew. I did not know that this was Feds, or to what extent. But I did notice things and came to realize that I had a "Leak" in my life—I knew that someone close to me was working for the "Authorities". Seeing as how I smoked marijuana (no drug dealing—and I was very non-social in my "smoking", preferring my own company: I was not a "constant smoker"), yet they were "Climbing up my Ass", and had more serious issues of their own, more serious "Issues" than mine (smoking marijuana is a misdemeanor), I was concerned to say the least. To top it off, like anyone, I did not like the idea of "Good Friends" betraying me, and Betraying Me for people who had "Priority Issues". In short if you have a Badge and a "Position of Authority" it does not "Negate Heavy Crimes" simply because you do, and all things considered I was clearly the "Lesser of two Evils" here. Obviously many people take positions in Law Enforcement just so they can get away with Breaking the Laws—that has been my Repeated Experience throughout my life, and had a lot to do with "Paranoia" on my end. It still does. Having said that it also "Sharpened my Abilities" over the years.

Having said that these Feds Misusing Psychology, "Evidence", and a wide range of Tactics, Fooled and Misused a lot of people, who I now consider Idiots in Retrospect, for Creating Dirt, and for Creating a "Fake-Profile". Again by Taking things Out-of-Context like two people in a Nasty Divorce, the movie "War of the Roses" perhaps works here. As it is, when you are Targeted by COINTELPRO, it all occurs behind your back, you remain unaware of it, so that they Can Get Away With Doing This, and so they are not exposed. You can not "Fight or Expose" what you do not see.

I quickly narrowed it down to three people, all three "Close Friends", who were on the "Inside" in terms of knowing me and my life. I shared with "These Three" things which I did not elsewhere. And we were "good friends" or so I thought. At any rate I knew that my "Leak" stemmed from one of these three, they were all "Close Friends" and so each was a likely candidate, and each a "Threat to my Comfort and Peace of Mind" for multiple reasons. One being that whoever it was, they were working for Corrupt People, and two the "Friendship Betrayal Issue" itself—to top it off, whichever one it was, was Ignorant, working for people who were Corrupt and more of a "Legal Issue" than myself. Whoever it was, their "Logic" was "Flawed", and this also made them a "Threat" to me, simply out of their Ignorance, providing Information to people who were not good—as an understatement. I knew that something was going on, and that it was not about me "Smoking Pot" but that it was about this Pedophile-Cop.

So I took this very seriously, and decided to find my "Leak", and remove them from the "Picture", my life. In order to do this I would have to do it to all three obviously. There were of course "Abuse of Friendship Issues" on my end in regards to this, as I would be doing this to all three, but I had suspicions about all three for various reasons, and in the end I could explain myself later, after having "Determined my Leak"—and quite frankly I was not "Safe" until I did so. Anyone taking the amount of interest in my life that I could "Feel" and "See to some Extent" that I was "Being Investigated", could not justify such behavior by my "Smoking Marijuana", so I felt "Threatened" and rightfully so. There was no "Legal Justification" as far as I could see, and I already knew some "Things/Secrets" about the Police Department so none of this led to a "Good Feeling" on my end. In the end I felt Justified, and had good reason for "Setting my Friends Up".

I decided to Set-Up my three friends. And I did. Unfortunately I "Set-Up" all three, all within a very short time period, and I am going to say within "two weeks". It was a very short time period, and as it turns out ALL THREE worked for the Feds. This Set-Up of all three was a major Tell for the Feds, and "Tipped them Off", as to what I was doing, as they watched me do this with all three and "All at Once". This was of course a major mistake on my end: it was "Sloppy and Stupid" on my end. The Feds simply decided to take my Set-Ups, and one in particular, the easiest to Manipulate, and make it "Work for Them". They simply took False-Information I gave, and I believe, in one case had on "Tape", and said it "Was True", when it was not and they KNEW IT WAS NOT. Later they would Manipulate and Program at least one person with one of these "Stories", the one I said and Sold to Biff, and then other people as well, later down the road. This was simply a matter of them "Keeping their mouths Shut", "Offering Proof" (Fake Proof), and Not Giving Context. The last time I checked, this is still called Lying, Manipulating, and Framing someone—for the sake of Character Assassination, and for Manipulating Others. Remember this was never a Real Investigation, this was all about DISCREDITING ME, and this is a matter of "Looks" not Reality.

So I fed all three Big-Stories and Lies, to find out which one was "My Leak". As it is, all three people I suspected were working for the Feds—as I would later find out. The Feds watched me "Drop Big Stories" on all three of their people, all three "Closest Friends", and all three being the only ones "Inside" in regards to me—their three Most Valuable Informants. They also watched me drop these "Bombs" on all three, all in row time wise—which was extremely stupid on my end. I did this all within a very small amount of time, and the Feds knew immediately what was going on. They knew I KNEW I had a "Leak", that I was trying to "Weed it Out", and they knew exactly what I was doing. They understood totally, and decided to keep this to themselves, instead passing these Stories off as True, and used one in particular the most, the one I said to Biff. As this "Story" was more damaging if you did not know the Truth, and as they had "Room for playing with This", for Manipulating This, as I would later find out. They also may not have had "Tapes" of the other Lies and Stories, although I consider this "Questionable and Doubtful", in the end the Feds simply picked the one that was the "Most Damaging" in regards to Destroying my Reputation, and the "Easiest to Manipulate".

Now all of this took place after the situation with the Pedophile Cop and his girlfriend being beaten and raped, but before the Murder which I will get to in a while.

Since they were Corrupt and Illegal, they also no doubt had my flat Illegally-Wire-Tapped, and had at least one conversation of me talking to Biff—Selling this Story to him, therefore one of the Big Lies and Stories, for "Parading Around" and Misusing BEHIND MY BACK. I base this on very good reasons as you will see—quite frankly this explains very much and it is the conclusion I eventually reached. I am sure that this is what they did, and what went on—I call this the "We'll make it work for us Tactic", and it is used by Corrupt-Liars. It is Used by COINTELPRO. It is used by The Clown.

Having said that, I said this one time, and said it only to Biff (again because it was a Tactic and Ploy on my end), unaware of my flat being bugged. As I am sure it was. This is a perfect example of how "Evidence" is "Not Evidence", how a Badge and Job-Title do not mean a damn thing, and how CONTEXT means everything. Quite frankly I am Pissed-Off at a lot of people, regardless of them being "Used", and regardless of this "Tape". Simply put, people who KNEW ME should have known better, and at the least asked at some point, which "Strangely Enough" no one did. Because these Agents knew what it was all about in the first place, and TOLD PEOPLE TO BE QUIET, so they could Misuse this, which they did. This is a Tactic Agents use to avoid being caught. Otherwise I WOULD HAVE BEEN TIPPED OFF to a lot of things—including these Manipulations behind my back, and they would have been Caught and Exposed: just like The Clown is here in Europe. No doubt what I was doing with their Three Informants made them "Paranoid" as well, no doubt they were not used to someone being this "Clever". As it is, to prove a point, I have figured it out many years later.

So they used these stories which they knew to be Lies, for Manipulating People which they did. They had a "Tape of the words coming out of my Mouth" and it does not get "More Convincing" than that, or Easier for Manipulating People—these Corrupt-Pigs must have been happy with their "Luck". If no one was told of why I was saying what I was saying, no one told the CONTEXT, they simply would not know the difference, and the Feds would get away with this Deceit—which obviously the FEDS DID: again I call this the "We'll make it work for us Tactic". And of course the only ones who knew the Truth of the matter were Me and the Feds, which is why they made sure I was not "Tipped Off" in regards to what was going on. They knew I Deliberately Lied, they knew why, and they Deliberately Misused This, by Selling it as True.

If I had known they would have been Caught and Exposed, and as it is many years later they are.

Thus with this Tape they could get away with anything and eventually did—it all starting with this Tape, and then Manipulating first a girlfriend and from there others. And I am convinced they had a Tape, in retrospect, because they Lied and Manipulated several people that SHOULD HAVE KNOWN BETTER, to the point that they were "Absolutely certain of a few Things" which they were NOT. They "Programmed My Girlfriend and Others", using this Tape as a "Center Piece" for a FAKE-PROFILE. I finally realized they must have had a Tape as everyone was "Certain" when not only was this Story not possible, but in fact I expected that if "Push came to Shove" and if things ever got too far, I would simply expose this for the Deceit it was, and people who knew me better would be able to "Attest to This". What a pile of shit. This is why I am certain they had a Tape, as in retrospect my girlfriend was "Convinced" and everyone else, there is no other explanation for her "Certainty" or Anyone Else's.

And to say that I am "Disappointed and Angry" at a few people in regards to this is a Total Understatement. A lot of "Friends" turned into Absolute-Idiots and A-Holes because of this, and this is why people were Told not to bring up any of this with me in the first place. As it is, I wondered at an "Absence of Themes", I wondered why these "Stories and Themes" I told to these three "Friends", did not "Pop Up" elsewhere. I can remember only one thing, a Staged-Event with Biff, far later in time, that centered around this Theme, and obviously it was so far down the road, and only one time, that I did not "Pick up on This". Having said that no doubt I had "Other things in my Life" distracting me as well. I have good reason for being angry.

And unless I am mistaken these Corrupt-Liars, The Clown and Her Buffoons, are still parading this B.S. around as True. Yet they know what the Truth is, because it is in their god damn files that way, and because they have this Exotic-Technology, and know this from "The Inside Out" anyhow. To top it off, it is because of this Clown, that I figured this all out in the first place, and she has no doubt as a result, gotten Herself in Trouble.

But back to Biff, my "Three Friends", and how having a Tape-Recording of something said, DOES NOT necessarily make it True.

I include a separate piece of writing from elsewhere, down below, which gives some further details and explanations about "Setting Up" my three "Close Friends". I go into more detail on this "Setting them Up" here. Rather than rewrite everything, I simply include a portion of a letter I had written for someone close to me. This letter also gives you an idea of my feelings as well, and the "Amount of wool pulled over this persons Eyes" (the person this was written to), never mind the amount of anger I have over all of this. And if anyone from my past reads this book, including ex-friends, and ex-girlfriends, now you know. As it is this is "Tip of the Iceberg" in terms of what these Corrupt-Pigs got away with: "Tell me what you don't Know". I'll help you out, you can not, because you don't.

A Separate Piece of Writing to Someone: a Letter

"First of all I was playing cat and mouse with the police (and others) because they were sticking their nose up my ass, and simply for smoking marijuana which is a "Misdemeanor", as in a "piddly offense". See the movie "Tequila Sunrise" for a better understanding, for a reference point (I will use movie references for you because this is an EASY WAY TO DO RESEARCH—Cops do this, so can you). For the sake of "Smoking Pot" they were keeping tabs on me, compiling information which is never good when you actually understand the QUALITY of these people. So I was well aware of these people (they were idiots), and keeping track of what was going on, and "Playing Games" simply to defend myself. In other words this was business NOT "Playing Games". To top it off, as I would later learn, this was about me knowing about this Pedophile Cop."

"A lot of these people—informants and undercover police were Total-Lowlifes. Simply because I was doing drugs at the time it does not mean I was one (a lowlife)—yet ironically enough the police and Feds, then and now—"Are your Friends". I find this ironic. On this you let your own "hatred" get the better of you—what else would you call it? You start throwing your clarity right out the window at this point, don't you? Simply being an officer or informant, does NOT make you a good person, and simply smoking pot, does not make ME a bad person, yet YOU GAVE THESE ASSHOLES the benefit of the doubt, while giving me NONE. What shapes your opinions and beliefs on this? Think I'M AN IDIOT? Think you are "INVISIBLE"? So what is the REAL ANSWER IN REGARDS TO YOU—put up or shut-up on this point. REGARDLESS, I already know the answers to questions I ask, it is obvious, these questions are FOR YOU. Do you know the answers, do you know yourself? Where is your head? At any rate the more I looked, the more I learned—picking up techniques and Tactics along the way. Having said that much was a given on my end, it is all psychology and "Mind-Warfare". It is common sense to a very large degree."

For my readers the part below explains my "Setting Up" of my "Three Close Friends"—parts of this letter have been changed a bit for a "Better Understanding", been made more clear, and names and "Direct Facts" have been removed.

"So I was keeping track of what was going on around me, people, police, others, connections, and "Webs and Tendrils". Again CONNECTIONS—interaction among people working for the police and the flow of information. Since you do not know I will tell you that you can follow the "Flow of Information" between people. IN SIMPLEST TERMS FOR YOU TO UNDERSTAND if I tell "Jessica" something and to keep it to herself, I can tell by "Anna's" (friends of each other) questions if the information has been passed on to "Anna"—if information has been "Leaked". In regards to police and informants, a "Leak" in regards to "Private Information", coming up somewhere else from a totally different person (no relationship) means that someone IS AN INFORMANT at the least (Readers: I would be watching for the "Theme" actually as they would have no other way of doing it "safely"). If I tell you something "Strange", then later "Biff" brings this up (theme-wise), it is a sure sign, a Red-Flag that there is a CONNECTION between the two of you—which would be the Police, as the two of you have No-Possible-Connection. In regards to my life at that time it would be a sure sign of police, and police connections. Got it? And do not give me some line of shit like "Well of course it is obvious", this letter is a result of what was not obvious to your THICK-IGNORANT-HEAD at the time, or even now. So keep your clever mind and bullshit to yourself, I am not asking "For your Reassurances", I am not asking you, I AM TELLING YOU."

"At any rate I began "Noticing Things" that things were "Amiss" on my end, I knew that I was being INVESTIGATED, not simply information being kept on me (keeping tabs). And at some point, I could guess why. I also knew that there were only so many possibilities "Threat Wise", as I had little contact with anyone on a "Deeper Level". There were only THREE PEOPLE I really had to worry about, all three "Close Friends". I also eventually realized that I had a "LEAK", someone close to me. And this was after this situation with the Pedophile Cop and his girlfriend, which explains why I began "Noticing some Things" as they were "Actively Investigating Me" based on me knowing about this Cop and speaking out about it. So what to do?"

"Simple. In order to find the informant, I would follow the trail of information. A leak is a leak, and in regards to "Close Friends" in my situation and circumstances, it would mean an INFORMANT. So I concocted False-Stories, LIES, for each individual to see what I got. If "A Story", the "Theme" of it, popped up elsewhere I would know that I have a "Leak", and depending on the "Story/Theme", I would know WHICH PERSON was guilty. But I was not the most brilliant here. I basically "dropped my bombs" on all three people at once, all things being relative time-wise. For the Feds this was a major "Tell". They knew I was trying to spot their people, to weed them out. These three people were Biff (who I was closest to, like a brother—and therefore the BIGGEST threat), Tark, and Kitch."

Now for my readers, the following has been edited and had material added as I am interested in explaining things here, but is basically the same and for the most part altered very little.

"I told each "Close Friend" a Different-Story, a Different-Lie. So I told Tark one Big-Lie and Story, I told Kitch another Big-Lie and Story, and I told Biff who I was closest to, the Biggest-Lie and Story, and because if "Push came to Shove" it would be obvious for the Lie it was—at least to others. Which is also why I believe a Tape was made of our conversation, because somehow this was NOT "Obvious to All"—therefore there had to be some

"Special Reason". Having said that Biff would be much more "Challenged" on this point as we were "Like Brothers" and I had given no indication of not trusting him, because I was not so stupid. Simple, and the Biggest and Juiciest Lie was reserved for Biff."

"I figured I was safe on this point as a lot of people knew me, knew the Truth of the matter, both in regards to what I specifically said to him, and the Subject and Theme itself—Duh! INCLUDING YOU. At any rate when it came out of my mouth I regretted it immediately, I had "Whored-Out/ Misused" my relationship with Kaye (I Lied about her in Particular) in doing so, but there was no putting it back in the bottle, and I had no way of knowing that until I did so. At the time it seemed like a good idea, and I felt threatened by my "Information Leak"—as the motives of those involved could not be good, and as the actual Truth was obvious anyhow. So then I waited for other people, to subtly bring up these various subjects, to find out who was my "Leak". The "Subject" would tell me the source. As it turns out, all three worked for the Feds by the way, but this is irrelevant to the story in one sense."

"But I totally screwed this up. Even now when I look back, I can not believe how sloppy I was. I basically brought this all up, with all three, at one time. Which is called BEING OBVIOUS, from an outsiders perspective. An outsiders perspective being the Fed's. So they knew, and knew right away. It was in this way that I would learn of a NEW TACTIC called "We'll make it work for Us". (And for my Readers this is what COINTELPRO is about.)"

"This TACTIC is about taking False-Information, and making it work for them—it is about how things "Look", not the Truth and Context of something. In essence taking things Out of Context, which is Lying when you know the Truth of the matter. Making it "Work for them Regardless". Since I "Stuck my neck Out", since they were Corrupt anyhow, they decided to use what I said to Biff, against me. And they did, behind my back, use this as a "Selling Point", after all they had THIS STATEMENT TO BIFF ON TAPE. At least I can assume as much as they clearly did have my Flat Wired at some point, and as "Everyone was Absolutely Convinced" of something NOT TRUE, later down the road—which is also why I could not later understand the "Theme of everyone's Harassment" as it was a million miles from the Truth (and later this lead to a Psychological-Snap—which ironically "confirmed" everyone's beliefs—having said that it is no wonder that I did). So I feel quite sure of this. Thus regardless of the Feds knowing the Truth of the matter, they now had something they could USE REGARDLESS. It was my mouth using these words, it would be easy to Misuse, which they did. I was not used to dealing with people who were TOTALLY CORRUPT. And although I said only a line or two in regards to Biff (his mind was thinking), this was all that was needed. So the Feds made this "Work for Them", a Tactic which I learned the hard way, and still pay for. It is called Corruption, and is Used by Corrupt-Liars."

For my readers that gives an idea, of what I was doing, and what it was all about.

Back to "Consequences and Responsibility from a Corrupt-Viewpoint" and an "Introduction to Programming"

To top it off, since everyone involved was "Told to keep their mouths Shut" I had no way of "explaining or pointing out" a damn thing—which of course the Feds did not want. So they Used this and Misused this against my girlfriend, and to get at me, and no doubt playing on her "Past-Issues" with her brother as well. I am guessing here, but I think it a safe assumption that they switched some of the "Content of the Tape" as well, so that she thought I spoke of Her in regards to Biff, when in actuality I spoke of Kaye—I spoke of someone who was out of my life, at that point, and living far away. I think they "Switched Names". I base this on the fact she was "Extremely Convinced" that something would happen that was not even possible—what she expected simply had nothing to do with me or who I am. And she seemed to think that what I said to Biff was in regards to her specifically, obviously (in retrospect)—which is and was ridiculous, never mind the Story and Lie I told Biff to begin with. I could be wrong about this, but I doubt I am wrong about the "Tape".

Which is why I picked the Lie that I had, for Biff, in the first place. HE might not know what was going on, but if push came to shove I could simply "Expose my clever Ploy" and my reasons why, never mind people who would "Attest to the Obvious" Lie of this Statement to Biff. This obviously backfired (an understatement). She should have known better, yet somehow she did not. To top it off she was studying and training in a Job related to Law. So I am certain a "Tape" was used as the "Center Piece" and other things "Built around This". They constructed an Artificial-Profile around this. I simply waited for her to explain her behavior to me one day, as we continued dating, she never did. In retrospect, for the moment, I am Pissed Off at her (Idiot and Jerk are good words here) and many others to say the least. As it is her "Flipping Out", got "Turned Around" and used for a "Center Piece" in addition to what I had said to Biff. Layer upon Layer, Manipulation and Deceit, for more of the same. Which is what Government Agents do, these ARE their Tactics. It is Text-Book-Procedure for Corrupt-Agents and COINTELPRO.

I call this Layering of Lies and Manipulations Upon More of the Same the "Burial Tactic". As more "Gets Piled On", the Truth "Gets Buried" and Farther Away. It is Standard Operational Procedure for Agents.

So what happened in a nut shell? It involves "Programming", and this is what Agents use, not local police. This is "Government Stuff" for lack of better wording, and it requires an Understanding of Psychology, and Training, to do this. If you are in this "Situation/Mobbing" it is Used and Misused on both the Public and Specific-Individuals. It is Used for "Steering and Controlling People's Perceptions" and therefore their Minds. It is in essence "Mind-Control" the "Controlling of People's Minds" through a "Programmer" who tells people, subtly and not so subtly, what to "See, Hear, and Experience", what "To Perceive and what they Will Perceive", who tells them "How to Think and Perceive Things", specifically "Events, Experiences, and Actions". The Programmer by doing this, "Dictates how they will Experience Reality", and "Programming" is in fact Modern-Day-Brainwashing. It is more "Invisible" and therefore more "Insidious", it is also Illegal. This is why Cults are watched so closely.

Thus people "See, Hear, and Experience Things", "Perceive Things" the way they are TOLD TO, not necessarily based on Reality—in fact, not based on Reality at all. Which is the point of Programming in the first place. Which is of course why much of the Public has no idea of "What is going On" in your "Mobbing" and can not figure out such simple basic things—which is why they are "Idiots on many a fine Point". Of course this is THE POINT OF PROGRAMMING in the first place, so that People, do not see Reality in the first place—otherwise most would not allow themselves to be Manipulated and especially in the way that they are. If people from my past actually understood to what degree they were Manipulated, and the amount of Lying and Deceit these Agents engaged in, they would be disgusted to say the least—and in fact would probably have "Hung these Agents" if they had known at the time. But getting back to the initial point, they were very fond of Programming, and this involves Government-Agent Subject-Matter, a Specific-Psychology and Training in it, for Misusing People. I have had a "Ton of It" here in Europe, Used and Misused on people around me and close to me, so at this point I understand what it is about, never mind what is going on.

When people are Programmed they "See Differently, Hear Differently, and Experience Differently", and this is all based on what the Programmer(s) WANTS, as opposed to Reality. And when we are talking about Government Agents who's goal it is to Create-Dirt, to Discredit the Target, to Debase their Character, then none of the Programming is "Positive" and it is all about making the "Target Look Bad" by "Altering and Programming People's Perceptions"—many times by Manipulating their Emotions. The point of Programming by Agents, like a Nasty-Divorce, IS TO GET WHAT THEY WANT.

They did this, Used and Misused my girlfriend, "Pumped her up on Fear", "Pre-Programmed her Expectations" to create unpleasant circumstances for her and me both. To "Scare Her To An Extreme" and to create a "Psychological-Snap" (P.T.S.D. Related) triggered by "Particular-Actions" and "Events". No doubt in some way, shape, or form, she was told such and such would happen—the "Fear and Expectations" built up. The same way someone might build up a "Child's Fears" in regards to scaring them for a particular event at Halloween, or a Haunted House. We had a situation where she "Flipped the Switch", turned into another person in a moment, I had no idea what was going on, and because she believed what she had been TOLD, and by strangers who could care less about her at that. I am sure the Tape didn't help any either.

She believed in a Job-Title, which said nothing about the "Quality or Type" of people with the Badges, never mind the Motivations of these people—they were a bunch of Corrupt-Punks with Badges, trying to create "Dirt", damaging material, because of this sick cop Meik and his "14 year old girlfriend". And in no small part they got away with it thanks to this Tape. I own what is mine, but I will not take on more than is mine. Thanks to her being Programmed, and a "Post-Traumatic-Flashback" (I'm guessing on this point) because of it, no doubt Past-Experiences with her Brother (Her "Family Life" on this Point could have been Better), her Flipping-Out, never mind the timing, I spent three days in shock afterwords, and this was all to "Discredit my Name" and to "Neutralize a threat to the Police-Department"—what a sick joke.

I am still Pissed Off about this, and am glad I figured this all out here in Europe actually. Otherwise a lot of people would have been personally pulling me out from their ass when I figured this out; I get Pissed Off just thinking about it. As it is when I figured this out here in Europe I had a near total Psychological-Snap, and called someone up to "Release my Anger", and was so loud other neighborhoods could no doubt hear it—literally, as I had the window open when I did so. To top it off, based on a personal experience with this guy Meik, later down the road, and after his "Fiasco", I now know HE KNEW what was going on with these Feds, and that he knew what they were up to and what they were doing—and because of him, which he knew. What a bunch of Lowlife Punks. We are not talking about Professionals here, we are talking about people who do not do their Job Properly, who are Corrupt, yet take a paycheck, while doing what they

WANT AND FEEL LIKE, based on what they can get away with. In regards to Meik, I am glad I am here in Europe, and so should he be.

They are following the Laws even less than many they are dealing with—this is acceptable? What ALL OF THESE LOWLIFES DID, from the Feds Involved, to the Police Department, to Meik, is ALL a lot more serious than smoking marijuana. I was a good person then and I still am now, it is just that people have no idea what happened THEN, and are even more clueless about what is going on NOW. I never changed, People's Perceptions did, and for all of the wrong reasons. Because someone CHANGED THEIR PERCEPTIONS, Altered Their Perceptions, for a "Desired Result": for a Desired-Result of the Programmers Intentions and Wishes. None of which was or is, based on Reality.

And regrettably later, I had one "Psychological-Snap" as a result of this "Situation/Mobbing" which "Fed their Theories", and was a direct result of the Mobbing—but other than this there was nothing in the first place, which is Sick and Ironic. And here in Europe these Idiot-Citizens are just the same: they see what they are Told to See, and they Experience what they are Programmed to Experience, through Programming and what the Programmer WANTS. So here in Europe The Clown gets to "Look Accurate" through PROGRAMMING, all the while Lying and Manipulating, and I get to be "The Bad Guy". It is a Sick-Joke, Farce, Lie, and Deceit perpetrated on Citizens, Individuals, the Public. Everyone remains Clueless except Me, and these Corrupt-Spooks.

This Clown, is literally a Psychopath, and these Citizens have absolutely no idea of what She has gotten away with. And much of what She has done, has been done to other people in our neighborhood as well. It is all a matter of "What She can get away With", and she acts with Impunity and with absolutely No Worry of Consequence—because of Her Agency and "Job-Title". By Her "Job-Title" alone She has gotten away with many things because of people's BLIND RESPECT FOR AUTHORITY. But the Citizens remain Idiots apparently, just like in America. But back to "Programming" and an eventual explanation of it, and examples, as people need to know how Sick and Dangerous this is.

In fact The Clown has done this here in Europe as well, with my girlfriend, it is a matter of "Programming" an Individual with "Expectations" in regards to a Subject which acts as a TRIGGER—in which they have "Fear"—it is hard to explain, but this is what Sick-Corrupt-Government-Agents do. They "Pump an Individual Up on Fear" and /or, to Create False-Expectations. It does not, and is not, always about Fear as I have found out the hard way, but this will be the "Example and Context" I use, initially. Basically the

"Programmed Individual" has a "Psychological-Snap" based on Fear, they get "Programmed into having a Post-Traumatic-Flash-Back" (some do—not all qualify for this) simply by "Events which remind them of Past-Events". It is a sick Manipulation of People, and both involved get Traumatized—my girlfriends because they believe something that is not True, and experience some "P.T.S.D. Flashback", and I because I have actually no god damn idea of what is going on. I got totally flipped upside down on my head and it is extremely confusing in the moment, and then some, to say the least. Once you understand it, is "Easier to Deal With", but it still sucks.

If you are Not Aware of what is going on, you will never figure it out, your mind will be damaged, and all because of some "Third Party". I hope this writing helps anyone else out dealing with this kind of Sick-Bullshit.

When there is such an "Extreme and Instant Reversal" in their Personality, you do not believe it, you experience disbelief, and figure there is something wrong with YOUR PERCEPTIONS. Minus these people being Programmed, there WOULD BE NO "EVENT". There would be no "Programmed Event" as I call it. As it is, both parties get Traumatized—again it is a Sick Misuse of Psychology and a playing on peoples Fears. Having said that, if they are not made aware of how they were "Used, Misused, and Programmed", and are Programmed heavily enough, they end up believing that "What they Experienced" was as they "Thought it to Be" (I'll explain), the "Event" itself, "Burning into their minds this Way". You the Target, experience Shock, Confusion, and Trauma, and remember it as a "Weird Event" to say the least. It is sick and evil, and also Illegal—this is why Cults are watched so closely. In the case of my girlfriend (in America) I simply had no idea of what had occurred, and after much time and "Recuperating from Shock", simply waited for her to explain one day what had happened—giving her "Her space on This", waiting until she was comfortable enough to talk about this. We continued dating, and she never did. Odd.

She never did, because she was Told Not To (A Tactic) by these Feds, and no doubt was "Convinced" that she knew the Truth, and in no small part based on what I had said to Biff, never mind Lying Feds Mouths. What I said to Biff, was a Lie and Manipulation on my end, for catching my "Best Friend". The "Tape" was Misused, and the Reality and Context of it was "Hidden and Conveniently Forgotten". So everyone got Used by these Feds, and later this "Programmed Event" with my ex-girlfriend got used for further Manipulating the Public and for Further-Building/Constructing an Artificial/Fake Profile.

All of this, if I had known back then, could have been easily exposed—as it was, and still is obvious, for the Lying and Manipulating it was and is. There are a lot of FACTS that expose this for the Lie and Manipulation it was, but what I offer here is enough, for anyone capable of thinking on their own. Having said that this is Not how my "Mobbing" started, but this was all Later-Used within it, and had a lot to do with what people Did, and Why (and I might point out I had no idea why until much later—and after my first heavy "Snap"). Having said that the same Tactic of "We'll Make It Work For Us" was later Used for getting me in this "Mobbing"—these Corrupt Agents again took something Out of Context so they could start this all. They also did this with one of my Co-Workers and got him in his own "Mobbing" on Information Deliberately taken Out of Context, obviously the work of "Real-Professionals", these Agents were absolutely Filthy. As an added note, they also got "Kaye" in this "Mobbing", because of my "Connection with Her". But back to Programming.

My girlfriend over here in Europe, had this three times with me (for this specific example—we actually have had this MANY TIMES), in regards to my "Drinking Alcohol", all within a week, and based on "Negative-Programming" (like there is a positive use) by these Agents, and her past with her ex-boyfriend (mostly, not entirely). We had a "Programmed-Event" three times in one week, compliments of The Clown, the "Trigger and Activity" being the Alcohol Drinking. Again they obviously Told Her "Something was going to Happen", built up her Fear and Negative-Expectations, which was a million miles away from the Truth. She had her head up someone else's ass, got Lied to and Manipulated, and out of Fear, actually created and initiated the whole god damn thing. She freaked out, as if "Things were Happening", and in the end "Experienced the Mess" she feared—it is Surreal. In Reality nothing has happened the way they think it has, and YOU get Traumatized with, and by, these Dupes, which happened to be girlfriends in these two examples. There are other examples as well, but these will do, as they both involve people who were close to me, and who should have known better, and in both cases they were "Programmed around Past-Events and Fear". It is a good example of the amount of Control these Agents can have through the Misuse of Psychology, Information, and Manipulation: through Programming.

I have been with this girlfriend ten years, ten years inside of this "Mobbing", and anyone in this "Mobbing" will tell you that "Relationships are Attacked", many no doubt destroyed, and I have experienced many upon many Programmed-Events "Through Her" and with her. "Deprogramming Someone" is difficult, but can be done. Having said that it takes time, and is a pain in the ass—you have to "Walk through the Event" with them, showing them the "Truth of the Matter", you have to "Restore common Sense". I at least now know what is going on, and it is still a Shock, and you still get

"Flipped on your Head" when it happens Initially (and always out of the blue), but having said that "Recognizing what is going On" is a big help.

What it is, is that two people experience the same event TWO DIFFERENT WAYS, when in fact they would normally experience the same event, based on knowing one another, the SAME WAY. This is because the Programmed Person is wearing someone else's "Viewpoint and Perceptions", which in turn "Colors and Shapes their Experience". They are not "Experiencing as they normally Would", as things REALLY ARE, and as only one Person's Perceptions has been altered things are still the same for the other person. Rather the Programmed Person is "Experiencing" through someone else—and in a way that someone else WANTS THEM TO, not as they USUALLY DO.

Which is what absolutely turns you upside down, as THEY ARE NOT NORMAL, and their REACTION IS EXTREME, never mind that YOU HAVE NO IDEA WHY IT IS OCCURRING.

There is an Abuse of Trust that exists and takes place between the Programmer and the Programmed Individual—which then is "Put Into" the relationship between the Programmed Individual and the "Person they are Programmed to React to"—the Target. The Programmer ATTACKS the relationship between the Programmed Individual and the Target, getting the Programmed Individual to stop Trusting the Target. This of course happens without the Target's Knowledge, and as it is Not Based on Truth, the Target has no reason to "Suspect Anything is Wrong". Keep in mind that these people are Programmed to "Get At" the Target. Keep in mind that the Programmed-Event, using the Programmed-Individual, is a Set-Up, made against the Target and Designed to make the Target look bad. Trust is an Issue here, and the "Trust between the Programmed Individual" and the other person they are "Programmed to React to"—the Target, is attacked and made questionable by the Programmer, obviously.

IN REALITY the Programmer is "Abusing the Trust" of the Programmed-Individual, the Programmer has Ulterior Motives, THE PROGRAMMER SHOULD NOT BE TRUSTED, and gets the Programmed-Individual to question the Friendship and Trust with the Target. Thus what is taken for granted normally by both parties, both people, is now "Questionable and Suspicious" by the Programmed Individual. Of course the Target, the person the Programmed-Person is "Programmed to React to" has no idea, and why would they—this is all going on behind their back. In essence the relationship between both parties, both people, is "Abused" by the Programmed-Individual for not Trusting, which they always did before, and because they "Trusted an Outsider" and allowed the relationship to be Used and Abused for some Sick Asshole, the Programmer.

To top it off, the Programmed Person has no idea that they have been Programmed, and will believe what they have experienced afterwords (it "Burns in this Way")—and be resistant to anything you say to the contrary. It takes time, and you have to "Walk them out of It", and you have to point out Facts which show the Truth of the matter. This is a pain in the ass, and can be initially frustrating. But it can be done.

In regards to my girlfriend here in Europe, she "Flipped the Switch", turned into another person and began "Freaking Out" and "Attacking Me" out of the blue, Out of Fear, because she had been "Programmed to Flip Out" at the sight of me "Drinking a Glass of Wine"—again "Programming based on Past Memories, Themes, Issues" and Fear. There was nothing unusual about this—I had been doing this for some time, during the course of my writing (On and off for many months when I did my writing, and to "negate shock" in regards to anything I "found" while rewinding on the past.). Yet she "Was Sure" that something was going to happen out of a "Failure in Trust"—none of which was normal, and all of which was Totally-Unexpected, as someone else "Put this into her Head". Quite frankly SHE TRUSTED THE PROGRAMMER who had convinced her "THAT SOMETHING WOULD HAPPEN" which was COMPLETE BULLSHIT, A TOTAL-LIE AND MANIPULATION—which is exactly what happened to my girlfriend in America.

I was minding my own business, quite relaxed, and working on my book at the time—all three times actually, and as typical I experienced the "Shock and Disbelief" that comes with an Instant-Personality-Reversal. It is Weird to experience and you get Traumatized; and I sure as hell was not expecting this three times in one week. No doubt, each time, got Added to the Next—building the "Fear Up" as there was no time to "Sort Things Out" from the first time, never mind the next two. As an added note, and you will note the irony here, this happened while writing about this very "Event" with my other girlfriend in America, at that time—The Clown, like many Agents, likes to use "Material from the Present" for picking "Events and Themes". So I was given the experience again, the "Gift of Giving" and able to make some mental notes this time around. Having said that my girlfriend was "Totally Convinced", and it was out of the blue to say the least. The piece below was originally in my "Tactics Section", and I feel does a good job of explaining Programming.

Programming: Illegal-Psychology, This is Why

I hate this Terminology and have from the beginning, but there is no "Better Wording"—I think of this as almost being like "Programming People" in terms of a "Computer Program", because in many ways that is what this is. This is PSYCHOLOGY USED BY GOVERNMENT, BY AGENTS, and it is Sick and Illegal. Programming is in fact Modern-Day-Brainwashing. Citizens Used and Misused for this do not get it, because it is too hard for them to understand, and you either need a degree in Psychology, or need to have experienced this Repeatedly, which I have, to get it. Either that or be "Observant and Questioning" which none of these Citizen-Jackasses are. It is subtle for many, and disgusting. For all intents and purposes this is "Invisible to The People it is used On". I hope to make people more aware of this, and I hope that anyone who has had this Used on Themselves, is helped by this. You have got to be sick to use this on people, and this kind of thing is a favorite of The Clown. Actually it is a favorite of Corrupt-Agents, a favorite of this "Mobbing/Situation", and a favorite of COINTELPRO, as they could care less about anyone else, and "get their rocks off" over Abusing their Authority, and GETTING AWAY WITH IT. This is True of the Feds in America that I dealt with, and the Idiots I deal with here in Europe. Thanks to The Clown, I understand this much better now. For these Agents, getting away with this Sick-Deceit, is like "Performance-Art".

This is for my ex-girlfriends and some others along the way. I hope this clears up a few issues for you, never mind others.

So this is a Misuse of Psychology, and there are "Different Forms" of this, as there are different Motives and Situations for using this, as well. The practice of this Psychology is Illegal as well, as it is Dangerous. This is what is Used by Cults, and as we all know Cults are watched and observed closely because of this "Programming" in and of itself. I have seen this Used, Abused, and Misused, in my life repeatedly—like it can be "Used for anything Good". It has been Used on Individuals in my life, to Create "Convincing Events" in which the person Programmed "Believes what they have Experienced" which creates an "Airtight Alibi", a "Witness" (to an event which exists differently in the mind of the Programmed Individual than Real Life), and a "Convincing Event" with a "Convinced Person" for an Artificial-Profile which the Agents SELL: It is a Sick Lie, Deceit, and Manipulation.

BOTH PEOPLE in these Individual-Circumstances, these Specific-Events, these PROGRAMMED-EVENTS get Traumatized and Used by the Agents. It is Insidious, Repugnant, and Evil—and it is hard to "Disprove" because the Programmed Individual, like any, believes what they have been "Told and Programmed With", and believe the "Event" itself, which "burns into their minds this way". The only way to "counter this", is to wait for later, for their mind to "return to normal", and then to go through it with them—in essence you have to "Deprogram Them", and doing this in the "Moment" or shortly after, does not work. I know I have tried. You have to "walk them out of it". Having said that if you do not know what is going on, what has happened, this is more difficult—the Individual who was Programmed simply ends up believing what was not True, and you are left wondering and waiting for an explanation. Never mind your own pain and confusion from the experience.

The first time I had this happen, that it was Used and Misused, was in America, and I simply waited for my girlfriend to bring up the "Issue and Subject" again she never did—so the air never got cleared. Having said that it never would have anyhow, because "Her Personality Reversal" was based on things being said behind my back, her relationship with "Others", that I did not know about. Unless I am mistaken, and I doubt that I am, she was Told to not talk about this with me. This was to protect the Agents involved from being exposed, and to Hide this. This was in response, by the Agents involved, in regards to a Pedophile-Cop who got his "girlfriend" beaten and raped, due to Negligence, and the fact that I knew about this. I have many more details about all of this, but they are not relevant here. In the end what this was really about was Corrupt-Agents deciding they did not like me, and using their Job-Title in any way that they could to "Get to Me". Quite obviously they did NOT give a rat's ass about my girlfriend who "Trusted Them".

Having said that, if you are a Target, these are the "Signs". The person you deal with will "Flip the Switch". There may be "Warning Signs" as their "Irrational Fear" builds up, as their Colored and Altered Perceptions "Re-Writes" what they are Experiencing. You the Target usually WILL NOT Recognize this, as there is no reason to—what is different here is THEIR PERCEPTIONS, not Reality. Thus they are "Experiencing things Differently" when things are not—thus you have no reason to "suspect what is going on" within their minds. Things, whatever it is you are doing that acts as a "Trigger", were never a problem before. When you begin to notice something "Strange and Odd" with them, no doubt you will feel it is your "Perceptions that are Off" as there is no reason to think that THEIR "Perceptions are Off". You are both experiencing things differently but as you know your Friend, and as you normally share the same Perceptions and Understandings, then "Your Perceptions" must be off. What is really going on is that there is a "third party" involved.

Thus you ignore what you are "Perceiving with Them", and write it off as incorrect on your end. Having said that, what is "Triggering their Behavior" is what you are doing, some Event that they have been "Programmed Around"—this is the center of it. Since you think your "Perceptions are Off" you continue with the Behavior, oblivious with what is going on with them—you have no reason to think otherwise, and NORMALLY they would not be "having these thoughts or contemplations" anyhow, that is rather the point—they have been PROGRAMMED TO. These "Contemplations" are NOT THEIRS and normally you would not even think, suspect, or expect that what you are doing is an "Issue". As it is Normal and the two of you share the same Perceptions and Understandings. Normally you would not even begin to suspect "The things they are contemplating, their Perceptions". But as it is the "Fear begins building around your Actions", and they eventually have a Psychological-Snap based on these, they instantly turn into someone else.

You as the Target, have no god damn idea of what is going on. There is nothing "Unusual about your Actions", but they "Flip-Out", and you the Target experience Disbelief, Confusion, Shock, and Trauma. You in turn wonder what has happened, and how it could have happened, what is going on, and the "Blame is Put On You" by the Programmed-Individual, as they obviously feel that you have done something that you have not. Their "Perceptions were Pre-Altered" to "Conform to a Particular View-Point", to the Programmers "DESIRED VIEWPOINT", in regards to a Particular-Event. The Programmed-Individual experiences the Event as they were "Programmed To", through the eyes of the Programmer, NOT through their own eyes, and thus they "Change into Another Person". They "See Things" as someone WANTS them to "See Things" in regards to something particular. They act differently and experience things differently than they Normally Would. Their Perceptions have been Altered, so that a Normal Event, is Perceived and Experienced differently than how it has occurred. To make it even WEIRDER, this is something that has occurred between the two of you MANY TIMES without a problem. Why would it be a problem?

This as an added point, is how the Public, is Affected and Controlled, as well, many times. In fact this is Used all the time on the Public, their thinking is "Whitewashed and Mis-Steered", and as I said "fear" is not the only way of going about doing this, it is a matter of the Programmer getting people to see things the way that they "Want Them To". Their "Thinking and Seeing", their Perceptions, become Altered and "Made in Line" with how the Programmer wants. In a way that is "Desirable and Beneficial" for the Agents, none of which centers around Truth. It is about them "Looking Good" and about making you "Look Bad". They do this with Specific-Individuals in your life, and the Public, Truth goes right out the window.

For you the Target, until they "Snap" everything is normal, as what you are doing is normal—for the Programmed Individual it is different—basically you experience the same Event, two different ways, and what counts is the "View-Point". Theirs has been Altered, yours not. What and how YOU ACTED, is Normal, and under Normal Circumstances, how the Programmed Individual has "Reacted" is NOT Normal, and they are having a "Psychological-Snap" under "Normal Circumstances and Behavior". WHO is acting weird, different, and why? What you are doing was never a problem for them before, why is it now? If this happens to you, do not blame yourself, it is bad enough as it is—I was in shock for three days in one case, and walking around on "egg shells" for days, in other cases.

They have been AFFECTED by an Outside Influence, and are different as a result. You have not been affected and there is no reason for you to be "different". Once they have a Psychological-Snap you are dumped upside down on your head. And in many of the cases I experienced, unless I am mistaken, and I do not think that I am, these Individuals experienced a Post Traumatic Flash Back based on prior events in their lives. It is as if a "Trigger is Activated" for a release of fear from "Past-Events"—and it is a ton of it. This will not always be the case, based on the "requirements" for this to happen, but the "Psychological-Snap" and Personality-Reversal will happen if they are "Pushed Enough" by the Programmer. It depends on what and how they are Programmed, the Programming itself being about "What and How" the Programmer wants something to be Seen, Perceived, and/or Experienced, and what "result" they want.

You as the Target are put into "Shock" to say the least. You have no idea of what has happened, your Friend has completely changed in an instant, and the "Blame" is put on you, as they have "Flipped Out" and "based on your actions". Further more you have absolutely no explanation for what has occurred, as what you were doing was never a problem before, and always normal before. There were never problems before, so what happened? You simply have no explanation for their "Psychological-Snap", their Flipping Out, and you never will, unless you know what is going on behind the scenes. They have "Acted and Reacted Differently" because of an "Outside Influence", they have turned into someone else because of this, over Behavior that is nothing new, because of someone that has Programmed Them, an "Outside Influence". You have no idea of what is going on, because it is not "Your Friends Behavior", but someone else's. And this was used on my girlfriend in America, and my girlfriend here in Europe. The circumstances were different, but nothing else was. In fact this has been used on different people at different points in my life, and in different ways.

Now there are "Different Versions" as well, and this is also Used, although it is a bit different, on the Public as well. For a simplified version, and not far off the base, what I have experienced with girlfriends is a "Smaller More Focused Version", and what I have experienced with the Public, is more "broad". But the processes are all basically the same, although usually not "Centered around Fear". This involves "Pre-Steering" and "Pre-Coloring" a person's mind, in regards to Specific-Events. Fear does not necessarily have to be an issue in and of itself.

It is the same you would do with a child who you are taking to the "Haunted House" at the fair, for "Kicks on Your End" (personally I would not—but to make a point), as you build up "The Fear and Tension" in regards to going, so that when you do the "Kid Totally Flips Out". This is a matter of building up fear, tension, and expectations, and it is sick, especially when the person involved has experienced Past Traumas related to the subject before. Imagine the "kids reaction", after Programming, when at the Haunted House a person "jumps out with a knife in their chest", and the kid had experienced the death of someone they knew in a similar manner—with all of the "Fear Pumping"/ Programming and Negative-Expectations, the Event of this person jumping out is experienced as something much more "Real and Traumatic" than it is. My example is not perfect, but I think you get the idea. Whatever the deal is, minus the Programming, the "Event" is experienced in Proper-Context and as a normal event, with the Proper Context in place.

In a nutshell, and without a Degree in Psychology on my end, this is Programming, and this has been used in my life repeatedly.

As another example I offer part of a letter that I wrote in September 2010, to someone I know, and who has very little understanding that she is also being Manipulated and Programmed herself. I wrote this shortly after a "Programmed Event" happened with my girlfriend. It is not as clear, but I had not sat down and written about this yet for my book, in regards to "Tactics and Techniques". I "tweak it a bit", but again it is all basically in original form. This also shows "How Ridiculous" the Programmed Person can be. My girlfriend was Convinced that I was "Getting Mad at Her" because she would not "Acknowledge that there are Spooks in Our Life", which if you consider I have a Book about these People, is a "Ridiculous Thing to Think". Quite frankly she should have Known Me Better, and should have "Seen the Facts". This is a good example of the Power of Programming, and the amount of "Sway" the Programmer has over the Programmed Individual. My "Girlfriend KNEW" and WAS "CONVINCED", of something absolutely NOT TRUE.

Part of a Letter: About a "Programmed Event and Person"

"Now they also do this, and have DONE THIS, WITHOUT EXOTIC-TECHNO, Misusing Profiling, Conditioning, Idiot-Citizens, and "Staged-Events", and by Lying and Manipulating. However it goes easier for them with this Exotic-Techno, less people are involved, few people know of this Exotic-Techno, less people know of what is going on, and I have less to talk about in one sense—I can not just go up to people involved and start talking about Exotic-Techno. Having said that there is enough "Exotic-Techno" being developed in the Public-Sector, that this "Machine and it's Capability and Misuse" can be "argued for" logically—quite frankly there has been a lot released in the news, and I have a collection of articles all pointing to this Exotic-Technology."

"But back to Maggy, machined anger, and "Staged-Eventing", building a FALSE PROFILE. Tricking and Manipulating Maggy into thinking something NOT TRUE, is True. And doing it by "Programming"."

"They tried convincing Maggy that I am "Angry, Tight, and on the Verge of Hostility" when I am not. Why? Read below. This will give you a REAL IDEA of what is going on, and a real idea of how PEOPLE LIKE YOU ARE MANIPULATED. You won't find this on FOX NEWS, you will find it from people like me WHO HAVE TO DEAL WITH IT. You were expecting Corrupt-Government-Agents to explain this to you? How would they Manipulate You otherwise?"

An Example of The Clown's Lies, Manipulations, and Cheating:
September 17th 2010: In regards to Me and Maggy

"I caught these Spooks making this game yesterday with Maggy—Maggy had been "Whitewashed" into thinking I was angry at her for "Denying this Agency's existence in regards to our Life" (stupid of her to think this by the way), instead I was getting irritated with her, because she would only Repeat what THEY TOLD HER TO SAY, and she would NOT LISTEN to the fact that I was EXPLAINING TO HER that these people were Lying about me being "Free on Tuesday", and WHY. (For my readers, I was trying to explain to my girlfriend that The Clown planned on cheating and "Keeping Me In" this Situation on Tuesday)."

"Instead Maggy was being Manipulated to start a fight, Programmed To, and at times "Machined with Exotic-Techno". She is easier to "Remote-Control", because she is Clueless and therefore easily Manipulated. Like yourself. If you let these people get away with Lying to you, then THEY WILL (For my readers: Having said that, both my girlfriend and the person I am writing to here, both have contact with these Spooks and both know of this Exotic-Techno)."

"This Whorbat (For my Readers: The name has been Changed), THE CLOWN, likes to Manipulate everyone around me, "Blame Me", and then

FLAUNT IT to be SADISTIC. Just like these Feds in America did on many a fine point. To rub things in my face, however as Whorbat has learned I am capable of figuring things out, figuring her out, and catching her. As a result she laid a booby-trap in regards to Me and Maggy, and to keep the Truth from getting out. The Truth of Whorbat's Lying and Cheating, in regards to "Holding Me In" yet again on Tuesday September 21 (2010). Which I know She will do, as She has been doing it for YEARS: let me know when this sinks in."

"I looked at what was going on and realized this Whorbat is Lying and Cheating and has every intention of "Holding Me In" on Tuesday. I realized what the bigger picture is, much to Whorbat's irritation. To make me look bad, "AGGRESSIVE AND ANGRY", to get people I love to stab me in the back (to isolate me), and to rub it in my face—in spite of the fact that I am NOT aggressive or tight in regards to this Situation, Whorbat, or others. Whorbat knows where my head is at, they know I am not tight about this, so the "Beauty of It" for Whorbat was to make me "LOOK" this way. BY MANIPULATING PEOPLE AND BY USING THIS EXOTIC-TECHNOLOGY."

"I have a book coming out about this Idiot and in light of this, I am relaxed. And in regards to any Citizens involved in this "Mobbing", they are Clueless-Idiots who couldn't think their way out of a paper bag. They couldn't begin to get what is going on here in my life—that would require them asking questions and thinking on their own for starters. Any "Tightness" is a result of this Techno, minus "Normal Circumstances", involving "Normal Emotions", with "Normal Emotional Levels"."

"Naturally I was getting Pissed Off at Maggy because she was too busy being a "Know it All", too busy being Used, Misused, and Manipulated, to even listen, it just was not for reasons that Maggy understood correctly. She had been TOLD I would get angry, just not the REAL REASONS why—she had already been TOLD AND SOLD it would be for "Other Reasons" which she then saw it this way, for these "Other Reasons" which were a Lie. As a result she "Experienced" things this way when it WAS NOT."

"This is called Programming—she was seeing "Actions" through "Someone else's Eyes", as something they were NOT, for "Reasons different than they really Were". She was "Seeing and Experiencing Actions" the way someone else WANTED HER TO SEE AND EXPERIENCE THEM, NOT AS THEY REALLY WERE. And I was trying to communicate to Maggy for the sake of US. For my Family so that she understood that this woman Whorbat "Planned on Holding me in this Mobbing", so that Maggy knew, and so that Whorbat would be less likely to Lie and Cheat, knowing that Maggy was watching."

"Maggy had been Deliberately Programmed, for my speaking to her about this in particular, in case I figured out The Clown's Plan for "Holding Me In"—thus it was a "Booby-Trap" waiting for me.

"Maggy had been "Deliberately-Misinformed" as in LIED TO, and Manipulated. My going to Maggy and speaking to her, to expose that this Whorbat was going to "Hold Me In" was predictable—of course I would, look at my Motivations for doing so."

"My getting angry, irate at Maggy (for MY reasons) was predictable, as my reasons were NORMAL. It was a Set-Up, a plan FOR BUILDING A FALSE PROFILE to sell to Maggy and others. Like YOU AND **** (name of other person) actually. To top it off my speaking to Maggy was about ending this nonsense and making sure that I can go free on Tuesday September, 21. To end Whorbat's "Clever Plan for Cheating" and "Holding Me In" yet again. How is that for frustrating? I am trying to speak out for my families freedom, and Maggy has been "Programmed" to be resistant and aggressive at cost to US. Talk about being Blind. I love Maggy but you get the point. Now look at yourself in the same regard, where my "Tone" is involved, you are equally frustrating and Blind. And yes obviously you both piss me off, for reasons neither one of you are capable of understanding."

"So I tried speaking to Maggy about Whorbat's-Cheating, Maggy had obviously been told to keep repeating "This Agency? Where? Who?", and had obviously been TOLD that I would get mad about THIS. She had been "PROGRAMMED" told to expect things that were not True, and for reasons that "LOOKED TRUE" but which were NOT."

"My own reasons for approaching Maggy were not what she expected, and neither was my anger for the reasons which she was Told. But both my behavior, and Maggy's, was expected, the whole thing a "Staged-Event", a "Programmed-Event", and it shows the level of Psychology being used here, and the amount of DECEIT, and Manipulating. I believe they had something also prepared with you along these lines as well, for when I called you, but seeing as how, they "Got Caught" the day before with Maggy, no doubt you were asked to play along different lines by the time I did call—you would know better than me, but it is a fair guess. If I am wrong on something so simple, it DOES NOT MEAN I am wrong about everything else, I am not, and at this point you probably realize this, as well you should. You were told to merely listen by the time I called, weren't you? But back to Maggy being Manipulated."

"If Maggy, OR YOU, knows of this Whorbat's-Plan to "Hold Me In", this Whorbat can not get away with it. Not the Plan she had in place at any rate: obviously same as always, she will use a "BACK UP PLAN". She plans on keeping me in anyhow."

"Maggy had been "Prepared", PROGRAMMED, to NOT LISTEN in case I figured out this woman's plan. She had been used as well, to get us into a fight, and I WAS NOT THE ONE WHO INITIATED being Aggressive. I initiated the conversation, NOT the Aggression. Maggy was actually, and SHE WAS "MACHINED INTO AGGRESSION" to top it off—her anger was much more than it should have been. She got Used. This is the level of Planning and Deceit we are talking about here. And "ironically" the plan was also "To Prove" to Maggy that I AM (this is to insult Maggy's intelligence behind her back) "Hostile, Tight, and Aggressive", when I am NOT. Talk about being blind on Maggy's end, and Sick and Manipulative on Whorbat's end. I have a book coming out, and in no small part it is about this woman, and Her mistakes and being caught, I have little to be "angry" about in one sense. This "Situation" is almost done, one way or another, and in light of that I am more relaxed than you know. If I do get angry, minus this Technology, it is for normal reasons."

"In the meantime this woman was and is trying to "Machine me Tight, Aggressive, and Angry" (I know it is this because I know my own feelings and can tell the difference—never mind that this woman is extreme), and has "Whitewashed" Maggy into thinking this, and trying to start fights through Maggy-to "Prove" that I am tight and aggressive. This is for a FAKE-PROFILE, an ARTIFICIAL-PROFILE, it is a TACTIC, let me know when this sinks in."

"Maggy took my becoming irate with her, and my growing frustration in talking to her (Maggy started getting aggressive first), as being "Tight and Irate" about the fact that Maggy would not address the existence of these Agents in our lives. Maggy thought this because SHE WAS TOLD THIS, AND DID NOT QUESTION WHAT SHE WAS TOLD. She was Programmed on this point like a particular computer program. Instead I was trying to expose Whorbat's "Corrupt Plans" for "Holding Me In" again on Tuesday, to Maggy. We experienced the same "Event" two different ways, and because her Perceptions and Viewpoint WERE NOT HER OWN. She started getting "Angry and Aggressive First" as a result of her own Programming, never mind that these Spooks also "Affected Her" with this Technology. She would not even listen to what I had to say, and on this point even, because "She was so Convinced". She had "Total Knowledge", "Absolute Certainty", because of these Lying Agents who could care less about her. They "Attacked Me", and Our Relationship, through her, and by Using her and Manipulating her. By Programming Her."

"THIS IS, AS YOU MAY NOTE, A MANIPULATION ON Whorbat's END."

"When you get that I have a BOOK OUT ABOUT THESE PEOPLE, get that they are caught, and get that I am relaxed in light of all of this, you see how stupid this line of "Thinking" was on Maggy's part. Brilliant on her end. She "wrote off", took my anger in regards to one thing, and tried to "attach it" to another. IT LOOKED THAT WAY TO HER BECAUSE SHE HAD BEEN TOLD BEFORE HAND (Programmed) THAT IT WOULD OCCUR THAT WAY, AND FOR THE REASONS THESE SPOOKS HAD TOLD HER—WHICH WERE A LIE. She had been PROGRAMMED. Just like Cult-Bullshit, and when you are non-questioning, like any Idiot joining a Cult, this is what happens. Of course they knew I would get angry at Maggy when she would not listen to the Truth of the matter—they just did not tell her the REAL REASON WHY. They understood why I would be angry, they just did not share this "Small Detail" with Maggy. It is called Lying, Programming, and Manipulating. She "Saw and Experienced Things" as she had been Programmed To, because Her-Perceptions had been ALTERED. They had been ALTERED for a "Desirable Effect" and one Beneficial for Agents, not based on Truth. And Certainly not beneficial To Her or Me."

"Now you are an outsider, and you can see how ridiculous this line of thinking is on Maggy's part. It is absurd but it is because she was PROGRAMMED. Just like you on many a fine point. Want a quick example, let's try the word "NORMAL". Are you F-ing stupid? Does any of this sound "Normal"? Could you even begin to figure out any of this shit if you were in my shoes? WITH ME EXPLAINING IT AND POINTING IT OUT YOU OUGHT TO BE ABLE TO. Where is your F-ing head? So get a good look at what has happened to Maggy."

"On the SURFACE it looked the way they Told her, the way they PROGRAMMED HER. There is nothing more believable than a Half-Truth and these Lowlifes do it all of the time. Who is the god-damned idiot here, me or her? THEY EXIST IN MY LIFE WHETHER MAGGY "ADDRESSES THIS OR NOT", it is a god-damned MOOT POINT. I have a book out, on these Lowlifes, and Maggy believed this is what I was getting irritated, and angry about? I do not "need" Maggy to acknowledge their existence, any more than I NEED YOU TO, I COULD CARE LESS. ARE YOU GOD DAMNED CLEAR ON THIS? There are a few other points here as well that will have to be addressed in regards to Maggy and some of the Lies she has been SOLD lately. This is to open your eyes by way of Maggy."

"How else could she believe something so ridiculous, such a ridiculous point, and line of thinking? They not only Programmed Her, they were "Mocking Her Intelligence" and showing me what they can get away with. Whorbat was "Mocking Her", a favorite of Whorbat's and part of

Whorbat's Profile. And it was DIFFICULT to get her to see the Truth of this, which is ridiculous—and because she was ABSOLUTELY CONVINCED BEFORE HAND. This "Programmed Event" "Burned into her mind that Way", "She Experienced it that Way" because she was Programmed To, and as a result, she did not want to "Let Go of the Programmed Version"—which had nothing to do with the Truth of the matter. It had nothing to do with Reality, and sure as hell had nothing to do with me. This is Sick, and both people get Screwed (and Traumatized). Her Perceptions were Altered in the way that they wanted, to "Start Something" and it happened because she listened to them, and they could care less about her. Obviously these Agents can NOT BE TRUSTED."

So for my readers this is Programming, the "Art" of getting someone to "See something Differently and Experience it Differently", and it is a Sick Farce. This is why Cults are watched closely, why it is Illegal, it is Sick. This is what was used on my girlfriends in America, and my girlfriend here in Europe, and if you are in this "Situation" they will use this on anyone you are close with, including Family and friends. Especially them as they have to be the "Most Convinced"—these Agents do not want anyone to "See what they are Getting-Away-With". You have to be careful, that being the case, as the person you are dealing with is not "Their-Self". In one sense it is like dealing with people on hard-core drugs, having a "Bad Experience". Wait for things to "Return to Normal", before trying to sort things out, before "Walking them out of the Process". You may have to wait a few days and you have to point out the obvious facts. Do not wait too long or they will remember it this way forever, obviously.

So that is the Corrupt-Form of "Responsibility and Consequences"

I have quite a few people I would still like to give a piece of my mind. They used this Tape for Programming my girlfriend, with already past/ prior "Family Issues", to get a "Look" that they wanted—I spent three days in Shock over this Bullshit, their Experimental-Head-Trauma-Shop. It was a Sick-Deliberate-Manipulation, and it explains what "Went on behind my Back" and happened to my girlfriend, especially since I have had it happen here in Europe so many times. What a bunch of Sick Bullshit, and no doubt she is still Clueless. Brilliant.

So from this the Feds (in America) began dragging my name through the dirt, which I had no idea about, because they were Hiding what they were saying from me, saying it all behind my back to Manipulate People—knowing full well that they would be Exposed by me if I knew what was being said. Which would have been easy for me to do, if I had known. And to this day, my girlfriend along with a whole lot of other people, believe and believed, something that is and was not True. This was the "First Framing" and came about to "Create Dirt", for Manipulating People behind my back, and as a way of "Neutralizing" what I knew about this Pedophile Cop if "Need Be" (so I can guess)—to Discredit Me, to make me "Look Bad". To top it off, the Feds began to build an Artificial-Profile around this, which at one point became obvious to me, and it was Their Fear that I had figured it all out, the Manipulating and Programming of my girlfriend and everything else, that led to them putting this Electromagnetic-Technology on me. To find out, how much I knew—for the sake of Protecting their own Asses. If you are one of these Feds, guess what, I have it all figured out, damn near to the finest detail. May your Agency give you a healthy dose of this "Mobbing" for being the Screw-Ups you are. This ought to be a Standard of Your Agency, for Screw-Up Employees like Yourselves.

The Murder

So the event that eventually led to my "Situation", and which can start belonging to other people as well: namely the Corrupt-Government-Agents who started this all. May your own Agency put you in this "Mobbing".

A bit of background information. I lived over a restaurant which looked out onto a large parking lot, shared by two businesses. On the other side, about twenty yards away, was a Kentucky Fried Chicken. One night while reading, and the business downstairs was still open, I heard an argument outside, a loud one. Naturally I looked.

An argument was taking place between what looked like a bunch of high-school students, two of them actually, but behind one guy was a group of about fifteen kids, obviously there out of curiosity and no doubt friends of the shooter. I figured I would go back to my reading, and if things got louder, and escalated, I would call the police. Well to some extent things did quiet down, until there was a gunshot. I hit the floor immediately in case there were more shots. I waited about five seconds and peered out the bottom of the window. All of the kids were standing there, apparently in shock, looking at the kid sprawled on the ground. A moment later they all took off running.

Well within ten seconds I had grabbed the phone and called 911. The dispatcher answered, I gave him the address and details, a shooting, what have you, and then he asked if the person was white or black. I paused a moment wondering why he would ask, but I knew every second counted, the kid in the parking lot was obviously dying, so I said he was black.

It was a big mistake, the ambulance took ten minutes to get there, and there was an ambulance station only around sixty yards away. Talk about disgusting. I got dressed and went outside, by the time I was there, a female employee from the restaurant underneath my flat was holding him up, it looked like he had been shot in the neck. Other employees, and people who lived over the restaurant, were outside as well. We were all in a state of shock, there was blood all over, the woman holding him up was sobbing. I kept wondering about the ambulance, thinking it would show up any minute, it did not. Unfortunately, none of us thought to run over to the ambulance station no doubt due to shock, maybe it would have made a difference. As it is the ambulance showed up ten minutes later, the kid died.

Obviously the 911 dispatcher was racist, and was responsible for the ambulance showing up late, there is no question. I began to speak out about this, again to undercover cops, and informants as I had no idea of what was going on behind my back in regards to Meik: of all of the Lying and Manipulating that was going on behind my back as a result. Eventually I spoke to one who lived above the restaurant, where I lived, and no doubt in very large part he was there to keep an eye on me. He became very angry and told me, "Mind your own damn business."—what a redneck. Clearly the police department was very nervous about me speaking out, and had something to hide. I do not know what the connection was, but clearly the police were protecting this 911 dispatcher, and no doubt for their own reasons. Obviously there was a connection, otherwise they would not be protecting a Racist 911 Dispatcher who was responsible, in very large part, for getting the kid killed. I think they may even have known this shooting was going to occur before hand. And everything in the Newspaper the next day was a Lie and Cover-Up of what really happened. And keep in mind as the Feds were "Secretly Investigating Me", they had a Wire-Tap on my Phone and all of this Phone-Call to 911 is Recorded somewhere, and a matter of RECORD.

So let me set up a few things for you here that were occurring at the time. I was dating Penny, an informant, my friend Kitch was a "Deep-Cover Cop" (the Feds eventually told me he was a "Deep-Cover-Cop" once my "Mobbing" started, along with the fact that he was bisexual—these Lowlife Agents love the sex stuff), I still had on and off contact with Eve (no sex though at that point, as I was dating Penny), and I knew I was being "Investigated", including at the restaurant (not the pizza place) I was working at. This was no longer a matter of the Police Department keeping an eye on me, this was a full fledged "Investigation", and I was the center of it all. To top it all off, the Feds knew, I knew, I was being "Investigated" (Keep in mind this was a Dirt Creating Mission—NOT an actual Investigation), that I knew about Penny being an Informant, and that I knew I was being "Profiled" at work. They also toned things down in regards to Penny and work so I would forget, so everything "Seemed Normal", so I would fall "Asleep" on these points. Which I did to a large extent—not entirely however.

Eventually the Feds sent Kitch, to ask questions, and "give me information", which was a laughable joke. Keep in mind that the Feds had a Recording of my telephone call to 911, and that they knew this 911-Dispatcher "delayed the ambulance"—the Newspaper said no one called for about ten minutes as I remember it, actually I called in the first TEN SECONDS. This guy came over to find out how much I knew, and to quiet me down, by giving a "logical explanation" for the late ambulance. As an added note, this guy Kitch, was one of the three I had "Sold a Story" to, a Lie (and that is all it was), for the sake of catching. I suspected these three repeatedly, on and off.

This guy Kitch explained that this kid was a drug dealer of some sorts, which I took with a grain of salt on my end. I was given the impression (although his mouth was saying something else) by Kitch that this kid was not much of one in the first place, if a drug dealer at all. To top it off this guy Kitch dealt drugs from time to time, was involved with guns to some extent, hung out with all sorts of shady characters, was black, and I pointed out the stupidity of his mouth. Simply put the same justification could have been used for him, but since he was a black Deep-Cover-Cop that dealt drugs from time to time, I guess that made it all different for him. Talk about dumb "Logic", what an Idiot. Of course I later found out that the Feds had sent him over, and had been sending him over for years, anyhow.

I had never heard of 911 never giving a damn before, and certainly not based on race. All this was new to me—but again it simply comes down to Job-Titles, you always have to look at the "Type and Quality" of person who holds the Job. Having said that it never occurred to me or anyone else there in the parking lot to run to the ambulance station, not to me and I am usually a man of action—until writing this book and this portion of my past. Now when I started looking around and sticking my nose into this, the following days and weeks, I realized something was wrong with everything here. I began speaking out about the 911 Dispatcher and the "white or black question", and the lack of response from the paramedics. I also looked at the newspaper the next day which said that no one called until 10 minutes later, which is and was an obvious Lie. This person obviously had some connection with the Police Department, which was Covering Things Up. Talk about a Total-Farce.

Actually there are at least 10 witnesses to the telephoning part that will tell you otherwise, The People in the restaurant downstairs had also called 911 and it sure as hell wasn't 10 minutes later. I eventually started speaking directly to undercover cops, who knew I knew about them, and one told me to "Mind your own damned business!"—which is weird because this guy was a 911 Dispatcher I was speaking about. So the police were covering the whole thing up, which makes me think they knew this shooting was going to happen. At the least they were protecting some Racist-Jerk who took a part in this boys dieing. Maybe he would have died anyhow, but we'll never know now, will we? In America we should all expect better than this.

I was making waves with the Police Department, and you have to be an idiot to think they were not nervous. And of course, as usual, they knew I knew what the situation was because I was speaking to their people. This is the same Police Department with an employee having sex with fourteen year olds, and then leaving "Them Alone" on crack-alleys with dangerous criminals. Think this made me popular? Last time I checked, Police Departments don't like getting caught for covering up a Murder, by a racist 911 Dispatcher—which is in part what this was. It is like the boy was killed twice. Looking back on it now I think the police knew this kid was going to be shot and decided to look the other way, and the Dispatcher was waiting for a call like this, was in on it with the Police or Feds. If that's the case this kid was even less popular than me with the Police. And think about this comment because it might have been ME if I had kept pushing the issue. Like I'm lucky in retrospect, never mind my Family.

Months later, after the shooting, the next Framing by the Feds came—as I had one prior to this next one, which I would later find out about, and put together here in Europe. It happened out in the parking lot at four a.m., and happened as a way to get me in this "Situation", which it did; and which I still live in over here in Europe, due to more Corrupt-Spook-Jerks. In my opinion these people are just as bad as the others in America, "Framing Me" has been a Common-Tactic to keep my "Situation" from ending.

The Framing which Started my "Situation/Mobbing"

So I was still living over the restaurant, the same one where the murder had occurred. Then came a van and a bunch of "Ninjas", at about 4a.m. in my parking lot. The Kentucky Fried Chicken parking lot.

The Feds watching me 24/7 knew I was awake. The whole thing was a Set-Up to put me in this Mobbing, or as I refer to it, Localized-Fascism.

I was awake in my flat, probably reading with no work the next day, when I heard a car door slam shut out in the parking lot (this is stealth?). People rarely parked out there, and not at 4:00 a.m., obviously I looked, never mind the recent murder. A bunch of people with black ski masks and dressed in all black piled out of a van, and ran across the street, after congregating for a moment. Across the street to where a drug dealers house was, hidden in shadows by trees and undergrowth. Now I have to give a few more bits of background info here, before I can go on.

A few important points here. One is that a guy from my university days, who I knew to be an undercover-cop at the time, was now here in my town and hanging out openly with these drug dealers, who lived in the house that the "Ninjas" ran to. These drug dealers were dangerous and preying on the homeless, I had seen them up close, passing them on the streets, along with this guy Orlin, whom I kept my mouth shut about. We would pass, and I would not even begin to act as if I knew him or recognized him. I was glad he was going after these people, they belonged in jail. As do the Feds who framed me, and The Clown.

Now the thing is back in my university days, he had been incredibly sloppy, and so had I. He pushed the boundaries of obviousness with the click he and I hung out with (small time hippie dope smokers), which was a bad thing to do in the first place, never mind around me. He liked to "Wave around his Invisibleness" like an Idiot. I was still learning techniques and styles of police, informants, etc, and was a bit underdeveloped in my "James Bondness". At any rate I too was sloppy, and made no point of being careful of who I spoke to. But not all people are friends, and not all friends, criminals, and cops are created equal. I eventually heard that he had been hit by a car, by a drug dealer (no one I knew), although he was quite fine. As sloppy as he was, it may not have been because of me, but it is irrelevant. I completely changed as a result, and you may note a reoccurring theme of me discussing things only with undercover police, and informants, when there were "Issues" to discuss. This is why. The last thing I wanted was someone to be dead as a result of me speaking out. My "speaking out" is not about a Vendetta it is about doing the right thing. I'll add a note here, "My Speaking Out" in regards to The Clown is at this point "Personal on my End" as well, as this Imbecile has Tortured me and my Family, and is indeed a Criminal and "Rapist" in the sense of the word any psychologist gets.

To top it off I made things clear to Penny, my girlfriend and an informant working directly for the Feds, and a couple of others working for the police that I was glad Orlin was going after these drug dealers. I made it TOTALLY CLEAR that I had absolutely no intention of making problems. I could care less what he was doing, I had no contact with him, he posed no risk to me, and he was not sticking his nose into my business. Quite frankly, with cops or informants, whether you are a dope smoker, or "Big-Time", these people are always documenting everything, and in short you have no privacy, everything personal is going into a Data-Base that anyone can read and Misuse. At any rate I was glad he was going after these drug dealers, and my attitude was "go-get-'em", I was never one to hang out with, or like the dangerous types. You may have noticed that I am against murder as a general rule...

In addition to all of this, I was being investigated by the Feds (COINTELPRO), and they were in charge of all of these informants, cops, what have you, that were in my life: I was under 24-hour surveillance. To put it bluntly, they knew I had ABSOLUTELY NO INTENTION of screwing things up for this guy, Orlin. They knew this beyond a shadow of a doubt. But then this kid got shot out in the parking lot, there was a Cover-Up to protect some Racist-Piece-of-Shit, and I was in the middle, once again of a nasty mess perpetrated by 911 dispatchers, local police, and Feds. So everything changed. Again, I'm just lucky that way.

Note the irony. I was all for Orlin getting these drug dealers off the streets, I figured they were coke or crack dealers, they sure as hell didn't have that whole "Passive and Mellow" university-hippie-dope-dealer vibe going. They were dangerous, and lived a "stone throws away". Yet eventually, as you will see, the Feds framed me by "saying" I interfered with a Federal Investigation, and that I did it because of this guy Orlin, from my university-days. To "get at him", like I had a perpetual grudge against the guy, which was, and is of course, a blatant lie. Actually it is ridiculous. They "White-Washed" the entire thing, pulled the wool over everyone's eyes so it LOOKED THAT WAY, in short they built a "faulty profile", a "FAKE PROFILE", an ARTIFICIAL-PROFILE to SELL to the Public: and it worked.

For my readers, and anyone else who may not know, this is a Tactic, "Fake-Profiling", it is the taking of things Out-of-Context, to fit an Artificial-Profile which gets SOLD to the Public who are Misused by this Lie, and "Proof"—Fake Evidence. Having said that, as it is not "Physical" rather "Mental" it is not Real-Evidence and can be EASILY FAKED by knowing Psychology, especially if you have the means to do so, and The People willing to do so. Thus it is NOT HARD EVIDENCE, it only "LOOKS LIKE" "Hard Evidence"—having said that not all is as it Looks, especially when dealing with people who deal in Espionage (to make a point), and "Not all that glitters is Gold.". Especially when dealing with Corrupt-Agents. IT IS PROGRAMMING PEOPLE, around a FAKE-PROFILE OF THE TARGET.

It is the Deliberate Misuse of Psychology, to get "Results" (usually actions) based on a persons "Psychological Make-Up" which can be Deliberately-Misconstrued to be for "Other Reasons". Like my example with "Maggy" in my "Programming Piece" from earlier. These "Other Reasons" being a part of a "Fake-Profile". In other words they are getting you to do things based on Predictable-Behavior, and for Predictable-Reasons, which they then say you did for ENTIRELY DIFFERENT REASONS. These "Other-Reasons" are part of a Fake-Profile, they have a way of "Gathering-Evidence", Actions which support this Fake-Profile. Actions which support the "Mental-Reasoning" they say you are "Practicing", Actions which THEY SAY is a reflection of your Psychological-Make-Up. Actually the "Psychological-Make-Up" they are selling to people, is actually very different from your own. They SELL a FAKE-PSYCHOLOGICAL-MAKE-UP OF THE TARGET. Everything is taken Out of Context. The Agents know this as in order for them to predict your responses to "Particular-Staged-Events", they have to know what your REAL PSYCHOLOGICAL-MAKE-UP IS. And in fact in most cases it is simply based on COMMON SENSE and how people would NORMALLY ACT under the circumstances.

It is the "Taking of things out of Context" to fit a FAKE/ARTIFICIAL-PROFILE which gets SOLD to the Public: the Public is Programmed to "See and Experience" these Staged-Events as they are "Told To" by the Programmer. Thus "They See and Experience" that this False-Profile, this Artificial-Profile, is True, when it IS NOT. The Artificial-Profile is the "CENTER PIECE" OF PROGRAMMING THE PUBLIC. It is all Lies and Manipulations, Layer after Layer, to Bury the Truth, and to Bury the Target.

BACK TO THE "PLUMBING NINJAS"

So their Framing-Me, their Fake-Profile worked, and in no small part because I had no idea what was being said behind my back. This is how Character-Assassination works, this is how Fake-Profiling works—if you do not know what is being said about you, you can not expose other people's Lies. This is a Tactic as well—Hiding what is being said so that they can not get caught at Lying. Creating a "Buffer-Zone".

I basically got "Framed" several times, in this manner, at the beginning. Over the years it has become far too many to even count. It is a Technique and Tactic Misusing Psychology, Information, Citizens, Context, and Programming. In the beginning, they went using a "Tape" of me talking to

Biff, which they did for "Convincing People", and the words were indeed coming out of my mouth—the simple matter of "Context" however had been left out, and so people had no idea that these things were not True. I did, the Feds did, no one else had a clue. In regards to my girlfriend, I suspect they also altered and doctored the tape so that it was "Specifically in regards to Her"—thus that way also, when she was "Programmed to Freak Out" this would all fit together—and "Look Great", and be Convincing, when in fact be the biggest Pile of Deceit on the planet. Again building an Artificial-Profile. So this is more common than I thought, and easy to do, at least if you are a Corrupt-Agent, you only have to be sick enough to do this in the first place: a rule of thumb for Corrupt-Agents is "If it is cheap and easy, but will work, then it is fair game". But back to building an Artificial/Fake-Profile and Framing Me in regards to Orlin.

So they Framed Me based on Orlin's and my "Past dealings Together" like it had anything to do with things in the present. They built a Fake-Profile out of this. Having said that I made a couple of mistakes along the way that only made it easier for the Feds, but I did not make many. They simply LIED, FALSIFIED EVIDENCE, TOOK THINGS OUT OF CONTEXT, BUILT A FAKE-PROFILE, OMITTED MAJOR FACTS, PROGRAMMED PEOPLE, AND DELIBERATELY MISCONSTRUED EVENTS, so that I "Looked a Certain Way" in people's eyes—what a Pile of Deceit. They did the same thing with me and my girlfriend and their Psychological-Trauma-Head-Shop experience, and later "Sold-This" down the road to a bunch of Ignorant-Jackasses, the average Idiot-Citizen. This is what happens when you give Corrupt-Jackasses Power and Authority.

But at the time of the "Ninjas" I was not thinking of this guy Orlin, in fact once I had spoken out to the "polices people"—like my girlfriend Penny and others, I pretty much put him out of my mind. That is because I had made my "Intentions Clear" in regards to Orlin and knew that they "Got the Message". They always had before when I spoke to Their People. Excuse me, but I was not expecting to be FRAMED on behalf of dirty-pigs, followed and harassed 20 years continuously, and across three Continents and in three Countries. You would have thought people would take Responsibility for people who were Criminals and total Screw-Ups. Instead I wondered if it was a hit by a rival gang. I didn't really think it was cops. I had never heard of the cop-ninja thing before, it was out of my realm.

Now note here this was all done out in the open, when there was all sorts of cover across the street-buildings, shadows, trees. Plenty to conceal both the van and the cops, from the street and any passer-bys, never mind other criminals. Also they could have done this much closer to the drug dealers house with greater concealment. Strange huh? Instead they chose a well lit parking lot with NO concealment and total lighting, totally out in the open, and outside of the window of someone they were "Investigating"—a "Known Paranoid". Gee that makes sense, doesn't it? At least it worked out to be "Convenient" in regards to getting me in this "Mobbing"—which was obviously the point.

So while contemplating things, the van had moved across the parking lot further away from their own people to DIRECTLY outside of my window. Talk about these Lowlifes setting me up. When I looked outside of my window again, there was the "Plumbing Van". It was complete with a phone number on the side. The image of people dieing came to mind if I didn't do something. I knew someone was still in the van, because I saw them move to the back, but not go out.

Like an idiot I decided to go out to the van, but ready to run for my life (I could get away quickly enough because of the buildings and corners) if need be, I more than half expected to. Having said that I would not have otherwise if I could not have, obviously.

Poised to run I went to the van and knocked, strangely enough, NO ONE answered. No one came to the door and said, "This is a police-operation (or Federal Investigation), mind your own damned business", or anything else for that matter—I mention this in retrospect .Yet as I now know, they were Feds (technically I suppose they may have been police—but the Feds Orchestrated this), and if it was not a SET-UP against me, why couldn't they do so? No doubt because the WHOLE THING was to Frame-Me, and they KNEW it was me, and if they had "Flashed a Badge" and said "Go Away" they would not have "False-Pretenses" for Framing-Me. They would not be able to LIE to their Bosses and say this "Guy did this", they would have nothing for "Selling a Fake-Profile" they would have nothing for "Deliberately Misconstruing my Actions", for taking Out-of-Context. So, still poised to run and a bit emboldened by the fact that no one came out immediately to shoot me, I knocked again. No answer.

I went back to my flat, one eye behind me, and then contemplated what to do. What ever was going on it was weird. I didn't know what was going on, what to do, but dead people kept coming to mind, along with the kid who had been shot out in this parking lot. My experience with the police was less than satisfying, never mind making myself a witness to a gang hit. And then what? The police were going to protect me? After my past experiences do you think I believed that? Would you?

As I was saying if it was a gang-hit that night with the Ninjas, do you think I would expect the police to protect me? Forget it, ask the kid from the

Kentucky Fried Chicken parking lot. So much for being paranoid. To top it off, my entire life prior to this, I had had nothing but problems with irresponsible police, so my own expectations were less than positive about this.

So the van had parked right out of my window and was complete with a telephone number which I could read from my Bedroom Window—which I would not have been able to do if the van had stayed on the other side of the parking lot. Imagine that. I doubt they had expected me to "Knock on Their Van" but they sure as hell obviously expected me to CALL on the phone. I called the number on the van. I expected it to be a non-existent number, instead someone answered, so that "People could Call"—and so that Feds could say I had "Interfered with an Investigation", and all "because of this guy Orlin". What a Lie and Farce. Imagine my surprise at the time, when a woman answered. Our dialogue was brief.

"What's a plumbing van doing out here at 4 in the morning?"

She hung up. I watched and waited not certain of what to do. Obviously her reaction was strange, never mind that someone answered at 4:00 a.m., and at a "Plumbing Business". But hey maybe the drug-dealers had an exploding toilet, and the Ninjas were a team of emergency plumbing specialists, at least that would explain why they were so obvious and out in the open.

Within a minute the van took-off to the original drop-off-point of the parking lot (where they had all originally gotten out), all the ninjas came scrambling back, hopped in the van, and it tore off. Obviously the caller had communicated with these people, and instantly.

Contemplation on my end, the possibilities, began.

What had I just witnessed? Near gang-hit or police stuff?

The fact that someone had been waiting to answer the phone, had me beginning to lean in the direction of police. On the other hand I did not consider this entirely out of the realm of an "organized gang" either. But if that was the case I had screwed up a police investigation, maybe they were going to do wiretapping. I knew Orlin was involved in retrospect. I add that because at the time of this "Event", he didn't even come to mind. I thought it was a gang hit, the telephone call brought other possibilities to mind. Having said that I was undecided, and worried that it may have been a gang that would return another time. I eventually decided to find out on my own, before letting the police know—given past experience I was not in hurry to have dealings with them. Another of my mistakes that the Feds made work for them, to make me look guilty, when in fact they knew what was going on.

To put it bluntly they figured I would call the number on the van. In part due to the recent murder in the parking lot months before, and the lack of response by the police, never mind them telling me to mind my own damned business. This is why the van parked outside of my window, I couldn't read the telephone number otherwise. Having said that a lot of people uncertain of the circumstances, and what to do, would have done the same, to try and find out more, before telephoning the police. This is not an unusual chain-of-thought. Having said that no one came to the door of the van when I knocked, flashing a badge, and saying "This is an investigation, mind your own damned business.". Imagine that. They did everything they could to Entice a Reaction, a normal one at that, whichever way didn't matter. What did matter is that I react in some way, shape, or form, so that they had something to work with and "White-Wash". To make fit an Artificial-Profile that they were already "Selling the Public" no doubt. I have experienced this repeatedly over the years, especially the last eight years here in Europe. As long as it "Looked Like" the Lie they Planned On "Telling and Selling" to everyone else, it did not matter: again Fake-Profiling, simply taking things Out of Context and saying it fits an Artificial-Profile being SOLD to the Public.

The next point and the most important is that they knew where I was at mentally. I remember saying to a few people I was glad this guy Orlin was going after these drug dealers because they were "Dangerous and preying on the homeless". I said it to Meik, I said it to Kitch, and I most certainly said it to Penny. I was also stating this for the record, after our "fiasco" at the University, I wanted the police to have a clear idea of where I stood on this matter in regards to Orlin. This last part in quotes I remember, and due in part because I repeatedly said it. I also, as you may have noticed, have "Protocols and Standard Operational Procedures" of my own—the big one here being that I speak to "Police People" in regards to "Police Business", thus I made sure the Police knew where I stood in regards to Orlin by speaking to THEIR PEOPLE. How god damn effective is it going to be for me to "Convey a Message to the Police" if I am speaking to Joe Dingleberry, as opposed to one of their people? How uncertain could the cops or Feds be on my Viewpoint? There was ABSOLUTELY NO UNCERTAINTY ON THEIR END. I said a couple of things later, "To find things out", which they then took Out of Context, but it is THEIR RESPONSIBILITY TO BE ACCURATE, and having said that they had no "doubts in their minds" and did this Deliberately. Obviously.

There was no "Uncertainty, Doubt, or Not-Knowing" on their end, they FRAMED ME, and they did so for a Pedophile-Cop, and a Redneck-Racist-911-Dispatcher. And I am the bad guy in all of this? To top it off, my kid and Family suffer TO THIS DAY, based on this Bullshit from twenty years ago and a bunch of Corrupt-Feds. You tell me, is there anything wrong with this picture? What do I have to do to end this Bullshit, "Appeal to Amnesty International", the United Nations, the President? These people belonged in jail, along with the Feds who started all of this.

Need more? Why didn't they move when I knocked on the window? Why park outside of my window? Why move further away from their own people with the van, never mind that a TOTAL-RETARD would have conducted the whole thing across the street on the same side as the drug dealers house, where all the concealment was in the first place. Why do all of it initially in a well lit empty parking lot where you are totally exposed? They were totally under a street light, and unless it has changed, to say that the parking lot there is lit up like a football field for a Saturday-Night Football Game, is only an overstatement by a slight margin. Why park outside the "Window of someone Known to be Paranoid", and better known by police "As being good at spotting their People"?

Furthermore the drug dealers obviously weren't home (and under 24/7 surveillance), the Feds sure as hell were watching me (that much ought to be clear by now), why park across the street in the first place, why move further away from their own people? Everything done was about EXPOSURE, NOT CONCEALMENT: starting off with the slamming of the van door to get my attention. To top it off, if you know anything about how these Federal-Jackasses operate (these in particular), I seriously doubt that they did not already have the house bugged/wire-tapped, as "Following the Laws and Procedures" is obviously not a big deal or problem for these Lowlife Feds in particular. They certainly had my Flat Wire-Tapped, and for "smoking marijuana". Having said that they Framed Me twice in a major way, including using "Doctored Tapes" or Tapes taken out of Context, they Altered Evidence, to convince my girlfriend(s) of several things in particular, and they Used one in Particular (never mind me)—obviously. I am not the Criminal here, they are and they belong in prison, especially after the Sick Deceit they have gotten away with. Never mind my three days of shock and having to wear other peoples Misinformed-Perceptions for years on end.

If these people were being careful, NOT DELIBERATE, Rainman, or Forest Gump could have done a better job. Boy-Scouts could have done a more intelligent job.

Remember, they moved to outside of my window, AFTER everyone had piled out. Why? Because of the telephone number. My calling the number constituted interfering with a Police or Federal Investigation. It is not why I called. It all came down to LOOKS, how they could make it look. It didn't take much did it? The situation was to "Entrap Me", to FRAME ME.

And here I am more than a decade later, and in a foreign country, still without any Rights, and suffering, due to a very similar type of Corrupt-Jackass (actually worse) in charge—a woman I call The Clown. And for what? Convicted Terrorists in Europe have received 10 years while I am nearing year 18 of this "Situation"—and if you think this is Freedom, then you are seriously mistaken. My Family is attacked as a result of this Bullshit—how much sense does any of this make? What a great big pile of Deceit. And they have "SOLD" everyone over here in Europe on some pile of Bullshit, from P.T.S.D., to "Paranoid", to "Split Personality", to "Borderline Personality", to it is "My Fault". Sounds like an Old Soviet-Union Handbook for Dissidents, doesn't it? It's no god damned wonder I am Pissed Off. The Spooks around here sure as hell haven't been forthcoming with REALITY. No one around here (never mind Idiots in America who took part in this "Situation"), Dumb-Citizens that is, KNOWS THE TRUTH OF THE MATTER.

I have Citizens here in Europe, twenty years after all of this began, taking part in Nasty-Bullshit, who have not the Faintest-Clue as to WHY they are doing what they are doing, for WHO they are doing things for, or WHAT they are doing. They do not even know that I have "Been Held In" repeatedly. THEY HAVE ABSOLUTELY NO IDEA HOW THIS EVEN GOT STARTED. WE ARE TALKING ABOUT CHRONIC-IDIOTS who allow themselves to be MISUSED in ways that are Reprehensible, Disgusting, and Flat-Out Stupid. And not one person who has taken part in this Bullshit has even bothered to ask ME. This is a perfect example of Fascism, "Localized-Fascism", Fascism IS when Government becomes Cult. After Hitler you would think that Europeans would have more sense. No doubt if I was a Racist-Pedophile-Cop I would be treated with more respect, never mind my Family. In the end it comes down to "Looks", having a Badge which "Connotates and Implies" "Goodness", as opposed to, and without actually having to be, "Good". It all comes down to a Stereotype over a Job-Title which in and of itself means nothing, because this says NOTHING about the "Quality and Type of Person" with the Job. People who are Irresponsible and Lazy-Minded simply look at the Job-Title, as I have found out the hard way.

But back to America, Feds, and Framing.

Again in retrospect I wouldn't be surprised if the drug dealers house was already bugged. Legality hasn't meant a damned thing to these Corrupt-Government-Agents in regards to my life, never mind a bunch of dangerous drug dealers. They have never admitted to any of this, that I tell you, except TO ME, in a variety of ways, just to piss me off BY LETTING ME KNOW WHAT THEY ARE GETTING AWAY WITH (here in Europe that is). So much for Justice. For all of this I have lost all of my rights as a Human Being and

Citizen, never mind my Family. So this is Justice? Ever wonder where all of those Tax-Dollars are going? Take a good god damn look because I doubt I am the only person this is going on with, or that this has happened to. Millions of dollars have been spent for the sake of a Pedophile-Cop, and a Murdering-Redneck, based on Corrupt-Federal-Agents. There's nothing like Justice and Responsibility. And because these Assholes were worried that these "People and their crimes would be exposed to the Public"—like these were minor violations here.

I was FRAMED by the Feds, and it was done to put me in this "Mobbing", which has spanned three Countries, three Continents, almost Eighteen Years, and god only knows how much money, but no doubt at this point it has cost the Tax-Payers MILLIONS OF DOLLARS, and all because of a bunch of Corrupt-Feds, who saw a Pedophile-Cop and Murdering-Redneck as "Less of a threat to Society" than a person trying to do the right thing. This certainly says something about the "Quality of these Agents". But don't worry, as long as you are white, and not a little girl, you'll be safe with them around. In the meantime they no doubt continue to Frame Innocent People who they do not like, Personal not Professional, and Falsify-Evidence so that they can put people in this "Situation" or jail for things they did not do. They are extremely confused about WHO is the Criminal and WHAT their Job is and Manipulated Ignorant People to "Go Along with Them". Consider yourself Exposed.

And so my readers know, what I mention in this book is only the tip of the iceberg; unfortunately I have "other issues" going on now as well. But I give more than enough that anyone with a brain, in America, or here in Europe and taking part in this "Mobbing", can begin to question things, see things more clearly, and in the end realize that what is being "Sold" is a Lie, Deceit, and Manipulation.

This is what passes for Responsibility in the "Justice-System" these days in America. Until American's begin cleaning up their Country, they can forget about Justice, and they can forget about a Government that is Capable of doing it's Job. Until there is a "War against Corruption" things will continue to go downhill. Corruption is a "Disease" and it is eating up everything in America. My life, and this "Situation/Mobbing" are good examples of this.

A Few Last Notes

As I understand it these people were also "Inadvertently" trying to take my life as well, before I was even in this "Situation", due to my "big-mouth"; as in I knew many a secret about the police departments "Mistakes". Mostly because they made them, and mostly because they made them where all could see: they were Idiots. Next time I will simply go to the Newspapers from the beginning. At any rate the Feds solution was to set up "Accidents", and "Natural Looking Events" that would take my life. For some people this is apparently their idea of "Responsibility".

"Reckless Endangerment" as a Tactic, as these Spooks here put it. In other words you may die as a result of Their Actions but "indirectly" so they do not look guilty. When it turns into dying, it is no longer officially "Reckless Endangerment" but Murder. Such as a mechanic, who supposedly changed my timing-belt, but then I found out the next year in a different city, at a different place, that he DID NOT. I KNOW the Feds were responsible because I had a nice conversation with a woman in the waiting room, who I knew to be an Agent, and felt was there because of me. I did not feel she was an undercover cop. What a coincidence.

As an added note when I bring up "Close-Calls" on my life here in Europe, with other people I know or speak to, it gets "Written Up" as "Paranoia" which is a Sick-Joke. It "Looks that Way" until they "Screw Up" (and they have already I might point out) and I am actually dead as a result. The Citizens somehow are unable to "Think or See" this far. No doubt due to "False Logic" (He's Not Dead Yet), and Programming. A False-Profile, with "Pieces taken from my Past" in which it is "Proven" that I am "Clearly Paranoid" when in fact NO CONTEXT for "These Pieces" has been given. Of course the "Context of these Pieces of my Life" is the important part. There are of course "Legitimate Reasons" for "Bringing up these close Calls", they were indeed "Sloppy but Hopeful Attempts on my Life" not simply a matter of being "Paranoid". In this situation the Spooks would say "Whoops" if something happens, all the while glad something did. So Citizens and others who are Programmed, are Stupid to think it is "Paranoia", and would feel differently if it were them, because they would see the Deceit of it, and realize the Truth. The Citizens are constantly Manipulated with "Evidence" that is TAKEN OUT OF CONTEXT. But back to my timing-belt.

This was BEFORE my Situation actually had begun. At any rate if your timing belt goes out, breaks on the highway (and I did a lot of traveling), you are screwed, to put it mildly. There were OTHER CIRCUMSTANCES as well involving my long bike rides out in the country (30–40 mile bike rides—think I was a chronic marijuana smoker?), and people nearly hitting me, deliberately. Clearly, killing me "indirectly", or "accidentally" was more than an option, it was being Implemented. Fortunately I live on to get this information out, people need to know how their Government is acting, what Agencies are allowing their employees to get away with. People need to know how some of these Government Agencies act—this "Situation", and what these Corrupt Agents get away with. Having said that I am sure that even these Agencies have no idea of how their Corrupt-Employees are acting, at least to what extent, or the Truth of the matter on many "fine points".

So, this is "My History": how things got started, how I ended up in my "Situation of Mobbing", what was going on, how I walked the high-wire, and why things continue on in a different Land and different Continent. The costs to Tax-Payers must be immense, never mind how ridiculous and hypocritical the entire "Situation" is. Simply put, when you look at what I have been through it makes no sense, and Agencies look like the Disgrace they ARE.

CHAPTER 29

A List of Government-Agent-Tactics which Have Been Used on Me which Can Help You See the World Around You More Accurately: A Guide to Spotting Corruption in the Media and World Around You

"She's not playing with a full deck, because she hides some of the cards for cheating."

—Mr. Lee in regards to The Clown

Tactics, Techniques, and Examples of COINTELPRO from my Life a List of Government-Agent-Tactics: Tactics used Against the Public

Lets start with the list of Government Agent Tactics before going on further in regards to this "Mobbing" and COINTELPRO, as this will help the readers understand better what is going on. As it is, I have already had to "Explain Several" in the previous chapter. These are my terms, my terminology, and if you look closely you will see these same "Tactics" Used by Governments and Politicians Across the World. These Tactics are based on Psychology, an understanding of Human Behavior, and the "Art of Manipulation"—thus there is a Hand-Book for these Manipulations and

Tactics. Being Aware of These will go a long ways to better understanding things happening in the World, things that you hear of in the News. These are Tactics used against the Public, and Masses. I do not mention them all, the ones I have observed and had Used on me in my Life, because I am on a Time-Limit here, as I have a group of Reckless-Spooks threatening me and my Family among other things. This list is an older list I compiled years ago, with a few recent additions made, but it is a good start. Some of the "Older Labeling", such as "Whitewashing" is really for the most part "Programming" as an example with perhaps a bit of "Difference in Connotation". If I were to properly update this, no doubt some things I would consider a bit differently, but no matter what these Tactics will only help others, and my "Assessments" are not far off the mark.

People should be Aware of these Tactics as they will help you see the Truth even within the News and Media, never mind helping people in my shoes. I wear a unique set of circumstances, that has given me a different set of eyes from most. Fortunately I have the person in charge of my life now (December 2010) and Her Buffoons to thank for these insights. As a result of The Clown and my "Spook-Education", I have been able to rewind on my Life and sort many things out from my past as well, including "How I got in this Mobbing" and "How Corrupt" the Agents back in America were.

Suffice to say for now, there was no book before, or observation of Tactics before, but due to sloppiness and stupidity of the person in charge and her Unit/Team of Buffoons, that has all changed. Again not all Government Agents are created equal and I will be getting to this later—time permitting. Seeing as how Exotic-Technology, Electromagnetic-Technology, is being used to interfere with my life, including with the writing of this book, I am going to be very brief and skeletal with this segment. So.

The "Rules of this Mobbing"

The Rules of this Mobbing, the situation is that there are "Stress-Tests" for a period of about one month. All sorts of Violations, Rudeness, Weirdness, and Crimes are Perpetrated against you for a period of one month, with an "End-Test". If you "Pass the End-Test" you "Go Free"—usually there is only a day or so afterwords, in which there may be a "Smaller-Version" of the same Test from the day before. At this point, if you "Pass the Big-End-Test" you "Go Free". If you want an "Example" of this Mobbing, see the Movie "The Game" with Michael Douglas, which is obviously a "Feel-Good Film" for Idiots who take part in this Mobbing, and was obviously written with a "Government Insiders" knowledge.

So if you are in this Mobbing, keep a calender and pay attention. You have a chance of Freedom, every thirty days or so. You go through several "Ups and Downs", a build-up to tests within this thirty days, and then a period of "Less Activity", and then it "Starts over Again", for the stretch of about thirty days. So it goes "Up and Down" in terms of "Stress" and Tests, until the "End-Test", which is in around one month. The idea is to "Attack what you don't Like". Within one months time you will go through many unpleasant circumstances, but if you pay attention, there is a "Build-Up" to many of these "Circumstances/Tests". This is to Build-Up the "Psychological-Pressure" and to make you tight, anxious, worried, aggressive, what have you—whatever it is about it is not about "Good-Mental-Health". So within this month you are "Tested and Profiled", and at the End-Test there is usually something "Extra-Nasty or Weird". If you get through it, you are Free.

In order to get through this Mobbing, and End-It, you have to go through a month of Stress-Tests, and at the Last Test and Day, "Pass". You can "Fail Smaller-Tests" along the way, the "Last One" counts the most. This means no "Losing your Mind", NO OVER-REACTING, no "Threats", yelling and screaming in anger, what have you. If you get through this without "Reacting Externally" (You can Flip Them Off in your mind like an American) then you "Are Free", or at least will be in a day or so. You can always just "Pretend to go Along", you do not have to actually like these people. Who would?

So the key is to keep a calender, and to "Figure-Out" when the End-Test is. You might "Fail One" before being able to Figure It Out. The key to this is to "Watch Everyone's Behavior", to watch the "Stress Events" as they Build-Up, and then "Fade Away" after occurring, and then the next Stress-Test Builds-Up. Find the "Rhythm" figure out from watching this calender and the "Signs and Behavior of Others", when the End-Date is. Also you may want to keep a Journal, without writing "Thoughts and Feelings"—do not be stupid and do this, merely "Write down a Record" of times, places, people, who are involved in this Mobbing, and whom you Interact or have problems with. Keep a record of all of the "Weirdness" that is going on in your life—a record and file: Facts. Also watch for "Themes and Patterns" as there may be "Hints of things to Come" if you are "Paying Attention". They may even hint at the "Last Day" what have you—through "Innuendos". I personally, mark the "Heavier Days" on a calender, noting what happened on that particular day, and even write down "Notes from Smaller-Days" to "Chart it All". I hope this helps others in this Localized-Fascism, this Mobbing.

COINTELPRO and "Mobbing"

COINTELPRO stands for Counter Intelligence Program. It is a Program designed to keep people, Citizens, and Groups, from acquiring "Power and Influence", and to keep people from "Speaking Out" against a Corrupt-System. It is a Line of Defense for Corrupt-People within Government: it is Designed to keep "Future Leaders" who are for Social and Government Reform from ever "Reaching a Point of Influence"—which would take away Power from Corrupt-Politicians, the Elite. In this way Corrupt People within Government are "Not Challenged or Threatened". In this way the chances of there being a Martin Luther King in the future, as one example, are about Zero. This is why this "program itself" is so corrupt, and the reason why most Agents who engage in it are so corrupt. COINTELPRO is a Government Program which is still "Active and Running", and which is Illegal and Non-American. It is for Corrupt-Lowlifes within Agencies who are "Protecting Other Corrupt Lowlifes" higher up the "Food Chain" in Government. It is a Program thought up by Corrupt-Minds for Corrupt Purposes. COINTELPRO can be "Summed Up" in one word: assassination. It is about Character Assassination, and "Physical Assassination" as well. Thus this one word totally sums it up. COINTELPRO when it was "Exposed in the Seventies" was known for both.

Agents took part in "Raids on Targets", within their own homes, and shot and killed many people Targeted by the Agents, in this manner. That is because this was the "Intent"—these were in fact, literally Assassinations, there was no other goal, there was "No Purpose" of actually "Catching People". We are in fact talking about "Government gangs", who act as any gang would and does, only they have the technology, means, and training of Government: and "protection" of Government because they are doing the Governments Dirty-Work. We are talking about "Government Paid Criminal Gangs" who "Target Citizens" on behalf of Government, for control. These "Gangs of Agents" a part of this COINTELPRO, doing what they do, are of course Absolutely Criminal and Absolute Lowlifes. If you need an "Example" the Agents who I dealt with in America, and The Clown, are perfect examples.

"Mobbing" is the "Next Level" of this COINTELPRO. It is where it is "Moved out of the Shadows" and "Put out in the Open" where all can see it, including The Target. This is to Indoctrinate the Public, to make them Compliant to Government, and to do the same with the Target, and/ or to Destroy the Target. All through illegal means. This "Mobbing" is Localized-Fascism, and is the "Ground work and Foundation" for New

World Order—when Government takes all of the "Power from The People" and gives it to itself. This is all preparation for "We the Government".

The "One Percent Tactic"

This involves single isolated events and putting them under a microscope, making them seem bigger than they are, for the Public—for Manipulating the Public and people. It involves taking a small piece of a picture (like a puzzle piece) and making an ENTIRE picture from it. Politicians running for re-election use this all the time in trying to defeat an opponent. Basically mistakes, what I call "growing mistakes" are made in ones life, things that people may do in life that they may not necessarily be proud of, but did regardless due to their youth (or life) and the learning that comes with it, are then put into a "Spot-Light" and harped upon. This is why I call this the "One Percent Tactic". In reality we are talking about less than one percent, we are talking about isolated events and then blowing them up to a much larger size, so that they become a major issue and for a while THE WHOLE PICTURE, one hundred percent.

The One Percent Tactic, blowing-up and "Magnifying" isolated portions of my life, and a few exceptional times, into a bigger picture makes this seem more than it is. The Feds could then run around behind my back, or in the future saying "I did this". It is a way of taking Isolated-Events from one's past and blowing them out of proportion, Lying and Manipulating, to create dirt on a particular person. It is a way of Manipulating others in regards to a persons reputation, it is a way of creating a "False Profile". And it is Character Assassination. In the end by the time these people are finished taking a few isolated-situations and magnifying them and multiplying them, everyone is in fear of you, or "Gossip-Mongering" over these "Magnified-Points", these Manipulated and Distorted "Facts"—I put this last word in quotes as the "Facts" are no longer "Facts" at this point. Note here the "key ingredient" for this Tactic, is a Lack of Context, not giving the whole Truth, while at the same time Lying—all of which is Used for Manipulating people's PERCEPTIONS of you.

Politicians make this a matter of habit when dealing with opponents during Re-Election, which I find very ironic, because with only a few exceptions they are all Corrupt and Filthy, and how Realistic is it to expect that Politicians would not make mistakes during the course of their Life anyhow. Everyone does, that is the point of Life—of growing and evolving, Life's "Learning Curves". To think that people, Politicians—or otherwise, have not made mistakes in their lives, is ridiculous. It is false and unrealistic, and a Manipulation by Politicians, a Manipulation of the Masses, and shows a "Sheep Mentality" as everyone "harps upon these" while denying their own "Growing Mistakes".

If you live in fear in regards to your life, do not experiment and "take chances with it", then you do not "fail, learn, and grow" as a result. So how realistic is any of this?

People are being lead through the nose with this Tactic, and end up electing "The Better Liar" as a result. Good people running for Election, get "Damaged" by this, as singular acts are "Magnified", and the "Masses" without putting things into Context engage in brainless-gossip as opposed to thinking of Context. Thus their Ratings go down, and the Nastier-Politician's Ratings go up. The Masses have got to look at Actions as well—including the opponent who is Deliberately-Manipulating "Past-Events" of the other Politician and "Magnifying them for Manipulation" of the Public, and the Public's-Perceptions. It is in this way that many times the "Greater of two Evils" is Elected, as for the most part these days, it is all about electing the Lesser of two Evils anyhow. Having said that the Public many times loses it's Head and Perceptions, and allows itself to be Manipulated by this Tactic. If a Politician is trying to Manipulate your Perceptions in this way, then how good a person are they—this is a "Low-Blow" and "Dirty-People" resort to this. The best idea is to not elect anyone who resorts to this, as they are trying to Manipulate You before they have even been Elected.

Basically a persons life gets tarnished and possibly ruined over what basically amounts to nothing, during the course of a life time. This Tactic is used to Manipulate and to get people to loose their own sense of Perspective, and therefore their own heads. WHEN YOUR HEAD IS NOT YOUR OWN IT BELONGS TO SOMEONE ELSE. And in this case it would belong to someone who is indeed not a Good-Person. Who else would engage in this Type of Behavior? It should be easy to see.

AN ARTIFICIAL-PROFILE

An ARTIFICIAL-PROFILE: An Artificial-Profile is very much like a Nasty-Divorce. Building and Constructing an Artificial-Profile, involves MANY THINGS AND PROCESSES though. It is what was Used on Me, by the Feds in America, and here in Europe. This involves Cover-Stories, Layers, Taking things Out of Context, and MORE. I speak of this in my Book, so I give a brief explanation here.

When people get a Divorce, and they "Hate one Another" and go to court, they do this. Lets say there is a house, money, possibly kids or pets at stake here, lets say all. This is what happens, they fight for "What they Want" which is all of it, and they engage in Building and Constructing an Artificial-Profile of one another. They "Decide on an Image" that they want to "Create" in regards to the "Other Person" and begin going through all of their "Material" that they have and own, to "Create this Image". This Image is "Inaccurate and False", it is in fact Character-Assassination, to DISCREDIT the "Other Person" so that they can get "What They Want".

They are in fact CREATING AN ARTIFICIAL-PROFILE. So what do they do? They begin going through "Videos, Pictures, Tapes, Photographs, Letters, Writing, Books, you name it", and begin SELECTING WHAT "FITS" this "Inaccurate Picture", this Artificial-Profile. They take "What will Work" and what all "FITS TOGETHER". They "Put it All Together", and ATTACH ALL OF IT with ONE THEME, or perhaps with a couple of Themes in Mind. They Omit Context for all of these different "Little Snap-Shots" and "Pieces", and instead "Tie them all Together", through "False-Context", connect them all together through False-Themes, Lies. They take all of these "Separate Pieces" accumulated over time, and String them Together, AS IF they are "ALL RELATED", and all one thing. And they say this in Court. Which of course "Looks Accurate" and "Convincing", but is a Lie, Deceit, and False. It is in fact a Manipulation, Character Assassination, Lie, and AN ARTIFICIAL-PROFILE.

So as an example, one is "Portrayed as Violent". So "PIECES" of Videos of the one person being angry are shown. Now there may have been a "Legitimate Reason", like being pushed in a swimming pool, spilled hot coffee, what have you, but that will be Lied about as well, and argued. It will of course BE TAKEN OUT OF CONTEXT TO "FIT" THIS FAKE-PROFILE. The point is "There are Pieces" of Video shown, and a "Story" to go with it. The "Story" is also False, a LIE, but they are hoping the "Snippets of Video" are "Convincing": and people are easily suckered by the "Looks and Sounds"—it is all very "Immediate and Real" to their senses. Now to top it off, other things are "Brought In" as well. Perhaps Photos, showing an angry or grouchy face, what have you. And recordings, like messages left on an answering machine what have you. ONLY PIECES "WHICH FIT THIS LOOK" ARE USED, and depending on how long they have been together, this can be VERY EASY TO DO. They could EASILY do the same thing showing a "Happy Person"—simply based on the AMOUNT OF MATERIAL.

Well this is very much what Government Agents do. Except they do very many other things as well, and have VERY MANY MORE ADVANTAGES. I get to this in my Book, but I will add a bit here. With Agents, they are first of all FOLLOWING AND MONITORING YOU 24/7 SO THEY "ACCUMULATE MATERIAL TO MISUSE" VERY QUICKLY.

The next thing is, like a "Nasty-Divorce" they take these things OUT OF CONTEXT. Taking things Out of Context is very important for understanding what they do.

The next thing is they "Arrange Events". These "Events" in my case are called "Programmed-Events", as things are "TAKEN OUT OF CONTEXT" before anything has even happened. These "Programmed-Events" are "Used" to "Prove" something that is NOT REAL. They are Used to Build and Construct an Artificial-Profile.

The "Actions of the Target" are Deliberately Misconstrued, to "Fit" this Artificial-Profile.

These "Programmed-Events" are REPEATED OFTEN, for "Much Material", "Much Lying", and is done for "LAYERING". In the case of the Public, the more this is "Done" the "MORE CONVINCED" they are.

All gets "Added" to an "Artificial-Profile". It is all to DELIBERATELY PORTRAY SOMEONE IN A "FALSE AND INACCURATE" LIGHT. It is all Character Assassination, and it is all CONTINUEOUS. This "Maintains Appearances". The rest of this is in my Book, but I think you get the idea.

Using a "Cover-Story"

An example from another Book I am almost finished with, involving a "Cover-Story" and Exotic-Techno. I along with Family-Members, have had this Techno used on us, constantly in a Negative Manner. As a recent example, yesterday, my girlfriend came back from the hospital (September 28, 2010), and was there in the first place, due to these Spooks, and their Exotic-Techno. Nothing beats the "Gift of Giving" unless you are in my, and my Families shoes. So what happened?

A couple of weeks ago her jaw, her lymph nodes were slightly swollen—by way of Exotic-Techno. Eventually it went away. This was the "First Cover Story", and a building up of Layers, of more to come (This "Layers" is a Tactic). Shortly after, perhaps a week, she began having "Heart Palpitations", "Heart Problems", and Discomfort, which in turn she began to worry about. Now the "Swollen Lymph Nodes" were Techno related, and the following "Heart Palpitations" as well. But from a MEDICAL VIEWPOINT it looked as if "The Virus" had gone from her lymph nodes to her heart, and infected it. To top it off, no doubt when my girlfriend began wondering at all of this, they were sending her "Scary Thoughts" ("Disastrophies" as I call this), and every time she "Bought into One" they raised the level of fear (Subtle Emotion-Pushing) she felt, by one notch. It turns out she was fine, but the point is it was all done with Exotic-Techno. They CAN AND DO MIMIC PHYSICAL AILMENTS, and use "Cover Stories", to sell this. The "Cover Story" is simply the "Logical Cause or Pretense", that which "Sells You" the rest of the Exotic-Techno Con-Game. Of course this applies to "Normal Life and Manipulations" as well; they do this without Exotic-Techno as well. They killed a woman's cat, that Whorbat was Pissed-Off at, using a "Cover-Story"—Innuendos were given for days before this "Event". Whorbat was Pissed-Off at this woman as she did not do everything for Whorbat "Correctly", she in fact did not "Let her dog bite my Kid", which is what Whorbat wanted: Apparently they had an "Agreement". A "Cover-Story" of the city putting out poison for the mice, and her cat "Eating a Mouse" was Used when in Reality these Spooks killed her cat. A Cover-Story is THE "LOGICAL PRETENSE AND CAUSE". It is in fact a Deceit, what Spooks "Hide Behind" to get away with doing things. A Cover-Story is about "APPEARANCES" to "HIDE REALITY".

The Layers Tactic

This is Standard Operational Procedure for this "Mobbing", Government Agents, and COINTELPRO. "Layers" is a Type of Tactic. "Layers" is a way of "Making something Look Credible" when it is not. This is used for Lying, Manipulation, Stories, and for Creating an Artificial Profile. One Lie gets built on top of another, in Layers, so that if someone decides to "Dig Deeper" they encounter yet another Lie, and if they are not "Paying Attention" and "Stop There", they STILL end up BELIEVING A LIE. It is way of HIDING THE TRUTH.

The same thing with Stories. A False-Event, a Story, gets built upon another which is also False. Very much like "Gossip". The "Gossipers" attach "False-Stories", Stories taken Out of Context, to "Other False-Stories", and these get "Layered" as well. It is a way of "Building an Inaccurate Picture" of someone—which is what "Gossipers" do. They engage in "Character Assassination", could care less about who they "Talk About", they are in it only for the "Cheap Rush" they get from it, and so they attach "Stories" TAKEN OUT OF CONTEXT to one another, and "BUILD AN INACCURATE PICTURE" of someone. It is done with "LAYERS". Someone "Might Disregard One Story" only to believe in another that is more "To their Taste". It is a way of HIDING THE TRUTH. Only Assholes and Government Paid Criminals do this.

Telling People "To Not Speak with the Target": Misusing "Facts", Hiding what is Being Said, Character Assassination

This is USED CONSTANTLY with Me. It is how these Corrupt-Jerks get away with so much. This is Standard Operational Procedure, and is OBVIOUS, yet people do not seem able to "Figure it Out". People are Told to "Not Speak With You" about "Things", to not even acknowledge this Mobbing, to keep their mouths shut. This is used every day as "Things are said Behind Your Back" to "Influence People" and to "Control Them", Lies, Gossip, Manipulations, are Hidden so that you can not "Dispel These". You can Not Expose People's Lies, if you do not know what they are. This was used repeatedly by Corrupt-Agents, for getting away with everything. THIS IS HOW COINTELPRO OPERATES. These Corrupt-Pigs got away with the Biggest Lies and Manipulations in this way.

Everyone got Used, Misused, and Manipulated and it all started with HIDING "FACTS" (Hiding the Truth and Misusing "Evidence"), HIDING WHAT WAS BEING SAID, so that I could not even begin to prove or point out the Truth otherwise. You can not expose People's Lies if you do not know what is being said in the first place. You can not "Fight" what you can not "See or Hear". So everyone from the Public to the Target gets Used, Misused, and Abused. Obviously "Hiding-Facts" from the Public, and "Hiding What Is Being Said About The Target", from the Target, is a Tactic. And I experience this CONSTANTLY, anyone in my "Mobbing" does. It is Used for Manipulating People, the Idiot-Citizens, the Public, and anyone you are "Close To". It is used for Programming People. This is Standard and a Given—it is Standard Operating Procedure for these Spooks and Government Agents.

TAKING THINGS "OUT OF CONTEXT"

This is exactly what it sounds like. THIS IS USED CONSTANTLY and is also Standard Operational Procedure for Corrupt-Agents and used on people in this Mobbing. This is Used with "Tapes", Staged-Events, Videos, Pictures, Books you read, Movies you like, you name it. ANYTHING that they CAN TAKE OUT OF CONTEXT, to "Create an Artificial-Image of You", THEY WILL (An Artificial-Profile). This is used for Programming People, Complete Idiots. This is what these Sick-Agents/Punks call "Work".

PLAYING BOTH SIDES OF THE FENCE

This is to "Confuse You" among other things, as when "You Look" you can see things "Either Way". This also gives "The Profiler" an idea of "Where Your Head is At".

Whorbat The Clown, "Bets on both horses so that no matter what she Wins". She will take ANY SUBJECT and then "Create a Look", Appearances, on "Both Sides". Things are "Played both Ways". She gets away with Incredibly Stupid and Obvious Manipulations this way. As an example, I was having "Sleep Problems" because of Exotic-Techno, NOT "Stress". In regards to this "Mobbing" I have "Elephant Skin", by the "Rules of this Mobbing" I can "Pass Easily", and Whorbat knows this. So I was "Sleep-Deprived Heavily" all of the time, and The Clown was "Pushing Me", FORCING ME; to go to the Doctor. My attitude was "Get Stuffed, why should I go to the Doctor for a "Sleep Problem" that is not mine?". Well as is Typical of The Clown, this Pissed-Her-Off. So she "Made Me Go" anyhow. So I went and the Doctor gave me sleep pills, of course. Well, Whorbat "Content" "Allowed these to Work". Of course this is all a Pile of Deceit, she merely need not keep me awake with Techno in the first place. And of course she was going for a "Look and Appearances". Well later, This Clown "Held-Me-In" on an End-Test saying I went "When I did not need to", that I was "Trying to Manipulate the Doctor for Pills", which was an absolute Lie, Manipulation, and Farce.

Once I had these pills, when I took one, these Spooks would not use this Techno on me to keep me awake, they would let the pill work. Naturally I felt "Relieved", it felt good to get sleep again, and these "Pills Relaxed Me" before falling asleep of course: which I enjoyed, why the hell wouldn't I? I did not want to go to the doctor in the first place, but once I did, I most certainly was not going to "Beat myself up about It". I made the most of it, which is Intelligent. So eventually, The Clown had "Tapes" of me "Talking about Enjoying One", SHE USED THIS AGAINST ME, and no doubt in regards to my Doctor. To "Convince Him" I was "Manipulating Him" and in REALITY these Lowlifes did not give CONTEXT. THIS IS HOW IT WORKS. What was I supposed to do, "Talk about Exotic-Techno"? The Doctor knows about these Spooks obviously they interact with him, but they got away with, and do get away with Bullshit like this.

SO THE CLOWN "PLAYED BOTH SIDES" AND SHE USUALLY DOES. This way, down the road, if need be, SHE CAN "PORTRAY ME EITHER WAY"—she has "BOTH OPTIONS".

THE MISUSE OF "FAKE-TAPES"

This is pretty much as it sounds, these are "Tapes without Context". In my case this is Used Often, and usually to keep people Programmed: there is nothing more "Convincing" then a "Tape" as it "Sounds Alive and Immediate" yet it does NOT give a good idea of "Context" in spite of what people may think. With The Clown, this has been a "Favorite" as this Idiot has Exotic-Techno which She Misuses constantly for "Making Fake-Tapes", what I call "Techno-Tapes". The examples I give below also give some good information about this Exotic-Technology and it's Exotic-Misuses.

Keep in mind that they are "Watching 24/7" and it is easy for them to "Acquire Tapes" and "Material" of anyone through "Normal Means": having said that a "Tape" does not always equal Truth. IT DEPENDS ON CONTEXT. And when Agents are Engaged in only Creating a Look, Appearances, they are not going to give Context—that is rather THE POINT. And in Whorbat's case, at this point, and for many years now, she has been RELYING HEAVILY ON EXOTIC-TECHNO, to do this. I have had many recent "Techno Related Events", Remote-Controlled Behavior, and I will explain a few of these here quick.

In these examples, as you will see, what you "HEAR ON TAPE" is NOT necessarily so. That IS THE POINT. This Exotic-Techno is "Undetectable" especially by way of "Tape"—we are talking about "Remote-Controlled-Behavior" here, and this is one reason I believe Whorbat will be in Trouble. You can no more "Avoid the Effects" of this Techno than you can a "Bullet from a Gun", and in this way The Clown has been Misusing this Techno for "Holding Me In". There is "No Passing" this way, which is NOT an Agency Standard, it is the Standard of someone who wants to "Keep Me In This Mobbing", so that I can not Expose Them—so The Clown has an Excuse for Controlling the Information and a reason for Interacting and Manipulating People, to Cover-Up the Truth.

So examples of Exotic-Techno used for Fake-Tapes.

On Christmas Day The Clown "Remote Controlled" a "Traumatic Fight" between me and my Girlfriend at her parents, so there would be "Fake-Witnesses". I did not say it was a "Remote Control Event" at the time to my girlfriend, because I thought it was the Last Day, and IT WAS. I GOT HELD IN FOR BEHAVIOR NOT MINE, AND WE ALL SUFFERED. One would have to be stupid to think "I am tight on Christmas Day", and unfortunately Whorbat had things both ways: If I said "Hey I am being Machined-Anger not mine", Whorbat would have "HELD ME IN", and because of "Fake-Witnesses" and "Tapes", and as it is The Clown "Held Me In" saying I "Over-Reacted" for Behavior NOT EVEN MY OWN. Either way there was no "Passing", and no doubt a "Tape" of this "Fight" was made for "Misusing on Others". If you do not have the Real-Context, it is simple to say it "Was Me", simple to get away with this as I did not "Speak of What was Really Going On" during the time this "Tape" was made. Because of course "I would be Held In".

Now the fight was "About no Sleep", and having said that it really WAS NOT: it was about Remote-Controlled Behavior. If you "Hear a Tape" you do not get the CONTEXT OR REALITY OF THE SITUATION: it "Looks and Sounds Real" it is not. Later I explained "What happened to my Girlfriend", only after I realized I had been "Held-In" on this day—I waited a couple of days to be sure. Having said that, there are "Many Days of Tapes" in which I had to "Play Along and Keep my Mouth Shut" because I could NOT SPEAK THE TRUTH OF THE MATTER. Whorbat has gotten away with many of these "Tapes" since September, her way of BURRYING ME, after trying to expose her in September which obviously Pissed-Her-Off. Typical.

So in January, Whorbat Misused Exotic-Techno to "Hold Me In", and for Fake-Tapes. She used here "Scripted-Dialogue". "Scripted-Dialogue" is when YOUR OWN WORDS GET REPLACED WITH WORDS THEY CHOOSE "AND THEN THEY LITERALLY GET YOU TO SAY". I'll explain. They turn down YOUR "WILL-POWER" WHICH IS ONE "PIECE OF IT". Then they REPLACE YOUR WILL-POWER, SUBSTITUTE IT, WITH A "WILL-POWER-PUSH" OF THEIR OWN. In this case Your Will-Power GOES DOWN, and then they make a "Push Of" THEIR OWN-WILL-POWER, which gets TURNED UP. They then make a PUSH FOR SPEECH. And in this case a "Script/ and Dialogue" is PREPARED. To do this they have to TURN YOUR WILL POWER DOWN. So that the "Will to speak Your Own Words" gets "Over-Ridden" by "Their Choice of Words". They also do this with your "Mental-Finger-Print", "Your Flavor"—which makes it more difficult to "Resist or Notice".

EACH PERSON has their own unique "Mental-Finger-Print" and when "YOUR BEHAVIOR, ACTIONS, WORDS" are "Sent to You" with your "Own Finger-Print" it is "Received as your Own". This is a way of "SNEAKING IT PAST YOU" so that you think it is "Yours". Having said that there are ways of "Picking up on This" and a "Major Tell" is the "Words and Actions" themselves. For instance of "Peculiar Behavior", I have "Forgotten the Last Day" many times: this is like a man in Prison "Forgetting his Parole Date". It is not possible. In regards to "Scripted-Dialogue" many times the "Subjects Themselves" come out of "Left-Field"—keep in mind that someone is "INVADING YOUR BRAIN" and DOING WHAT THEY WANT, which is USUALLY NOT WHAT YOU WANT OR ARE ABOUT. This is FORCED IN MANY A WAY SHAPE AND FORM.

So in January, this Clown, "Scripted-Dialogue" in which I spoke of this "Mobbing" on the last day, for a "Quick Moment" (Impulsed), and for this, I was "Held-In". Of course this was Their Choice, not my own, as I knew it was a Last-Day. Again a "Tape" was made, so that if I Spoke-Out to "Others", the "Tape" would be "Proof"; and of course because the Real-Context of Remote-Controlled-Behavior and Scripted-Speech, was "Hidden and Omitted". This Clown in this way, keeps getting away with saying "I Failed a Test", when I did not: and has Fake-Tapes, "Tapes without Context", to "Prove I Failed". Talk about Bullshit.

With this Techno it can be DIFFICULT TO CATCH MANY TIMES if you are "Distracted", having said that you can always catch it on a rewind by "Looking at your Words" or the sense of it all. If you are "Consciously Paying Attention" it is much easier. So in January this was done, and I was Held-In. And it "Got Pushed Out of my Mouth", IMPULSED OUT, as I was Busy and Distracted, and I knew right away what had occurred. There are MANY COMBINATIONS OF EXOTIC-TECHNO TACTICS THEY USE IN CONJUNCTION WITH THIS.

In the end, it is all Remote-Control. Every "Action, Thought, Thing" has an "Identity Marker" and they simply "Enter this into their Techno" to get that Desired Action or Event—it is as easy as Pushing a Button. Many times you are "Left Wondering" at what you have said or done, your head reeling, but it has been done to me so much at this point, I recognize it quite well.

For my readers keep in mind that this Mobbing and "Passing It" is about "Passing Stress Tests". However in my case none of my "Failing Tests" is about "STRESS": it is all about TECHNOLOGY AT THIS POINT.

So lets get to February 2011. THIS WAS NASTY. This Lowlife Whorbat wanted that I "Speak to my Kid" about all of this Spook-Stuff, WHICH IS ABSURD. She knows that I AM RESPONSIBLE about this and that "I Keep my Problems away from My Kid". I told this Idiot to "Stick It in regards to This" as they were "Pushing for This" with Techno. So they "Machined Me Angry", and naturally I "Said many Things"—which is Totally-Sick on Whorbat's End. Having said that I was able to "Cut it Short" but not after "Saying Things". I in fact got "Held-In" on this day, and again Whorbat had it both ways: as I did eventually say to my Kid, that I was "Not Myself" which stopped the "Techno-Attack". It sucked for both of us, and again "No Chance of Passing". This was ALL DONE WITH TECHNOLOGY, it never would have happened otherwise. This Whorbat is sick. And THEY FIRST MADE AN "EMOTION-PUSH" OF ANGER, THEIRS I MIGHT POINT OUT, AND THEN "SCRIPTED-DIALOGUE". When it is done this way it is almost impossible to "Catch the Words" and "Stop Them". So I got Held-In. And if you realized how much I have done to "PROTECT MY KID FROM KNOWING ABOUT ALL OF THIS", you would get this. Whorbat is Sick, and wanted it done: and again SHE USED FORCE LIKE THE RAPIST SHE IS.

Naturally a "Tape" was made, and Misused on people I know, to convince them I am "Stressed-Out" about this Mobbing—a complete Lie, Deceit, and Manipulation, but Idiots Involved in this Mobbing would not know the difference, and DON'T. I'll let Her Agency "Sort This Out".

The "Temptation Tactic", and the "Sleep Tactic"

Now the next Tactic is the "TEMPTATION TACTIC". Informants working for the police use this all of the time. A very basic rule of thumb for informants, you can count on it. First of all when you are dating an informant (local police or otherwise) or become "friends" with one, they engage in a very typical behavior of trying to get you to hang out with Their Friends, these friends are of course other informers, and part of one "click". Basically, if you look at behavior and motives you are talking about, a gang. The irony is that informers think they are better and different on this point, this is very debatable (or if you are me, not debatable at all), and as a general rule even local police do not like informants: POLICE do not sell out their own. Informants do. A little wake up call perhaps for a few informants out there who think they are part of the privileged few; if a cop had to choose between an informant going down or a fellow cop he would pick the informant every time. Informants are looked at as a source of information, but as total Sell-Outs as well, and in the end expendable. Cops almost never sell one another out, informants do all of the time, it is a matter of habit for them. Cops look down on informants.

In their own eyes they are special and important. Everybody thinks they are "Special" due to a pat on the back and a bit of "Compartmentalized Knowledge". This is Idiot Mentality. To quote Mr. Lee: "The Thinking Stops where the Ego starts". Each level gets Manipulated by the level over it. Each level gets used.

Once an informant gets you to start hanging out with their friends, they start playing you for a fool. In ordinary life friends speak to one another, but with informants they are sharing everything you say or do with each of them, with the Collective so they can better Manipulate you. They pool their information about you. Think you have a "good relationship or friendship" with someone new you meet through this click? Forget it. All of the information is being shared and pooled, including what you like and don't like about other members of the group, with those members, so they can each better Manipulate You. In short everyone knows what you think about them, so they can Manipulate You and push your buttons. Having said that, look at the "Quality of Friendship" you are getting, these people are Lowlifes.

Now having said all of that, the "Temptation Tactic", "When in Rome Tactic". If you are not already targeted by the police, you may well be at some point due to these informants. Regardless, given the ignorance and cheapness of informants, they will use this nonsense on you in regards to sharing information, even if you are legal and dating "one of their own". Betrayal is like breathing for these people. So when in my "Situation" (or if they do not like you anyhow—misery loves company) these people as individuals and as a group will encourage you to Break Laws with them. Seeing as how you are "THE TARGET" these informants are "protected", you are the only one who has anything to lose. When under "Investigation" you will be encouraged to Break Laws in "designated areas", where surveillance is active, as in video and microphone surveillance. Also the more witnesses available the better, especially when they are all informants—and especially when they are all Breaking the Laws.

Now this is before, shortly before, my "Situation" began, and when it did "begin", it began at this Job.

As an example from my own life, at a restaurant where I worked, at the end of the night quite a few people got high at the back of the restaurant. The boss, also a dope smoker, looked the other way. I was encouraged all of the way by this group of "restaurant misfits" (try informants) almost daily. Having said that, I did not like to mix business with pleasure, and for an extremely long time did not engage in such behavior. In the meantime a "moving truck" had parked in a parking lot across the highway about 60 yards away. It remained there indefinitely, "unused". Again it comes down to "When in Rome" and eventually I broke down on this point myself (around the time the Feds started having my boss smoke in front of me out back of the restaurant after closing hours) on and off, all of it documented by the Feds across the street in the moving truck.

Now as usual when something does not work Tactic-Wise, the Feds will try DIFFERENT APPROACHES. Thus when I often time bring up a Tactic it leads to others. In this case mentioned above, the "Sleep Tactic" or "Attrition Tactic" was also used. They simply wait for you to FORGET about the dangers of your situation (I knew I was being "Investigated"), through time and persistence, as they wait for you to become "Desensitized" to this. The Feds knew I knew I was being "Investigated"—actually they were waiting and hoping to get something to Frame Me with which is why I put this in quotes. Now this was at the restaurant prior to my "Mobbing Situation" starting, it actually began at this restaurant and NOT due to smoking somebody else's marijuana with The People bringing it to the restaurant—and informants at that. At any rate I was the TARGET thus everyone else felt they had nothing to worry about, and were ENCOURAGED by the Feds to Break the Laws in front of me AND TO ENCOURAGE ME TO DO THE SAME. THEY were also encouraged to Break the Laws in front of the Feds, by the Feds. So they were using "Temptation", the idea of "When in Rome" to get me to Break the Laws, which I eventually did start doing with them, and once my boss started doing this.

The other Tactic used was the "Sleep Tactic", waiting for me to "fall asleep to the dangers" of what I knew.

Now having said that, these Agents knew I KNEW I was being Investigated, and they knew this because I had told other people outside of work (one was a Deep-Cover-Cop-Kitch), people I did not work with—"friends" in my Life who worked for the Feds. This is obvious, as they Changed Tactics and Techniques at the restaurant. Specifically as one example, they had put an expensive beer at my work station, these Feds hoping that I would take it, as it was out of place and did not belong there in the first place—thus it would not be missed. I wouldn't have taken it anyhow, so it is a moot point in one regard. I mentioned to Kitch that I was being "Investigated and Profiled" at work. It was here that they began engaging in the Tactics I mention above, as they "Switched Tactics", and this is turn was a "Sign" that Kitch was involved in some way as well, once they did. Having said that, I mention this now, for good reason, as this is relevant to the next Tactic I mention, and proves a point later.

Now there are a few major points of irony here. First of all EVERYONE IS BEING DOCUMENTED by the Feds in this Situation, and these circumstances in particular at the restaurant. Including a Fed smoking dope as well. These people at the restaurant encouraged by the Feds, WERE STILL BREAKING THE LAW. If you know anything about the Feds, when doing an investigation of this type, they INVESTIGATE AND PROFILE EVERYONE THEY HAVE CONTACT WITH, who has contact with the "Investigation". They, from THEIR point of view, would be stupid not to. They want to know where any possible information leaks are for "Damage Control" and more importantly want to know people inside-and-out for being able to MANIPULATE THEM AND THEIR PERCEPTIONS. Thus by Breaking the Laws for these Feds you put yourself in jeopardy, and give them control, legal control over your life. YOU GIVE THEM A LEGAL PRETENSE FOR INVESTIGATING YOU. What are you going to say if you get to a court room because they tell you to "jump" and you tell them to "stick it"? What will you say, that you broke the law because the Feds told you to do so? The Feds will deny it, and say you Broke the Law and we can prove it.

Now in my case I was known for smoking marijuana, and known for NOT smoking all of the time (I liked to bike 30 miles a day as an example), and known for being "Perceptive by law Enforcement", for being "Paranoid" by others (friends), and for having a "Big Mouth". Which is ironic, that being the case why am I in this "Situation", and why did I not go to the Newspapers to begin with? Obviously I should have. Having said that The Clown is saying that I will "pay", once I publish this book, that it will make waves and piss people off—having said that, everyone can "Stick It". People like The Clown are Responsible in the first place. There is no excuse for Corrupt-Assholes and an Ignorant-Public. Quite frankly The Clown should pay as well, and not just by me. You will note I was being "Diplomatic" back then, and you will note what it got me.

The "We'll Make it Work for Us Tactic

(Readers will have to excuse the Extra-"Personal-History")

You can and will end up in this "Mobbing" if you get involved with someone who is in it, are dealing with Corrupt-Agents, and they do not like you, and or can "Prove you are Breaking the Law"—I put this last part in quotes because this can simply be a matter of "Making it Look that Way", "Context". It is best to not have contact with these people, these Agents, and this "Mobbing" even if you are a Law-Abiding-Citizen. In fact because of me and Idiot-Tim (a Fed—and a marijuana smoker as a point) at the restaurant, one of The People, at the restaurant, "Bryan", ended up in my Situation of "Mobbing" before I even did. When he did, I knew something strange was going on, just not "what". As it is people kept trying to get me involved in "Gossip about Him" which I did not do in spite of the fact that everyone else was. I like to think this says something about my character, especially since I was the only one not to do so.

These Feds, and specifically Idiot-Tim, did not like him and said that he "Interfered with an Investigation" by "Tipping me Off" which he did not do—I already knew I was being "Investigated" and at work, and these Feds knew this ALREADY because I had told people outside of work that I knew, who worked for the Feds: SEVERAL PEOPLE. In fact they had "Changed Tactics" because of this. So they knew the Truth, they simply Lied about this, Hid Information, and took things Out-of-Context to get this guy Bryan into this "Mobbing"—which they did. They Manipulated Everyone on this point to get people to think he had "Tipped Me Off" in regards to this "Investigation", when he had not. Which is "Why" the Feds "Hid This" from me. These Feds were actually CRIMINALS, who got Him and Myself, into this "Mobbing" by Lying and Manipulating the Public, simply by HIDING CONTEXT, BY HIDING THE TRUTH. And they Manipulated a lot of "Good People" by doing this. These Agents were, and no doubt still are, Filthy.

And as an added point it WAS THE FEDS and THEIR SLOPPINESS THAT TIPPED ME OFF. They simply "Dumped this slight Detail" and Responsibility on Bryan. PERSONAL NOT PROFESSIONAL, part of the Profile and M.O. for Corrupt-Agents. No doubt it "Looks Better" in their Reports and Files this way.

In my speaking to "Idiot-Tim" about Bryan, I implied, and perhaps for reasons of my own even said that he had "Tipped Me Off" (at this point I am not going to "Dig Deep" in my memory—it is irrelevant in one sense—the Feds knew I knew anyhow), which of course he had not, and of course which the Feds KNEW, and no doubt Idiot-Tim as well, seeing as how he was one. As usual I was trying to get information from someone (Profiling Idiot-Tim) and these Lying-Feds simply used what I said to "Get what they Wanted".

Having said that, REGARDLESS, I can say whatever in the hell I want, especially when dealing with Lowlife Pieces of Shit, for reasons of my own—which I clearly had, and IT IS THE FED'S GOD DAMN JOB TO BE ACCURATE. All of this seems to have been lost in regards to everyone, including my girlfriends Eve and Penny, never mind "Good Friends" and "Family". If you were not there when these TAPES and or "Other-'Evidence'" was "made", if you were not given Context, if you did not know the Bigger-Picture, and what was going on at the time, THAT IS NOT MY FAULT. Neither is the fact that you did not look at the "Quality of The People" in charge. That was the RESPONSIBILITY OF EVERYONE WHO ALLOWED THEMSELVES TO BE USED BY THESE CORRUPT-AGENTS. These Agents simply made things "Work for Them". They did this with Bryan first, putting him in this "Mobbing" because they clearly did not like him, Real Professional I might point out, and then Me. It is called being Corrupt and Filthy.

They liked to use a Tactic I was not familiar with, and which CORRUPT AGENTS use, the "WE'LL MAKE IT WORK FOR US TACTIC". Thus REGARDLESS, they took the few sentences between me and Idiot-Tim, knowing the Truth already, that I knew I was under "Investigation" and that I knew because these Feds had been Sloppy, and dumped this on Bryan. Having said that, at the time there were many obvious things going on in my Life, and I had my own reasons for saying what I said—it is irrelevant, that the Feds knew the Truth, but "IGNORED IT", AND OMITTED IT, is not. I was "Finding things Out" and Profiling, they merely ignored the Truth, and Used what I said against Bryan based on Personal-Reasons, not Professional-Ones.

These Agents were in fact Totally-Corrupt, and the ones Seriously-Violating and Breaking the Laws. You can not Falsify Information, and Frame people because you "Feel Like It". I could have done, and still can, do a better Job, of following their Agencies Rules, and as a result have made no mess for this Agency, never mind others. As it is this is why these Agencies have Protocols and Rules, so that their Agents do not make messes for their Agencies. What they did, what they got Citizens to do, was Criminal. And these Idiot-Citizens were doing things BASED ON LIES in the first place. This puts a lot of things into Context.

It is simple and easy to use this "Tactic" and a child can use it, having said that you have to be very Corrupt to do this, and especially in the way that these people did. They were able to get away with this because people did not ask questions and quite simply did not care. My "Tape with Biff", may have made it EASY to Convince People that I was "Someone to not Like", but NO CONTEXT strangely enough was given, to do this. It is called Lying and Manipulating—and with this they started a whole "Chain and Campaign" of Deceits, Manipulations, and Lies. Against Bryan, and then Me.

So according to these Crooks, this "Mobbing" started for Bryan, for "Violation of Laws", and again because I was being "Investigated" (Remember they were only trying to Create Dirt)—in Reality they did not like Him, and did not like Me, and Framed both of us for "Interfering with an Investigation" which was a Lie in regards to Him and Me. These Agents were PERSONAL in their "DECISION MAKING" not PROFESSIONAL, and simply went sticking people in this "Mobbing" based on who they did not like, NOT based on FACTS. They Lied and Manipulated FACTS AND EVIDENCE to get people in this "Mobbing". Because they were Corrupt. They knew the Truth of the matter in regards to Bryan, and then later Me. They also did this to Kaye, a friend of mine who I loved very much, simply because of her past friendship with me—again not Professional, and no doubt they had "False-Pretenses" a Cover-Story for her as well. They would have had to, because quite frankly she was a very gentle person, with no "Cause for Mobbing". Like there is "Good Reason" for Mobbing.

The point is, even being near these Agents is Dangerous, they will Investigate you and Profile you ILLEGALLY, and if you are dumb enough to Break the Laws you can easily end up like "Bryan". Having said that "Bryan" did not screw up and should never have been "Mobbed" in the first place; Bryan got "Mobbed" because these Agents Lied, provided to the Citizens LIES, and these brilliant Co-Workers got Used and Manipulated. These Agents DID NOT PROVIDE CONTEXT. In regards to "Screwing up an Investigation", and "Breaking the Law in this Regard", he did no such thing, and they knew this and so did I—I just did not know any of this at the time, what was going on with him. If I had, this Lying and Manipulating would have been exposed, and this is why, like with me, they kept THINGS HIDDEN.

These Feds took "Information", "Quotes of things Said" which they knew was NOT TRUE and simply said it was True, in regards to Bryan and Myself. Thus the "We'll Make It Work for Us Tactic". This is Corrupt and Illegal, and based on this they Manipulated EVERYONE. These Agents are willing to look the other way in regards to "faulty information" if it is what they are trying to go for in the first place. Last time I checked, being "DELIBERATELY SLOPPY" is called LYING, MANIPULATING, and FRAMING SOMEONE. Again that "being the case", if they should try to "weasel out" they are Responsible for being Accurate. Having said that they did everything, all of this, Deliberately.

These Agents were merely Criminals with Badges, and Filthier than anyone could have imagined, including myself. I did not realize how Corrupt these people were, and for years as Innuendos and other things were made, thought that these people were merely "Mistaken on Points and Facts" (never mind that I did not get a lot of Innuendos in the first place). Of course over here in Europe, that all changed with The Clown and Her Buffoons who are very Corrupt, and once I did a "Rewind on my Past",

I began putting the pieces together. In fact what I did not see then, I see now. I was "Forced into figuring things Out" and "Forced into figuring out Who and What I was dealing With" as a result of this Clown. Including the Reckless Endangerment of Citizens, for Psychological-Snaps.

In short I was profiling a guy from work Idiot-Tim, who I am certain was a Fed, and he Deliberately took things the wrong way, Out of Context, and got Bryan Mobbed. When you know what the Truth is, and take something said to the Contrary, it is still called Lying. When you Manipulate People with this, not telling them the Truth, merely "Selling Looks" it is still called Manipulating Facts and "Evidence", and Lying and Manipulating People. Things Taken Out of Context, especially when you have "a hundred statements to one contrary one" is still called Lying. This Idiot-Tim hated Bryan, and in retrospect the Feds obviously did, they "Had it in for Bryan" like they did for Me—and the worst that could be said about the guy is that he was a dope smoker. The worst that could be said about these Feds, is an extremely long list—and words like Framing, Dangerous, Lying, Manipulating and Tampering with Evidence, using Citizens for Reckless Endangerment and Acquiring-Dirt, Filthy and Corrupt, come to mind. If the Citizens Used knew the Truth about these Agents they would be Disgusted and Angry to say the least. It is a good thing it was almost twenty years ago.

Actually the Feds in charge had it in for Him, but since Idiot-Tim was one, he was a part of this. He simply wanted something to "Use Against Bryan" to get Him in this "Mobbing"—which is what he later did with me as well. This Manipulator also ended up getting me "Mobbed", and again by taking something he already knew to be False—the other Feds certainly did, and Using it as Truth anyhow. Thus Lying, Omitting Facts, Manipulating, Framing: NO CONTEXT. Simply because of PERSONAL REASONS AND DISLIKES in regards to the Target, not Professional Behavior, or even Legal. In fact they had already done something similar in regards to my "Trying to Catch, Biff, Kitch, and Tark".

Seeing as how I spoke to both Bryan and Idiot-Tim out behind the Restaurant during slow business many times, and that the Feds were busy "Conducting a Dirt Creating Mission" and watching all of us at work, no doubt there were Tapes of these conversations as well. No doubt these were Misused as well, if they "Needed To" at all in the first place. Having said that, it is the conversations I had outside of work with other people working for the Feds, stating that I already knew that I was being "Investigated", that were "Omitted" to get this guy Bryan in this "Mobbing". And at the point in time that this all occurred, "I was not Asleep" in regards to my being "Investigated". I can "Safely Assess" that these Feds knew this, or they can claim "Incompetence and Ignorance", no doubt whichever one would get them in the least amount of trouble with their own "Bosses". Thus they said "We'll Make it Work for Us".

In regards to me, and my "Mobbing" which started by "Talking to Idiot-Tim" at work.

The funny thing is, I was trying to find out about these "Plumbing Ninjas", and whether it was the police or a gang, so I could "Know and put my mind at Ease" one way or another, based in no small part, because of the Recent Murder in the parking lot. The other thing was, regardless of me knowing about this guy at work, Idiot-Tim, of being in "Law Enforcement", and Because Of It, I wanted to know about this "Plumbing-Ninja Thing", so that I could take a more "Formal but Indirect Approach" in notifying the police if from Idiot-Tim's reaction, he felt that it was a Gang-Hit. I most certainly was not going to "Rely on speaking to Him" to do so—I felt this was my thing to do anyhow. And I had no idea the "Plumbing Ninjas" were related to my so called "Investigation". I would telephone from a payphone, and do it anonymously, which I had done in regards to other things before involving violence (as an example I was passing a house one night and heard angry voices raised and glass breaking). I "Pushed the Issue" and judging from his "Reaction" it was indeed "Police-Related", having said that I also "Blew smoke up his Anatomy" on a few points, to "Confuse the Issue". I said, and I believe I started off with "I think I screwed up a Police Investigation" which was a statement of fact, it indeed was a possibility just not one I had made up my mind about yet—which was the point of speaking to this Idiot-Tim. This among other things "Got Taken For Fact", when these were issues of "Finding things Out" and later "Confusing the Issue". Which was Typical Protocol on my end—and I might add the Feds knew to analyze things CAREFULLY in regards to me. I most certainly DID NOT DELIBERATELY screw up a Police Investigation or a Federal-Investigation, and even to people in this Agency (the higher ups), this is clear—minus Edited Files, for people covering up their own Asses. And in this regard I made it totally clear to begin with—so these people knew anything "To the Contrary" was to be so, simply to "Muddy the Issue"—again "Blowing Smoke".

So again as with Bryan, they said "We'll Make it Work for Us", and went with "Looks and Sounds" as opposed to Reality and Truth. They sure as hell can not claim to be "Sloppy".

Having said that, the Plumbing-Ninjas were all part of a "Set-Up" in regards to me in the FIRST PLACE, as I have clearly pointed out. This was a "Stroke of Luck" on their part in one sense, talking to Idiot-Tim, as it is exactly what they WANTED, and no doubt they knew I may take this approach as like anyone in Law-Enforcement I had a habit of getting information this way, and they knew I had "Issues over Murder". They also knew that I had a "Habit" of speaking to people who worked for the Police, for finding things out. They no doubt knew I would go "Looking Around" after seeing these "Plumbing Ninjas"—especially when it was in regards to my parking lot and neighborhood, due to the Recent Murder of this Kid. All of this on the Feds end I might point out, to "Cover-Up" this Murder and the Racist-911-Dispatcher, and to "Neutralize Me", by putting me in this "Mobbing". This is Corrupt, not Professional, and involved covering up Lowlifes and Misusing a lot of people to do so.

There's nothing like having your priorities straight.

And once I was in this "Mobbing", once it started, it was Implied and Insinuated that this was "Why I was In"—and when I brought this up a couple of times, the Truth of the matter (that I did not interfere with an Investigation), they "Whitewashed it and Buried It" by "Bombing me with Traumas" and saying it was a "Lie on my End"—they sure as hell didn't want people figuring this out, including me, and they did this to shut my mouth up in regards to the Truth. And at the time I even "Brought up Orlin" as I figured out this had something to do with him—but again I got "Bombed with Traumas" to "Shut-Me-Up" by Idiot-Colleagues, Total-Know-It-All-Citizens. As it is these Agents had already "Whitewashed and Pre-Disposed" everyone (Citizens at work) in regards to this "Ninja-Event" and Orlin—"Saying that I would do so, "Interfere", by calling the number on the Van" and again they knew what I would probably do, my Real Motivations, and I did.

I was in up to my ears in this "Mobbing" and had no time for "Figuring this much Out" (what I know now), I did not have a chance to speak any further of this, and in fact I thought for a very long time that everyone was simply "Mistaken", that they "Misunderstood" in regards to the Feds. And again later I thought they were simply "Mistaken", in regards to the situation in which my "Girlfriend Freaked Out", had a Psychological-Snap, and which the Feds years later "Brought Up"—instead I was a "Sucker and Dupe" like everyone else. Instead this Re-Occurring Behavior on the Feds part was of course Deliberate—the whole thing a Dirt-Creating-Mission, COINTELPRO. As at the time I was clueless that this was being spoken of behind my back, never mind other "Issues and points as Well". Everything "Remained Hidden" or was "Buried Quickly". And when I got to the Truth of some of these points, over here in Europe, I was "Livid and Enraged" to say the least. I had absolutely no idea of some of the things being said behind my back at the time. Things make a lot more sense now.

And the chances of the Feds not knowing these things are ZERO, as they "Obviously Re-Wrote" things, took things Out of Context Deliberately. They had and have a "Habit and Profile" of doing this, it's a part of their M.O., they did this Repeatedly. In regards to my situation with my girlfriend who they Deliberately Programmed for a Psychological-Snap, based on taking things Out of Context in regards to Biff, and Misusing Evidence—a Tape, these Agents did this. Then later with Bryan my Co-Worker they also took things Out of Context and did not tell everyone everything they knew. And then with me and the Plumbing-Ninjas, which was a Deliberate Set-Up as well, and in which they again knew what the Truth was, they merely Used things Out of Context yet again. And they put a friend of mine, Kaye, in this Mobbing as well. Corrupt-Filthy-Pigs. They understood the motives involved here, and in fact Manipulated some points here, based on their understanding. They simply "Added on Layers" of Deceit and Manipulations. This is how Government Agents do things, COINTELPRO—this is all Text-Book Procedure. I have observed this in America, and here in Europe as well. In fact I have seen The Clown put one person in this Mobbing because this person Tino "Screwed Up" doing The Clown's Dirty-Work, and The Clown was Pissed-Off.

In retrospect there could be no doubt that they KNEW I would "Go Snooping" to find out what these Plumbing Ninjas were about, and WHY, because of the Recent Murder obviously. They also were "Investigating Me" at the time that it happened, and they knew that it was "Undesirable Event" for me and that the last thing I wanted to see was more. Thus I would go snooping around, and with people involved with Law-Enforcement. This was all a part of my M.O. anyhow. So they had some "luck" but they were "hoping for it" as well, and based on Psychology. When they got what they obviously wanted, thus the Set-Up with the Plumbing-Ninjas, which WAS ALL FAKE, they simply "Ran with It", knowing I had done no such thing, knowing I had NOT DELIBERATELY INTERFERED WITH AN INVESTIGATION, and regardless of anything said to the contrary by me, "As I mixed things up to Deliberately Cover my Intents"—which is what I did with Idiot-Tim, or so I thought.

I had made my INTENTIONS QUITE CLEAR, REPEATEDLY, THROUGH VARIOUS PEOPLE WORKING FOR THESE AGENTS, IN REGARDS TO THESE "DRUG DEALERS" LIVING NEARBY—and a onetime statement of "Blowing smoke up this jerks Ass", could hardly render these Idiots "Incapable of Knowing the Truth". This Set-Up with the Plumbing-Ninjas, which was OBVIOUSLY FAKE, rather proves this. When you have so much "Evidence to the Contrary" taking one single thing and "Running with It" shows the real intentions. DUH! Any Law Student can figure this out, never mind a Lawyer.

I had already seen one kid die in the parking lot, bleed to death from a gunshot, and they clearly tried to "Entice a Reaction" in the first place, never mind that if the "Situation with the Ninjas" was real, they would not have taken such blatant obvious risks. This was about shutting my mouth up in regards to the Murder in the Parking Lot and the Racist-911-Dispatcher in the first place. They can Kiss My Ass. Who are the Criminals here?

And again it is the FEDS RESPONSIBILITY TO HAVE THEIR FACTS STRAIGHT, and what did they really believe when they had a hundred contradictory statements to one? Well?

And in the meantime, The Clown, who is Responsible for me figuring a lot of this out, is threatening to kill me with this Exotic-Techno, and Traumatizing my Family, because She is just as Corrupt, and a Psychopath, and Caught. The difference between Here in Europe, and America, is that this time I know what is going on: this Agency Punk Clown is Caught. She has also, "Set Me Up" in some similar but stupid ways, simply for Manipulating People I have Contact With, and again with "Tapes"—and again Out of Context. This is simply a matter of taking "Snap-Shots" of Life, and building an "Incorrect-Picture" again like a "Messy Divorce", simply to Manipulate the Perceptions of others. Quite naturally, She has "Taken these Tapes Out of Context"—taking the idea from my book, and to "Rub salt In". An Idiot can do this, She proves my point in Multiple-Ways.

Having said that The People She has Manipulated have not "Brought this Up", any of the "Lies and Stories", anything from "Manipulated Context", because they have obviously been TOLD NOT TO, which again is a TACTIC, and as a result remain Ignorant as to how they have been Manipulated. It is this easy, and when people are "Told not to speak about Things", things never get "Sorted Out" which is rather the point. And in the meantime they have been "Programmed with this Proof". This is a "Tactic" for creating a "Buffer-Zone" a "Layer of Protection" against the Truth, for the Agents, as in if you do not speak of it, it can not be Sorted Out.

And in regards to all of this, no doubt it simply gets down to Individual Agents, not Agencies. Having said that I will find out down the road in 2011. This entire "Situation" was Created by Corrupt-Agents, and goes on due to Corrupt Agents: and it started because of Two Cover-Ups. As a point here, if it had been the President he would have bled to death in the parking lot simply for the color of his skin. For this I lose almost two decades of my life? And why is my Family Targeted? Who are the Criminals here, and where is the Justice in this? And it all started with the "We'll Make it Work for Us" Tactic.

Again Cops and Agents do what I was doing, for finding out Information, it does not make it True when they say something merely for the sake of Profiling, or Lying to "Catch Someone", and it did not make me using these same Techniques for the same reasons True, when I said certain things. This was merely about MAKING THINGS "WORK FOR THEM" BASED ON THEIR DESIRES AND AGENDA. This was about Hiding the Truth, Hiding Context, so that they could "Manipulate Looks and Appearances". When I did a Re-Wind of my Life, based on my experiences with The Clown, I found this "We'll Make It Work For Us" Tactic, in quite a few places.

Again not all Government Agents are created equal, and some are more Corrupt than others. Those who are Corrupt, being Personal not following the Rules and Protocols of their Agencies merely using their Job-Title and Position within their Agencies to DO AS THEY LIKE. Obviously those who are Professional have a different standard, and do not Misuse their Jobs and Power and Agencies for Personal-Reasons. The person I am dealing with now, here over seas, The Clown, is a good example of someone who is both Corrupt and a Psychopath—as I have come to learn, Psychopaths are attracted to Jobs of Power, for the "Protection and Perks" they get with the Job-Title, never mind the Abuse of Power and Fulfillment they get out of it. At some point we have to realize, and ask, what exactly are the requirements for being a Government Agent? The one I am dealing with now, is Lying both to The People around me, and I can say quite certainly—will Lie to Her Own Agency, to compensate for the Truth. She has violated the boundaries of her own organization, and I base this on eight prior years of dealing with several Intelligence Agencies with different people in charge, including Her Own Agency.

The bottom line of my own Situation is that I knew of a Pedophile-Cop who got his girlfriend beaten and raped due to his Negligence and Stupidity and I basically witnessed a Murder outside of my window, knew of a Racist-911-Dispatcher, knew the Police were Covering It Up, and spoke out about it. As it turns out the Feds were also involved. This is really what created "My Situation". But it was in trying to "Identify my Leaks" and in trying to get answers from someone I knew to be in Law Enforcement, at work, about these "Plumbing Ninjas" that really got the ball rolling. Having said that the Feds had already decided to get me into this "Mobbing", going to great lengths to FRAME ME so it could occur. So not all Agents, or Spooks, are created equal, not all are Professional. This Tactic, "We'll Make It Work For Us", is what started everything. This has lead to Manipulation of People sometimes even putting them at risk, has led to this "Mobbing" for Two Different People, and almost two decades later it still goes on, and again I have a Corrupt-Dangerous-Agent, in charge of the whole thing, at that.

The "Buffer Zone Tactic" and/or "Insulation Tactic"

Idiot-Citizens working for these Spooks think they know all about you, when they are being Lied to their face. Just like the restaurant situation, and informants, they are only told what they need to know to get a "job done". They are used by the "Level" over them. These Citizens get used as the "First Line of Defense", doing things for the Agents without actually knowing WHAT it is they are doing, WHY they are doing it, or for WHO they are doing it for—the "Quality and Type" of people running the whole thing. Thus they are TOLD to "Deny Everything", if something comes up, which in turn protects the Government Agents. Thus the Citizens are a "Buffer Zone", "Insulating the Agents" from getting caught. All sorts of people act as a FIRST DEFENSE for these Agents, trying to discourage you from speaking out or acting out against these Spooks and Agents, and based on what they do NOT know. That is rather the point of Speaking-Out about these Agents in the first place, to let the Citizens know what they do not, to let them know they are being Used and Manipulated. To let them "Know what they don't know". In this case Lowlife Informants are the same, if they do not try to discourage you, they can end up in legal trouble, or just like you (In "Mobbing"). Thus Citizens are a "Buffer-Zone" and act as INSULATION for the Spooks, a First Line of Defense.

The Small Fish Tactic

One day at the Public Library while there with my kid, these Spooks, The Clown, had someone Masturbate from behind a bookcase about seven feet away. After much gnashing of teeth on my end with my girlfriend, we went to the police who later found him and identified him. Well by me in part obviously. They came to the door, showed some photos and asked some questions. Now anyone in my "Situation" will tell you that whatever "Laws" these Spooks follow enables them to Interact and Manipulate the Police. It is behind the scenes, but obviously there. Naturally I wondered why these jerks would let one of their own be caught.

Obviously I was being Profiled , but they had never sacrificed one of their own people before for the sake of Profiling. For this reason alone I never pressed charges. Something stunk, and these people are more aware of the Laws than myself. Eventually, years later, pondering this (and with insinuations dropped by the Spooks), I realized that this person had been caught for some big crime, and was being used to accomplish a goal for these Spooks. In other words he either "Comply with these Spooks", or go to jail. Having said that if I had pressed charges, if he had gone to jail, no doubt it would have been for less time than if the Spooks had put him in jail. Thus he was willing to take a chance and make a sacrifice—for details (Of how I know) I won't mention, this guy obviously had "no choice". The other thing is they no doubt would have pulled strings for him so he got less time, or in all likelihood told him this without any plans of actually doing so (What is he going to do, complain that he didn't get a fair deal with the Spooks for pulling out his penis at the Public Library?).

Thus THE SMALL FISH TACTIC. Some dupe, sucker, and pawn, takes the blame, while the "Big Fish" get away. It comes down to P=B-C (the profits equals the benefits minus the costs): Motivation. Some people don't have a choice. Any Investigation stops there, you have someone else taking the blame. You have someone taking the blame for someone ELSE'S actions, it is very believable and very effective. Behind the SMALL FISH is another level, the BIG FISH: in my Life these Bigger-Fish are Agents.

COMPARTMENTALIZING

This is used by the military based on Human Behavior because it works. It is also used by the Government, Cults, Politicians, the Elite, and anyone with any sense, with secrets to keep. The idea, this System, works by only letting people know enough information to get the "Job Done". It is Manipulation involving Layers, and Knowledge and Information. This is Used ALL OF THE TIME with these Ignorant-Citizens. Thus People, Ignorant Citizens are "Told to be Quiet" and to "Say Nothing" in regards to the Target—it is a Tactic, to keep people from finding out Information which would Expose the Agents Lying and Manipulating. Only Agents, The People at the very top really know what is going on. To quote Mr. Lee, "The difference between a sheep and a dog is power and knowledge, not blindness." Everyone is being Manipulated for The People at the top, of any given "Food-Chain". Wanting to believe otherwise will not change this. If you want to know what the "Layer at the Top" is about you have to figure it out on your own, Independent-Thinking based on Observation, Logic, Deduction, Motives. You have to be a Detective.

THE BURIAL TACTIC

The Burial Tactic is how Agents Bury the Truth with Lies. It is Layer after Layer of Deceit, Lies, and Manipulation. This is done with both the

Public and Individuals. The Truth gets Buried, and Farther Away. I have "Covered" the Public Version elsewhere, this is how it works on Individuals.

I experienced this Repeatedly, and it is one of their most Common Tactics. It involves what I call "False-Logic". This is what they use for covering up the Truth when you speak out to people, and tell them what is really going on, or what has really happened. In this case it is really a "Formula for Lying". In regards to any subject, they make "Layers and Levels of Lies", it is a long string of False-Logic, and the way it works is this. They get a person to "Buy into one Lie", which is of course Incorrect—again it is False-Logic, they "Bend the Truth". They then get a person to "Buy into a Second Lie", built on and based on, this First-Lie. To this they will add one more in the same manner, and then one more yet again. At the "End of it All", the person they have spoken to is so far removed from the Truth, they can not see it anymore. This is a form of Brain-Washing, Mind-Washing, Programming, and it is Used and Misused all of the time. It is a part of "Control Another's Mind".

Talking to people then becomes "Difficult", showing them the Truth that is, as you have to "Peel Back" each Layer and Lie, it is one Long-Chain of Erroneous-Thinking, with a "Conclusion that is Absolutely False". So people end up believing what is not True. To "Top it Off" they may add "False-Profiling" and Staged-Events to make this "Extra-Convincing" and to make this "Air-Tight" and "Absolutely-Believable" to the Citizens they are Lying to—as the Citizens "See Actions" which seem to fit the Layers of Lies these Agents have Sold. For these Idiots it is too much to comprehend, or perceive. So many times you may scratch the surface in terms of Truth, going two or three "Layers Deep" saying well "Why is this then?", and in Reality if you are going to reach where their heads are at, you may have to go five or six Layers Deep. For "Each Layer" the Truth gets distorted, and "drifts" further and further away. To top it off, you have to figure out what is really being said behind your back, in order to get to the bottom of it all. They will do this with anything True you say, the Truth in spite of being obvious Drifts away from the Citizens. Thus the Truth gets Buried.

Programming, Whitewashing, the Immunization Tactic, the Burial Tactic, and Creating an Artificial-Profile from a Staged-Event

Again I'll use an example from my life. The Spooks as a way of burying the Truth will go to people you know, informing them of a "Staged-Event" that has been "arranged for you", telling them how you will react—"Predicting How Things Will Happen", "Predicting You", to "convince people" that they are good at this and "know you". They do not give the Whole Truth of the matter though. This is based in part on the profile of the Individual/Target (me) and really, almost entirely, just COMMON SENSE. Almost all of the Nasty-Situations/Events I have been in and "reacted to", are "no-brainers" in terms of what to expect.

That is to say how I have reacted to 99% of these Events/Situations is how anyone else would have reacted. When people, Citizens, are engaging in Nasty-Behavior they Normally would not do, because they know it is wrong and would bring about a "Nasty Reaction", then what is different here? Thus the Behavior is Normal, the Idiot-Citizens PERCEPTIONS in regards to your "Reaction" is NOT. This is a matter of "Programming" and Brain/Mind Washing—the Citizens have changed, "their Perceptions, Thinkings, and Actions have changed", NOT YOU—they are for all intents and purposes CULT. Having said that being free of my "Situation" involves NOT REACTING to very unusual circumstances based on "what you know": which if you think about it does not involve Real Life, being realistic, or being normal. As an example I was at the park with my Family one morning. There was another couple with a child of similar age, and two old ladies approaching.

Now from the beginning I suspected something about these two old ladies (probably mid-sixties) as they seemed to be debating which couple was the "correct one" as they approached. I quickly realized this and decided to "Pay them no Attention" hoping they would choose the wrong couple, for whatever bit of nastiness they were to play out. They did, they chose the couple nearest them. The small child (2–3 years old) tripped and fell on the pavement, judging from the screaming, severely hurting her/his self. At that moment one of the old women standing nearby said something incredibly nasty to the couple in regards to the kid falling down, the kid was still screaming in pain. Obviously what the old-bat said was Nasty.

This also gives you an idea of how utterly and completely nasty people can be in regards to others simply because they have been "Convinced" and TOLD TO DO SO. These two old bats took, and take, no Responsibility for their behavior because someone else TOLD them to do it. IF THEY ARE NOT RESPONSIBLE FOR THEIR OWN BEHAVIOR, THEN WHO IS? They were not forced into this at gun-point, they still had the Ability to think of what they had been TOLD, and before they took Action. There was no "Intense Psychological Pressure" put on these people to do so, they did so because they were Non-Questioning. Of course they were "Manipulated and Programmed" what else would you call it? This is Kooky, who else thinks this way?

The entire situation was an absolute disaster for the Spooks. The mother of the child, as well as the father, let the two nasty old bats have it. The father quickly "toned down", to compensate for the mother of the child, who was in the faces of these two "wanna-bees" (In Fascism any sense of shame or guilt for ones actions, is dispersed amongst the Collective) yelling at the top of her lungs, thoroughly pissed, and from what I have experienced in my own life, no doubt for very good reason.

The entire scenario is extremely unusual, and involves behavior no one would admit to Publicly, to bring about an unusual over-the-top-reaction on your end. These two old bats would never say to anyone they engaged in this Behavior normally. The mother of the Child, like anyone would of course say "You're damn right I did."—as would most any parent. This right here should give "Citizens" cause to think—but they do not think this far. In short, in real life, Extreme-Circumstances bring about Extreme-Reactions—this is NORMAL and typical for all. This is a matter of Energy, and the Law of Cause and Effect in regards to people Interacting with one another—what I call Human-Physics.

However in my "Situation" any "Extreme Reaction" is labeled as "Unusual" and an excuse and pretense for DELIBERATELY-MISLABELING-YOU, Psychologically, for the Public. It gets added on to an Artificial-Profile, and offered as "Proof of This"—this is of course simply a "clever" Manipulation. In short it is REVERSE-PROFILING; that is arranging "Circumstances", then taking things OUT OF CONTEXT, to provide and "Prove" a "False-Psychological-Profile" to the Public. It is flat out Lying and "Framing You", to "Look Good" for the Public (the Spooks obviously) and to gain the Public's "Support" for making more of the same. They are building and constructing a FALSE-PROFILE for the Public. It is sick, and Conspiracy—Conspiracy is Illegal for a very good reason.

What were these two old bats expecting? This is why people in "Real and Normal Life" do not act the way that they do inside this "Situation" because they know they have no right doing it, and that a Normal Person will respond unpleasantly to them. DUH!!! THAT IS REAL LIFE—and to say that a Person/Target responding to you unpleasantly, for Behavior on your end which you have no business making, is "Unjustified" or a sign of "Sensitivity", or sign of "Ill Mental Health", is a Blatant-Lie, Stupid, and you have to WANT to play the "Role of Victim" to believe it. You have to be stupid an Idiot, Ignorant, to believe it.

ACTUALLY, LOOK AT THE COMMON SENSE OF THIS—IF YOU BELIEVE TO THE CONTRARY IN REGARDS TO THE TARGET THEN YOU ARE OBVIOUSLY "PROGRAMMED" AND BRAINWASHED—WHAT ELSE WOULD YOU CALL IT?

Quite frankly if that is what you believe for making behavior you normally would NOT, for reasons you already understand, then you are not a "Victim", but you are an ASSHOLE. This is what "Society's Understood

Boundaries", Mores, are about in the first place. No one needs to tell you that you should not be making that behavior in the first place, it is understood, and understood based on common sense. Your "Making this Behavior" because "You were Told To" makes no sense, and is SICK.

So, to get to "Immunization" again. As I spoke to my girlfriend who was, and still is, in league with these Sick-Jerks, I realized many things about this whole Sick-Scenario. The entire scenario was to Debunk the Truth, make me look bad, and therefore "IMMUNIZE" themselves (the Spooks) in regards to me speaking out to others. In the FUTURE people would be much less likely to listen to me as a result, because this process shuts off people's, the Public's, Eyes and Ears to the Truth. Thus the Spooks would be IMMUNIZED to future exposures of themselves, IMMUNIZED from the Truth, in the future. That was part of the Plan, and has been used on me countless times (they update it occasionally), it is just that in this case, the Spooks really screwed up—giving me a lot of insight. It is about their "Looking Good" and making you "Look Bad". And when you are the Target it is hard to believe what is wrong with People's Perceptions, and to see that they have been Programmed in such a way.

The Spooks wanted to label me as a "Mental-Case", an "Over-Reactive-Whacko". Instead, at least two of their people found out what "Normal" was. Again most parents would have Reacted this way, as what these bats said was ugly, inappropriate, and being yelled at was Justified (obviously)—this is why most people do not do things like this, it is UNDERSTOOD to be Incorrect Behavior, and Understood that there will be Repercussions. It is also Understood that this Behavior is Unacceptable by Society to begin with—Life's and Society's Understood Rules, Mores. But because of Agents Lying and Manipulating them, and telling them what to do, they do these things anyhow. Having said that my girlfriend out of "Programming" had agreed to go along with these Sick-Agents and later "Told me on their Behalf" that the couple over-reacted, the woman in particular. Yeah sure, Naturally. I could go out and do this right now to any of these Brainless-Jackass-Citizens, and get the same Reaction in an instant—how hard is this to figure out?

You have to be an Idiot to believe this Bullshit.

Well needless to say the two old women screwed up, and realized this right away as it was the man who was the Target and who was "supposed to react". In this case the man was trying to restrain the wife in this "Unusual Event"—he toned himself down from the beginning after initially giving them a piece of his mind. In this case as with all others, my girlfriend had already been PRE-INFORMED of the set up, this being one of the very reasons why: to prevent a reaction on HER END. It looks bad for Business if both are the same way, and proves a point against the Spooks which is what this "Event-Gone-Awry" did. This tends to show the reaction is NORMAL when two people, one of whom is working for the Spooks, reacts the same way—thus they "Pre-Inform" the other party, in this case my girlfriend, to avoid looking obvious. It looks bad for Spook Business and if you are not starting to get the gist of these people by now, they are all about LOOKS, looking good for the Public: NOT actually being good. How things "Look and are Perceived" can be and IS Manipulated, NOT the Truth itself. This is a matter of Programming the Public.

So the WHITEWASH TACTIC. I spoke to my girlfriend about the "Event". She eventually said that the mother "obviously over-reacted", feeding me her Pre-Fed Spook lines, based in part from her guessing what she was supposed to say, as this is what they had "Fed Her" in regards to ME. Their Propaganda and their Sales-Pitch, to the Public and my girlfriend. Everyone had been PREDISPOSED TO A PARTICULAR LINE OF THINKING, namely the Spook-Line-of-Thinking, and had been spoken to BEFORE the event. This is to steer everyone's line of thought in a particular direction, away from the Truth. It is really to keep people away from INDEPENDENT THOUGHT, from thinking on their own. It is a form of MIND CONTROL. It is to keep people from seeing the Truth. Almost ALL CITIZENS if it had been them, would have had the same reaction—minus Gandhi and Jesus. People, Citizens, become Delusional when they have been Whitewashed, and some how believe that the "Behavior is NOT Normal", and that they would have acted Differently—like they are Gandhi, and like the Irresponsible-Behavior is that of the person Reacting as opposed to the person "Initiating It" in the first place. What a bunch of god damn Idiots—OBVIOUSLY THEY HAVE BEEN "PROGRAMMED", what else do you call this?

The simple Truth of the matter is that the mother acted the way most people would. The two old ladies acted in a way that was anything BUT NORMAL, put bad energy out, and reaped the rewards. THE TWO OLD LADIES STARTED AND INITIATED THE WHOLE THING NOT THE COUPLE WITH THE KID. The mother of the child, given the circumstances, reacted normally, in spite of what any Spook says (by the way THEY KNOW THIS is a normal reaction given the circumstances) to the Public, and at any rate "her reaction"—if she had been in my "Situation", does not warrant or give cause for Illegally-Harassing her and following her around for years on end (this is old Soviet Style B.S. of stigmatizing an enemy of the state). She in spite of a "reaction" is not mentally ill, not dangerous, not illegal, and not even acting Out-of-Context given the circumstances. She is also not "initiating nasty unusual problems" with others, starting problems, OR engaged in Conspiracy, which is in and of itself, highly Illegal and a Crime. This is, the last time I checked, called Real Life, the mother's reaction that is. Having said that the Spooks are not about describing their SET-UPS from a Real-Life-Perspective. If they did there would not be something to WHITEWASH the Public with in the first place. As usual, many of their SET-UPS have Multi-Purposes, and involve Multiple-Tactics.

So I brought this up with people, as it was a very good example of the Manipulating going on, by these Agents.

So the "Selling Point" was "over-reacting", and anything I brought up to the contrary, was "Debunked" bit by bit, until the Truth of the matter which was in everyone's faces was no longer believed. The Truth got BURIED by every Lie to the Contrary. This is the BURIAL TACTIC. In short for every point I brought up, they brought up several counter points, which were built on Lies. Thus I would respond with the Truth for each of their LIES/POINTS, and the Truth of the matter would keep drifting further away, being Buried by a long string of erroneous thinking: Faulty Logic—one lie built on top of the other. In the end you are left fighting off so many Lies, and "Small Points" the whole initial point is forgotten by the Public. Thus the BURIAL TACTIC. In the end the original point is forgotten, and the Simplicity of the Truth is ignored. It is these Smaller Lies, that end up being looked at and taken as "Truth". People end up being deaf and blind. They end up looking at the smaller pieces of the puzzle, not the bigger, more obvious, and Final-Truthful-Picture.

THE TRAUMA TACTIC

This is based on a Government Program, MK-ULTRA, which still goes on, but which does not "Officially Exist" anymore. The name has perhaps changed but it is still going on which is Bullshit. This involves creating very unusual situations, shocking events, for the sake of creating Trauma and therefore Compliance. This usually involves people acting bazaar, unfairly, and unjust among other things, and this is the "Lighter End" of it. It is Surreal and a Total-Deceit. The person currently in charge of my life The Clown, relies and has relied on this Tactic for years, and it is the only thing she is good at. Having said that it is easy for anyone to do, most are just not sick enough to do this type of thing. When you are a Psychopath and Rapist (again this involves Power-Issues), this is no problem.

This is used to break a person down mentally, and to gain and enforce Compliance—thus controlling another's Actions and Thinkings through what amounts to Torture—which is exactly what it is, the "Delivery-System" is merely different. In this case the "Shock and Trauma" is the "Physical-Event" of Torture. As a result my hair has started turning prematurely white. How effective this Tactic is depends on you, I still have two middle fingers up for the person in charge. She is a Corrupt-Piece of Shit, and a Non-American wearing the "Title of American" without actually being it. As an added point The Clown relies on this Tactic heavily because of a severe lack of intelligence on her end. After multiple Government Agencies, with multiple people in charge, she has been the sloppiest of all. If Her Team of Agents, The Buffoons, were more Honest they would attest to this.

In short she uses Force to compensate for a lack of intelligence on her end. Since writing that many years ago I realize it is more than that, again this comes down to a Psychopathic Personality Type. She gets "Fulfillment" out of Abusing Others, and harming others, she gets Fulfillment by Violating People. Thus She uses it Excessively and Dangerously.

The person in charge before her, from the same Agency had created no mess, and was close to ending this "Mobbing" I am in (I was also near ready to ending this) without creating any "Mess" or Book, without looking bad or obvious. Unless I am mistaken the person before her cared about the reputation of the Agency and his Job, and in getting things finished. This woman (The Clown—2009) has created such a Mess, I am now schooled on Government Tactics, have "Re-Wound on my Past" and writing a book about it. This is the result of Holding Me In and "Bombing me and my Family" with Traumas—including killing one of our pets to Traumatize my Kid (Family), and this is many years after "Passing the first Time"—as in I should have been free. She is using the Agency for her own personal nonsense, her own "Desires and Fulfillment". If Her Agency "Takes sides with Her" because of "Agency Affiliation" then they can kiss my ass.

For another comparison, she is like any dirty cop within a police department abusing her authority without the police department knowing about it, while at the same time making the police department look bad as a result. Obviously Her Superiors don't know what is going on. She is also currently trying to keep this book from being finished and printed, for the sake of Her Own Ass.

This says everything about this Type of Person, they are cowards who can not take Responsibility for their own actions. When dealing with people who use the Trauma Tactic, keep in mind these are Weak Minded Individuals, who pick this and use this Tactic, because they figure it would work on themselves, and because they Lack the Intelligence to deal with things in any other manner. People Using this Consistently with you, are Corrupt and Dangerous, and no doubt "Psychopathic" as at least in my case "What is being Done" is Dangerous and Gratuitous. Only people who are sick, engage in this non-stop, and it gets down to an S.S. Officer-Mentality.

Again simply being in this "Situation" does not automatically determine which Type you have, you will have to Look, Analyze, and Determine this on your own. Once this "Situation" starts it obviously has a "Life of It's Own". So when dealing with this Tactic keep this in mind, this is what you should be looking at. As one last note, you can file a complaint against any police officer that over-steps their authority, at the police station, because the moment they do it is no longer "Their Authority"...SPEAK OUT AND BE ACTIVE, IT IS THE ONLY THING THAT MAKES A DIFFERENCE. And for my readers, yes I am filing a Complaint through this Book. I have my "Own Protocols and Methods" in place, just like in America, and I am following these. But very soon things will be taken care of in one way shape or form. The last thing I plan on doing is keeping my mouth shut for the sake of this Psychopathic-Jackass. Again you have to try and determine what is "Outside of this Situation"—what is not "Within Rules", and how much is even possible to put up with.

The Reckless Endangerment Tactic

The Reckless Endangerment Tactic is another favorite of the Spooks. This is used to create injury, or worse, and I was given "Reckless Endangerment" one day by the Spooks as I contemplated what to call this Tactic. They will try to "steer you" into a particular course of action that results in a chance of injury or death. Usually the circumstances are so Manipulated in the first place, that the chance of being hurt is very high. This is to get you hurt, and at the least to shake you up: to "Put Psychological Stress" on you. Never mind "Lucky Coincidences" on their end, of you Dying. Last time I checked, if that happens it is called Murder. Usually when you bring this up with people around you, they do not get how close things were, and laugh it off as "Paranoid" because they were not wearing the shoes. Both my Kid, and I, have had close calls. As is typical, Citizens "Having to wear your Shoes", would consider what you say "Less Laughable" if they did. And it also depends on "Who" is doing this to you: in my case with The Clown this Idiot is hoping for the worse and hoping to get away with it, by "Claiming it was an Accident" and "Misjudgment".

These people are incompetent, so always take this seriously, their job means more to them than your life and you do not know these people—there is no way to know how serious they are in "Taking Chances". As a very good rule if they are doing this, then they are Arrogant which also translates to "Sloppy". It also means that they do not treat the Issue of Serious Injury and Death seriously, again a clear warning sign, and "Tell" of the type of person you are dealing with. In one particular case, due in very much part to Exotic-Techno, my kid narrowly avoided a "situation" that could easily have resulted in death, at the very least an emergency trip to the hospital with very little time to do so. The "situation" could have resulted in a crushed windpipe, or a broken neck or both. As it is it resulted in serious injury. To top it off I spoke out about this to my girlfriend and parents, what had occurred. In typical Arrogant-Idiot-Style the woman in charge, The Clown, arranged for these circumstances two more times, with other peoples kids to show "What she could get away with"—to "Mock" what I said to my parents and girlfriend. The first kid was hurt, but at least no chance of a damaged windpipe (although crying heavily), the second kid ended up with an extremely bloody mouth. Again typical Arrogant Behavior of The Clown. She likes "Showing Off" for "Building Her Career", in reality it is part of Her being a Psychopath—the Citizens around here are absolute Fools and Idiots.

In all cases, when it happens to you, it is more than "Paranoia".

Stigmatizing

This is another common Tactic. When the Feds accuse you of something you have not done, or is not True, they dump a ton of Innuendos and Accusations, and Counter-Arguments, on you and down your throat, so as to make you "Feel Guilty" regardless. They also will do things to you "Based on these Accusations"—"Punishment with Accusations" to CONDITION YOU. They use the Public-Idiots for this—which is funny in a sick way, as the Public will "Help" because they are already convinced, and this in turn is used to "Further Convince Them". As they "Get their Hands Dirty" they become less questioning, they become Indoctrinated.

In my case I did not even understand the reasons for the Public's Behavior (especially at work) until many years later because the Feds HID what they were Accusing me of so they could get away with Lying and Manipulating. The Public in turn thought I was "Merely being Deceitful and Clever" and was Convinced to the point of acting on this. Eventually I had a "Psychological Snap" in regards to one of their people, and responded with the same Type of Behavior they were engaging in—once I did the Public became Totally-Convinced (As if there was doubt in the minds of these Idiots) that everything they had heard was True. In reality the Feds were going for this "Snap" in the first place, and I and these Idiots got used—fortunately for these Stupid-Citizens I am the only one who had to pay, other than the person I had a "Snap" with.

And all of this started by these Sick-Feds convincing a girlfriend of something that was not True, Programming her, to the point of her having a Psychological-Snap with me, which is a god damn understatement. And again, in Her case, these Feds were going for HER having a "Psychological Snap", and this was a matter of taking something I said Out-of-Context to someone I suspected of working for the Feds (I was trying to "Catch a Leak" in regards to a "friend I suspected of being an informant), and "Pumping Her Up" full of fear (and I suspect an altered tape), "playing on past traumas" from her life, and then they "Rewrote everything Afterwords". Like everything else, this Lying and Manipulating was Hidden from me, so these Assholes could get away with it. They then took this Bullshit, and Built a Fake-Profile around it. It took me years to sort this out, never mind my ex-girlfriends "Psychological-Snap", and when I did sort all of this out, here in Europe, I was in a Rage over the whole god damn thing. Atomic-Disgust is one of my wordings for it. In the end, thanks to The Clown, I got it all sorted out though as I finally decided to start looking and began putting pieces together. They used all of this for "Stigmatizing Me".

This Stigmatizing is to Convince the Public, and is a Sick-Misuse-of-Psychology—it is to "BRAND YOU" so that you "Look the Part" of the Accusation, and React to the Accusations. Actually this is not a matter of "guilt" which is on the inside, it is a matter of putting you on the defensive externally and continuously, in regards to a particular subject—Accusation, to burn the behavior into you—to CONDITION A GUILTY REFLEX, so that you "Seem Guilty" to the Unsuspecting Public. It is a matter of "Being Branded", of having an Accusation(s) being burned into your "Hard-Drive". This is used to make you seem guilty, when you are not, and for winning over the Public, who are used in the first place, to help create this. Remember these Spooks like to recruit people who are not conscious of what they are doing. In this case they Use and Misuse the Public to Condition you, so that the Public will become "Further Convinced"—these Citizens are Idiots. It is so you are "Conditioned Into Acting Guilty" WHEN YOU ARE NOT. If any of you see this French Interview with the guy from the DoD, you'll see this with him. When he brings up the "Accusations of being an Axe-Murderer" it is obvious that this is what was done to him. It is obvious that he has been "Branded and Stigmatized". It is also equally obvious that what was portrayed and perpetrated against this man was a Farce and Lie, a Manipulation made for Corrupt-People, and by Corrupt-People. And I might point out, because he was trying to expose Corruption.

First they Break it, then they "Fix" it

This, as most of these Tactics, can be observed in mainstream news. In my "Situation", many times the Spooks claimed I had a problem I did not have; convinced the Public this was so, through False-Profiling, Reverse-Profiling, and "Authority-Waving" ("Hey we have a badge, so what we say must be so"—believe me, question authority Folks), and then set about through Lying and Manipulating the Public, to Create the "Problem" they said I had. In other words they Sold the Public on a "Problem" I did not have, then turned around and Created it through Conditioning, Psychology, and Manipulation, and using these Idiot-Citizens to do so. They started it all by naming the VERY PROBLEM THEY WERE GOING TO CREATE. Then they turn around, after Creating the Problem, and say "We are going to FIX This". The whole time the Public remains a bunch of Ignorant-Jackasses, "The road to hell is paved with good intentions" being the general rule of thumb, and Motto here. So they Deliberately Create a Problem, then turn around and say they are going to "Fix It", making more Manipulations and Problems along the way. Of course in order to "Fix It" the Agents claim they have to begin Setting Up Staged-Events, to help you overcome "The Problem"...In other words they then begin "Setting Up" more of the same thing. "Fixing the Problem" being a matter of being so "Shell-Shocked", Thick-Skinned, Psychologically-Damaged, and/or Traumatized that you are "Non-Reactive" to these Violations of You and Your Rights. In this case as with all others, only the Agents know what they are getting away with. So first "They Break It" and then "They Fix It".

Defiling and Debasing

This is Used to Desensitize the Public and Citizens. It is used with the "Rumor Mill Tactic"—they are basically the same, although this Tactic here that I speak of, Debasing and Defiling, involves Actions as well. It is to get people "Desensitized" in terms of "Speaking and Acting against the Target".

This is done initially, slowly and gradually, as people become "Acclimated to their own Filth", the Filth they are asked to Do, and Engage In.

Now in eighteen long hard years I can tell you that all Corrupt-Agents love sex stuff. This is one of the favorite subjects for Defiling and Debasing. This is to drag the Target down, and is also a result of Agents using their Job as a "Shield", a Buffer-Zone, an excuse for their own Lowlife Mentality, and Perverseness. Any Agent that tells you it is a matter of "Job Only", is full of crap—they have many options for subjects, and a "Handbook of Ideas" no doubt as well. Young people in Nazi Germany killing Jews had the same Mentality and Psychology: many if you look, smiled for photographs completely oblivious to the Reality of "Who and What" they had become. People are always responsible for their Actions, especially when what they are doing is a MATTER OF CHOICE.

Many Spooks love the sex stuff as a Tactic, because it also Debases and Humiliates the Target. Sex as a general rule is intimate, and meant to be so, but not with the Spooks. Exposing ones sex-life is often Used and Misused for creating Shock and Trauma, and for pushing one into a Psychological-Snap, as it is a major violation of any Normal Social-Boundaries. Especially when it involves complete strangers discussing your sex-life, in front of you, without there being any way for them to even know of it. This is enough to Push anyone "Over the Edge" and was used with me repeatedly. This was done to me based on Lies and Ignorant-Jackasses, and done to get a Psychological-Snap. Anyone with a braincell, never mind a brain can figure this one out.

Now in my regards they also Lied about my Sex-Life for Manipulating others, which is "Twice as Sick", as if there are even words for this.

Your private life becomes Public, open for gossip and examination, to Humiliate and Debase you. Further more they did this in America, getting my Co-Workers involved in this based on Lies and Manipulations, and it took me many years to figure out why they were doing this (Far after I had left my Job). In fact what led to one "Psychological Snap" I had was coming to work and watching one of my Co-Workers impersonate me having sex, from an hour before with my girlfriend, in the same sexual positions and "Sequence of Events". My girlfriend at the time, who was working for the Agents did not of course have to witness this, and was kept "oblivious" by the Feds in regards to much of this, minus what I told her. These Idiots took part in this Lowlife Behavior based on what they thought they knew, as opposed to the Truth. Again Lies and Manipulations are built on more of the same, and added to an Artificial-Profile which these Agents "Sell to the Public". It is to a very large extent the "Burial-Tactic" in this regard—Layer after Layer. Quite frankly knowing what I know now, and understanding why these people did what they did, I could really "Knock their teeth In" to say the least. They all thought it was no big deal and funny, in fact that was the "Attitude of most of The People", most of the time: "It was hilarious".

Having said that many times I observed them do this to "Bryan" (Debase and Defile Him)—who got in this "Mobbing" first and because of these Corrupt-Agents, not because of something he actually did. The funny thing is, as people came up to me and encouraged me at work to engage in "Spiteful Gossip", Debasing and Defiling, I refused to do so. I would NOT because of my own Character, because this was Ugly-Behavior, yet later people chose to believe the worst about me—because of what they "Heard

About" and were Told—and because these Corrupt-Assholes provided no Context with "Evidence" which they Misused. Isolated-Manipulations for reasons these Citizens had no idea about: like these Corrupt-Feds were going to share this with the Public. These Citizens were quick to believe what they were Told and were Idiots. As far as I am concerned, they still are.

"Controlling the Agents Actions"—False Pretenses: Traumatic-Bonding

The Spooks will try to convince you that you have power and control, over their actions. By doing this or that, what they want, you can avoid such and such consequences. This is a flat out lie; they have your life planned out, a basic map of "events", in terms of the current day, the next few days, and coming weeks. To some extent things, Their Planning, may be altered by you and your actions, there is no doubt, but as a general rule, they have "Events" planned and will go through these to some extent or another regardless. Thus "False Pretenses". They will many times try to convince you that they gave you an "Event" due to your Actions, which ever ones they insinuate are the "Cause" of their own Actions—as if they are "Reacting" to something you have said or done. This is a lie to Manipulate your Behavior, to make you feel "Responsible and Guilty", and to make you Compliant and ass-kissing, do not believe it. In other words they tell you that you received this "ball-kicking" for a particular action or reaction on your end, when in fact most times it was Planned by them all along. They do this to try and convince you that you have CONTROL over what "Traumas", Nasty-Events, they give you, when in fact you have NO CONTROL over what they decide to do to you. If you believe you do, you have become Delusional, Manipulated, "Bonded with your Torturers", and Controlled at this point. This is what they want, and it is to get you to "Traumatically-Bond" with them—this is SICK, and used by Sick-Corrupt-People. Good Agents use Good Psychology, Healthier Psychology, not this.

Another point and reference: False Pretenses is also another name for "Framing and Lying" in reference to building a "Fake Profile" for the Public, and in Misusing Citizens.

"Being Good to You"

Occasionally the Spooks will do something nice for you. This is to see how grateful you are, and broken down you are Psychologically. In general if you are constantly ball-kicked and abused by someone, being grateful for a single nice act by that someone, shows that something is wrong with you. The Spooks are measuring your response, and checking for the Stockholm

Effect. They are checking to see if and how "Bonded you are with your Torturers". This is all it is, a test to see how broken down Mentally you are.

The Four Rules of Thumb

We'll make it work for us.

If it works its good.

First "us" then the Target. This is about the Agents "Looking Good", about Manipulating Public-Perceptions.

Attack what the Target does not like.

These are the "Basics of COINTELPRO" and of this Mobbing, which is the "Next Level".

The "Speaking to Others First" Tactic

Keep in mind that in part what you are going through is a Psychological-War with the Spooks, in regards to the Public. The entire "Situation", the Deeper Layers, are Constantly Hidden, the Public is given the Idiot-Version of Reality, along with those around you whom you interact with—"friends and Family" who are also not worth a rats ass usually. Of course those around you, depending on their questioning nature, may be given a bit more of the Truth than others, but as usual, no matter what "Level they are On", they to are Manipulated, go figure. Keep in mind that these Spooks use the "Burial Tactic" and may have seven or eight "Layers of Deceit"—never mind Staged-Events which they can "Write Off" any way they want, and which is never beneficial to the Target. This is the whole point of these Staged-Events, to "Deliberately Mislabel Your Actions" and then offer them as "Proof" in line and keeping with an Artificial-Profile. Appreciate the friends and Family who are more questioning, do not let go of the ones who do care, in spite of their "Involvement" for these Jackasses.

Now a constant Tactic used as a form of Mind-Control, and Damage-Control, is the Speaking-First-Tactic. This is simple, and is what kids in kinder-garden do, you would be surprised at how effective this is, and how stupid people are. As this implies the Spooks like to "Speak First" in regards to all of their planned nasty Staged-Events for you, to "Pre-Steer", "Color", Whitewash Peoples Perceptions. TO PROGRAM PEOPLE'S MINDS. In reality this is to keep people from seeing what is really going on, to keep people from questioning things, to steer peoples thoughts in a particular direction (MISDIRECTION), and to keep people from Independent Thinking and Thought. This is to keep people from thinking of OTHER POSSIBILITIES. This also leaves people Dependent on, and Predisposed towards Liars with Government Badges. People stop thinking for themselves and simply begin RELYING on these Agents to TELL THEM "what this is", or "what this is about", as people "Buy Into" their Manipulations and Deceits and stop thinking on their own—they become Reliant on these Lying-Jerks for the Truth. Eventually people stop questioning these jerks and START BELIEVING THEM WITHOUT QUESTION. Which is a dangerous when the Agents involved are Corrupt. This also means that for all Intents and Purposes, The People/ Citizens that they have contact with have now reached a point of being "Officially Brainwashed". This is what "Mind Control" is, the ability to "Brainwash" through different means than what is "Conventionally Thought Of" and "Typically Thought of", the "Stereotypical-View".

Programming through "Staged-Events" and an Artificial-Profile, are VERY BIG PARTS OF THIS.

Simply by Speaking-First these people many times "Defuse Your Speaking Out" before you have done so. Based on Staged-Events, as an example, they know what they are going to do, know what they have been doing, and know many times what your reaction will be—and based on "Normal Behavior" given the Circumstances. THEY ALSO KNOW WHAT YOU ARE GOING TO SAY, in regards to the Truth—and have "Prepared for This" as well. Take a look at the couple at the park, and the two old bats—by Speaking First the Spooks had already convinced people of "What and Why" things would happen a certain way, BEFORE things had even happened. And of course this was based on how normal people in normal life would react, not a "Psychological Profile" of a "Mentally Ill" person. People were already "Predisposed and Whitewashed", PROGRAMMED, with a certain View-Point, and one which HAD NOTHING TO DO WITH REALITY. Having said that these two old bats picked the wrong couple and proved many of my points. If it had been ME, or anybody else, the reaction would have been the same. And as it is THIS COUPLE got "Labeled as Over-Reacting", simply to "Control-Damage" at that point: they got Labeled with everything that these Spooks were ready to dump on me. So I got to watch this "Perspective" from an "Outsiders View-Point"—it was all a great Big Deceit. And as you may notice, they do this by "Speaking First".

The "Attack what you Do Not Like" Tactic

As this implies they attack what you do not like. Do not let them know how much something bothers you, and they will drop it more quickly, as they are only interested in doing things to you, that "You Do Not Like". The whole point of this Tactic is to obviously Piss-You-Off. Many times, all of the time actually, when you are angry, they will try to get you to tell them things out of anger, and try to get you to say things about people out of anger which they will of course use later. Regardless of Context, and Regardless of whether what you have said is True or not. They will also use this Tactic repeatedly in regards to various Themes, to get you to have a Psychological-Snap, or to Condition You in regards to certain "Subject Matter", Themes.

Psychological Mislabeling

This is another thing constantly Misused. The Spooks sell a "Mental Disorder" in order to "Win Help and Support" of the Public—having said that "The road to hell is paved with good intentions". This Label is what they choose to try and "Explain your Behavior" with many times. Often a "Center-Piece" for an Artificial-Profile. In reality through Lying, Manipulating, Reverse-Profiling, False-Profiling, it is what they are trying to Condition for "Looks", or are using to sweep everything "under the rug" with. For people who are so Blind, these Citizens sure like "Looks". It is their excuse, and distraction, from what is really going on. This is an old Soviet Union Style Tactic of labeling a "Dissident" with a "Mental Disorder" to Stigmatize, Color-Perceptions, Debase, and to Desensitize the Public in regards to you, the Public's Actions, and what you have to say. It is for "Programming the Public". In short it is to make you "Look Crazy", to "Stigmatize and Label You" like an object—the way people feel about someone with AIDS, to turn peoples ears off ("He said what? Oh well he's nuts anyhow."), and to get people, the Public, to "Help" with "Your Problem". Of course things are taken OUT-OF-CONTEXT, like your Actions, Reactions, and Words, for this Psychological-Label to "FIT". It is a way of also ISOLATING THE TARGET.

The Re-Write Tactic

Once the Agents are "Caught and Exposed" for having done something, they will run around trying to convince people that what happened was for reasons other than what the Target has said. They will try to "Re-Write" things, Program People, "After the Fact". This is ridiculous and stupid. Take a look at the two old bats at the park—how hard is it to see the Truth of the matter, yet these Lying Agents got away with it, many at least somehow thought, that how the "woman reacted" was unusual. Which is a Lie and Farce. Yet they eventually "Re-Wrote" it. This is a matter of Programming People after an "Event" has occurred, which is "More Difficult" for these Agents, as people have "Already Experienced Their Own Perceptions" in regards to what the Agents are trying to "Cover Up".

On occasion I have "Returned their Behavior", to some of the Citizens to prove a point, and to "Wake Them Up" from their Programming. When they got angry I rubbed their "Thinkings and Perceptions" in their face, which further pissed them off. I pointed out their Ignorance. And at times, from these One-Time-Events with me, these Citizens were almost ready to attack. Having said that the Agents had to "Cover some Points Up" afterwords, they had to "Rewrite Things". Usually by "Fitting my Actions" to an Artificial-Profile—having said that if you point out the "Obvious with these Citizens" when you give them a "Taste of your Experience" this becomes more difficult for the Agents.

And to think that I go through, and have been through Thousands of "Things", Traumatic-Events. These Citizens are Brainless-Jackasses, and the Psychology Used is the same Used for creating Cults and Fascist Movements obviously. How else would you explain this Contradictory Behavior, this Nasty Behavior, this Unusual Behavior and Warped-Perceptions, this Cult-Behavior? Well?

Having said that, this is what The Clown is promising to do down the road here, once I begin to speak publicly, to our neighbors, Non-Corrupt Field-Agents, others living around us, in regards to The Clown and Her Buffoons, and what the Public does not know. She promises to "Re-Write" things, and is "Insinuating" She has already Programmed People in regards to my "Speaking Out" to neighbors, the Public, shortly down the road. Which is Ironic, but sick: she is a Dangerous Psychopath, and Her Buffoons working for her are Dangerous Idiots as well.

What She has done, is a very long list of Crimes and Manipulations, involving and against my Family and the Public, and quite simply this Show-Off, The Clown, has gotten away with too much. She has Tortured me and my Family, and not even "Ended this Situation" and based on Her own Agencies Standards—She is Holding Me In, and when this is the case, then this is the Mess She gets. She has done all sorts of things to people living around us, unknown to them, and is threatening to take my life as a result of "Speaking Out" which is funny in one sense—I have two fingers up for this Imbecile and Her Buffoons, it is not my fault you do not like Americans, that you only like "Wearing the Title" and the "Color of the Money". Having said that if this Agency does not take care of things properly, it will be worse. I never should have been in this Situation, never mind being "Held In" and by a Corrupt-Idiot that gets away with it, by Lying to the Public, and even other Spooks around here, who have no idea that She is Corrupt. I know for a fact that other Agents around here have no idea of what She is getting away with. So, as things are now, March 2011, The Clown has already "Re-Written" some things, and thinks "She will get away with More". In the end though, this Clown and Coward will be Exposed.

Isolating the Target

This is done in a variety of ways, this is to Convince You the Target that you are "Powerless and Helpless", and can count on no one. Believe it or not many things in spite of this "Mobbing" are still there for you, and function regardless. They will try to Convince You that You are alone and have no friends or anyone you can rely on. Given the Ignorant-Jackasses I dealt with this is True to a point, again "Look and See" to find the Differences amongst People and Circumstances. Not everyone or everything is "Created Equal"—at any rate if this were REALLY TRUE they would not have to "Sell This" to you, this is to DEMORALIZE YOU and to get you TO GIVE UP AND STOP TRYING. Believe it or not these people can loose, they can be fought, they can be caught, and at heart they are cowards engaging in Behavior reserved for Dumb-Post-War-Nazis.

Two and Three Layers

This is Standard. The Agents, Feds, Spooks, what have you, have basically two or three "Levels" that are "Standard". There is the "Public-Version"/ the "Public-Story", the Agent-Version which is the Real-Level, and if you are not paying attention, there will be "Your Version and Story", that is what they try to "Sell-You". Everyone each gets a different "Version and Level of Reality"—everyone is "Compartmentalized" and the only REAL LEVEL is the Agents-Level. They are the "Ones Organizing All" based on their "Own Motivations".

What "Motivates Them" as I have found out the hard way, depends on the "Quality of Agents" how Corrupt they are. No matter what this "Mobbing" is difficult but not all are the same. In my case, like everyone else I could not believe that these people were Corrupt, I was ignorant and thought somehow that these people were "Merely Mistaken" in doing and saying what they were. Later, much later (years) I found out some of what they were saying behind my back and yet still they remained "Ignorant"(Because these Spooks had Exotic-Techno I knew they could not be "Mistaken") and they "Continued Doing" their nasty Corrupt-Bullshit. Eventually over here in Europe due to The Clown who is obviously Corrupt, I started doing a "Rewind" and found out everything many years later, to what extent these people "Were not Mistaken". To say I was Pissed-Off is an understatement.

Now this Agent-Level, is the Level you want to be on if you are going to understand and see things around you Clearly. In all likely-hood you will not Convince anyone you know of, of this Real-Level, but Trust Me this is where you want to be. When you are here, you can catch these Spooks and Agents at their own game and they become more predictable. They have to be careful, and they can be caught.

Many times the "Standard Recipe" is: one for the Public, one for You, and then one for the Agents—of course the Agent Level is the real level.

Misusing "Facts and Evidence", Hiding What is Being Said, Character-Assassination, and COINTELPRO

You can not "Fight what you do not See or Hear". This was USED REPEATEDLY by Corrupt-Agents, for getting away with everything. THIS IS HOW COINTELPRO OPERATES.

If you do not know what is being said about you, how can you explain or tell the Truth about anything. These Corrupt-Pigs got away with the BIGGEST LIES AND MANIPULATIONS in this way.

First Manipulating something I said to a "Close Friend" who I was "Trying to Catch" because I suspected him (And he was like a brother), then taking this Out-of Context (Which they did because they knew what the Truth Was), then taking this and Manipulating a girlfriend with it to get Her to have a "Psychological-Snap" (She was Programmed) based on Her believing this was True in the first place (Taken Out of Context to an Extreme), then taking this and convincing people at work that what I said to this "Friend" was "True" and that what happened with my girlfriend was AS THEY SAID (Building an Artificial-Profile/my girlfriend was Used for This), and Setting-Me-Up and Framing Me, and then getting these Idiot-Colleagues to ACT ON these Lies and Manipulations, which in turn pushed ME into a Psychological-Snap (several actually) after a very large time of Abuse and Traumas. This in turn got Used again for this Artificial-Profile, which is what these Feds wanted in regards to my "Psychological-Snap"—they were "Pushing for This". They were pushing for "Manipulated-Dirt", Artificial-Dirt, so that they "Were and Looked Airtight": for the sake of their Own Asses.

Everyone got Used, Misused, and Manipulated for a bunch of Corrupt-Pigs, and I got Fucked more than anyone on the planet, never mind those around me who got to deal with my "Psychological-Snaps" and Moments of Disgruntled-Charm. No one, and I mean no one had a god damn clue as to what was going on. And it all started with HIDING "FACTS AND EVIDENCE" (Hiding the Truth and Misusing "Evidence"), HIDING WHAT WAS BEING SAID, so that I could not even begin to prove or point out the Truth otherwise. You can not expose People's Lies if you do not know what is being said in the first place. You can not "Fight" what you can not "See". So everyone from the Public to the Target gets Used, Misused, and Abused. Obviously "Hiding-Facts" from the Public, and "Hiding What Is Being Said About The Target", from the Target, is a Tactic. And I experience this CONSTANTLY, anyone in my "Mobbing" does. It is Used for Manipulating People, the Idiot-Citizens, the Public, and anyone you are "Close To". This is Standard and a Given—it is Standard Operating Procedure for these Spooks and Agents. It is COINTELPRO.

"Programming People"

In Layman's Terms "Programming" is the "Art of Gossip" and getting people to accept it as "Real", and then getting people to "Act on this Gossip", "To make Actions because of It". Programming is in fact Modern-Day-Brainwashing. See the "Previous Chapter" for more.

False-Objectivity

This happens as a product of brainwashing, it is a "Looking from the Outside In" and "seeing things and believing things" from an outsiders perspective, not your own, and in regards to things which you experienced and know, yet they did not. I had this happen in America, not knowing or understanding several things, and having been "Traumatized and Tortured" to the point that I no longer had any sense of things that had happened. I simply "Accepted other people's Viewpoints and Perspectives" which were False. That would be a given as they had been Programmed on many a fine point. I having had my Trauma Enema, began trying to "Reconstruct Things and Events" from Their—Perspectives, no longer "Knowing or Relying on my Own" like I was not there. Between how Surreal Things were, and the Traumas, I was no longer even a "real person" and was certainly not myself. Again the Misuse of Psychology and Traumas, and in the end "Brainwashing". This was was the Low-Point of my "Situation" and near the beginning. As a result I said a lot of things, trying to "Rebuild Events and Reconstruct Them" based on what Other People Believed who were not there, and who had no idea. And as if I was not there, as if what They-Thought must be True, and from "This Point" I began "Building on This". I call this False-Objectivity, it is the product of Hard-Core Traumas and Brainwashing. You accept things from the OUTSIDE-IN, as opposed to the Inside-Out. You are not "telling people how things were", people who were not "there", when you were, you are "Accepting what you are Told" as if it happened the way they said it did or believe it did. This is False-Objectivity and it is the result of Traumas and Torture.

CHAPTER 30

Mobbing: the "Next Level" of COINTELPRO this is Not an Investigation: COINTELPRO—a "Dirt Creating Mission": The Bigger Picture: The Creation of a "Fascist Movement" and "MK-ULTRA" for Creating Citizens Compliant to Government: also "Reversed-Legality" for Illegally "Creating Dirt" in Regards to a Target

> "You gotta be responsible for your actions. If you want to give me your life, I´ll take it. And I'll put it in the deck and I'll deal it, you dig? And if you don´t like the way I deal it, then don´t give it to me. You dig what I´m saying? But when you give it to me then it´s mine. I deal it anyway I want".
>
> —CHARLES MANSON (I USE THIS AGAIN AS IT FITS PERFECTLY)

For my readers, this chapter could have been in better shape, however I am dealing with "Difficult Circumstances" on my end, and so rush this a bit. Having said that I feel that the material is good, and the Information within should be Shared and "Released". With that in mind read on.

Well whether you are familiar with this "Mobbing", this type of "Situation" (Localized-Fascism) or not, it ought to be clear that what I am going through is not an Investigation. An Investigation constitutes finding and verifying facts, many times resulting in prosecution, for crimes committed. This "Situation" is reversed. It is "REVERSED LEGALITY". It involves two parts: one Type of Psychology for the Target (that would be Me), and another Type of Psychology for the Public, for the Citizens. This entire situation involves "Removing Laws" and Committing Conspiracy, to get people to Break Laws and to Commit Crimes, both in regards to the Target, and the Citizens who get involved on behalf of Government Agents. This is a "Situation" where Crimes are Committed to "Acquire Dirt" for prosecution: this means that the "Target is too Clean or Piddly" to Prosecute Otherwise and that the Target, through Illegal Means (Conspiracy) has to be "Pushed into Psychological-Snaps" so that there is something. They Push For Psychological-Snaps, so they can have Manipulated-Dirt, Artificial-Dirt, as a result.

Again it is REVERSED-LEGALITY. Having said that one gets into this "Mobbing" on False-Pretenses, BECAUSE OF LIES, in the first place. I base this on knowing several other people, including a co-worker, who were in this "Situation" for reasons the Corrupt-Agents invented, NOT FACTS. And in the case of this Co-Worker, like me, what they offered as "Proof" looked and sounded believable, however it was not True. With him, as with me, they merely Hid Information, and Lied in spite of knowing the Truth, to get him in this "Situation". These Agents were Filthy-Corrupt-Pigs, and took things Out-of-Context in regards to him.

All of this is COINTELPRO Techniques and Tactics. COINTELPRO was all about how Agencies did and could break the Laws in regards to Targets, how to "Set Them Up", how to Frame Them, how to Manipulate People with Misinformation, it was all a "Book and Program" on how to Manipulate Everything, for Destroying a particular Target. THIS IS WHAT WENT ON WITH ME, from day one, after I spoke out about this Pedophile-Cop. COINTELPRO was and is, a long term "Dirt Creating Mission" against a Particular-Target: which is what I have been through.

This was NOT an Investigation—it was a long slow campaign to "Discredit Me" through Character-Assassination. I was being "Investigated" for smoking pot? The People involved in this, from this Lowlife-Pedophile-Cop (who knew what these Feds were doing and why), to this piece of crap Racist-911-Dispatcher, to the Police Department, to the Corrupt-Lowlife-Agents involved in this all "Defending these Sick-Bastards", they can kiss my Non-Pedophile, Non-Racist-Ass. And in regards to several girlfriends who let themselves be Manipulated and Used in spite of Very Obvious Facts, I could wring your necks in this moment just thinking about this Bullshit.

THIS IS WHAT COURT IS FOR, and two of you were involved in getting Jobs involving LAW. Brilliant. Back to business. (As an added note for readers: I have seen repeatedly that many of the Field-Agents who get involved are "Clueless" as well as to all of the details—the Real Facts.)

In regards to the Target, the Target is eventually "PUSHED INTO" Breaking the Laws, by REMOVING ALL RIGHTS AND LAWS WHICH PROTECT THE TARGET, and therefore through Traumas, Torture, Duress and Stress, PUSHING the Target into "Psychological-Snaps" which results in this Behavior many times. This is about attacking the Targets Mental-Health and Destroying It. TO MAKE THE "TARGET LOOK BAD" and to "CONFIRM" everything "Bad" being said about the Target in the first place. It is about Using this Psychology to "Create the very Person" they said they had in the first place—all things being Relative here. These Psychological-Snaps are not the "Product of a Healthy Mind" and are a direct result of what is being done IN THE FIRST PLACE: Traumas and Torture, to Create "Snaps".

Given the Circumstances, this Behavior is Normal in one sense, what would "Happen to most People", and if people think this is a result of "Normal Mental-Health" on the Targets part, then what is this all being done for in the first place? If "Snaps" are the Desired Result of the Agents, then they will eventually happen, and will happen to anyone being Targeted—but what "Kind and Quality of People", Agents, does this to a person? And these Psychological-Snaps, the Product of Negative Psychology-Traumas and Torture, are what the Agents involved want: for Documenting and "Giving Control" to them , by Creating Manipulated-Dirt. They know that this is "Normal given the Context", based on Psychology, which is why they do things this way: it is DESIGNED THIS WAY INTENTIONALLY.

Until these Agents CREATE PSYCHOLOGICAL-SNAPS, THE ONLY THING THEY HAVE IS LIES, DECEITS, MANIPULATIONS, THINGS TAKEN OUT OF CONTEXT, AND "PROGRAMMED-PEOPLE" WHO ARE "FALSE-WITNESSES"—they do not have anything until these Psychological-Snaps occur. Which is the POINT of the Agents: they "Need Psychological-Snaps" to "Look Justified" and so they can actually have something to SELL TO THE PUBLIC, "That can not be Disproved". THAT IS BECAUSE EVERYTHING THEY HAVE UP UNTIL THAT TIME IS A MATTER OF FALSIFIED EVIDENCE, LIES, DECEIT, AND MANIPULATIONS. And as always they SELL the Public a Different Story. As it is, these "Snaps" are not as they are "Portrayed" either, but they become "Center Pieces" for the Fake-Profile.

Keep in mind that this Government Program at it's core is Illegal, does not "Exist"—it is a Conspiracy, and these Government Agents can claim anything they like. Once they PUSH YOU into "Psychological Snaps" and Reactions/Crimes that come about as a result, they can Lie about the "Circumstances and Events" themselves so they are not Implicated. This is of course Bullshit and Illegal. Having said that this depends on the Individual Agents involved as well, some are more Corrupt than others. How Dangerous they are, and how hard they "Push You" into Psychological-Snaps, depends on how Corrupt they are, and as I have learned how Psychotic they are.

Having said that the Citizens in all cases are Breaking the Laws and being Documented as well. And the same "Rules of Thumb" apply here as well. Everything they do for these Agents, can be taken "Out of Context" just as with the Target, and be Misused down the road if need be. What the Citizens fail to realize is that YOUR RIGHTS are THEIR RIGHTS, and what they agree to do to you, can be done to themselves as well. These Citizens are Idiots. Again this is a Documenting of Crimes, as much as it is the "other elements" which I will get to here. And in the case of the Citizens they are taking part in Conspiracy as well, which is what Terrorists do, never mind that the Government Agents and Citizens alike are "Terrorizing the Target". I'll leave that for anyone reading this book, who has taken part in this nasty bullshit before, to ponder this. In the meantime you may want to see if you are on "The No Fly List"...

This is just a "Creating and Gathering of Dirt", Manipulated-Dirt", and it is done THROUGH ILLEGAL MEANS as it is the only way to "Get It" in the first place. Manipulations, Lies and Deceits, Staged-Events combined with Programming, Traumas and Torture, and Fake/False-Profiling, and more Lies and Deceit. If they really had anything on the Target they wouldn't have to Lie in the first place, that is what Laws are for. Further more these Agents have, and are, Breaking all of the Laws (Never mind the Constitution), in fact they are in violation of International-Law and god only knows what else. Amnesty International would have a field day with these folks, especially the ones I am dealing with now. With one major-exception inside of this Situation, I have for the most part lived clean by comparison (a Psychological-Snap on my end brought about by Intense Harassment/ Violations by the Spooks and their Citizens). On the Internet this stuff, this Situation is called "MOBBING" and "Street-Theater". In Europe it is called "Mobbing" as well, and usually is known for taking place only at work. Unless you are me. I know of two people here (in Europe) who have been Mobbed at work. I knew of five people back in America "Mobbed" like me, in my "Situation". It is really FASCISM, Closet-Fascism, and I call it flat out ugly. It is also Non-American, and against Democracy.

The Psychology of MK-ULTRA and the Psychology used to Create Fascist-Movements: Dissidents and the "Old-Soviet-Style"

In many respects what I go through resembles a formerly Secret-Project known as MK-ULTRA, done by Government Agents, except this "Situation" is designed for the Masses, as much as it is for the Individuals/Targets who are "Targeted" by this nonsense. In short this "Situation" of mine has two parts: one is to "Make Compliant Citizens", to "Traumatize and Shock the Target" (in this case me) in order to get the Target to comply with other peoples wishes (Which is really Government: Government Agents are Orchestrating the whole thing), and the other part is to Indoctrinate the Masses, the Public, so that their hands get dirty in regards to acting without Morals or Ethics towards another Human Being—and again all at the bidding of Government Agents. This "Situation" in it's entirety is about Creating Compliant Citizens in regards to Government. Which is what Fascism is, it is about people who are Indoctrinated, Compliant, Obedient, Subservient, and Blind, in regards to Government. People do what they are told by Government (Agents) and WANT to do it, regardless of what it is they are doing, or harm to others—just like the Nazis, Nazi Citizens, with the Jews. People behave like people in a Cult. What these Citizens fail to realize is that YOUR RIGHTS are THEIR RIGHTS. These people lose their minds, which go right out the window—because they have been Indoctrinated and Organized into a "Small Fascist Movement", by Government Agents, and these Citizens are too Stupid to even realize it. It resembles "The Wave" (movie), or the movie "Die Welle", because it is, and is that way by Intent and Design. Like any Cult, these People are Programmed, and Programmed to BELIEVE what they are TOLD, and to DO what they are TOLD—the Psychology Used here is Illegal, and is why Cults are watched so closely. Yet the Government is Doing This, and Programming People, it's Citizens, in regards to itself. And that is what is happening.

These Idiot-Citizens act like "Cult" and are too Stupid, Programmed, and Brainwashed to realize it, never mind "Looking at the Bigger-Picture". Fascism is when Government becomes Cult. You can not have "Fascism within a Democracy" and claim that it is a Democracy.

Now for those of you who do not know of this MK-ULTRA Program, it "was" (and IS) a Government Program, investigating and Implementing Traumas on Selected-Victims, for the sake of "Mind-Control", for the sake of Brain-Washing, for Controlling people's Minds and Actions, for breaking their Will, to get "Desired Results and Responses" by those doing the

"Programming", the Torture, the Brain-Washing and "Mind-Control". This MK-ULTRA Program was eventually officially exposed, and a number of victims received "Financial Compensation", no doubt a piddly amount, for their "Participation". The American Government, people involved with this Program actually, destroyed the Files and Evidence of all of this MK-ULTRA Program however. No doubt to protect their Sick-Corrupt-Selves—we are literally talking about the behavior of Nazi-Experimentation on the Jews, except this was done on people in North America. This is a matter of "Record", and in spite of what Lying-People's mouths say, IS STILL GOING ON, and in fact HAS REACHED A POINT OF "PUBLIC-IMPLEMENTATION". It is in fact being used on people like ME, people in this "Situation". And it is being done by very Corrupt-Agents. You have to be Sick, to want or enjoy doing this to anyone.

The difference here between MK-ULTRA and this "Situation", one of the major differences, is that this "Situation" involves the "Psychology of Creating Fascist-Movements"—quite literally. I am sure that just like with MK-ULTRA which involves Torture, Shocks, and Traumas for the sake of Controlling a person's Mind and Actions, this "Psychology of Creating Fascist-Movements" also has a Government-Name for this Program—YOU CAN BET ON IT. Actually this part applies to the Public who are "Used" for the Idiots they are, and the MK-ULTRA part applies to the Target. Notice that the key point is that EVERYONE must obey the Government Agents, everyone in my "Situation" IS obeying Government Agents and their portable, local, personalized-propaganda, and Localized-Fascism. Whether it is the Target, or the Public-Citizens, all "MUST OBEY" THE AUTHORITY OF A BADGE—GOVERNMENT AGENTS. You have to be blind and stupid to not see that this is Fascism, which apparently these Idiot-Citizens are.

So in Reality, this "Situation" has TWO PARTS and TWO TYPES OF PSYCHOLOGY INVOLVED, which compose it. One is the Psychology for "Mind-Control", MK-ULTRA, the use of Traumas and Torture for COMPLIANCE which is used on the Target, and the other component is the Psychology for Creating a Fascist-Movement, using Programming, for Controlling the Public—this Government-Program for doing so has a Specific-Name, I just have not heard of it. With these two parts combined, we have my "Situation". And to top it off, Dirty Techniques and Tactics are Used from COINTELPRO.

So all must "Obey Authority" and be "Compliant to Government" is the theme. As it goes on, Government Agents gather "Manipulated-Dirt", on all involved, for Misusing and Manipulating down the road if need be: and this includes "Helpful Citizens" as well.

At least that's what The Clown keeps on telling me, that I must "Respect Her and Obey Her". That and "I am the best in the Business" which is in reference to Herself. Judging on this Agency's "Reputation", maybe she is right, but it is because of Non-Professional-Idiots like Her that they have the "Reputation" they do in the first place with many people—especially with the other Agency back in America. Perhaps better "Psychological Screening" of employees is a good place to start, never mind the Corrupt-Agents who started this mess in the first place. Both Agencies need better Employee Screening, and better "Profiling of their own Agents".

This Agency in Europe will have to excuse me, but for now I am wearing the shoes. Having said that for my readers, as I will get to, I will cover Agencies and Agents (or not do to time constraints) in a separate piece, as I feel that not enough people understand that this does indeed come down to "Individuals" within Agencies, the "Quality and Type" of person with the Job, and not just Agencies themselves. The Clown, like everything else she "Taught Me", taught me this by "Accident" as I learned of the extreme Corrupt-Lowlife-Agent. And I am, for all reading, "Wearing the Shoes". I may be in the midst of this weird "Government Program and Bullshit", but having said that She is not even following the Rules in regards to This-Mobbing or Her Agency. And I have seen Agents here in Europe and in America who obviously did not have all of the facts in regards to Me, or who they were working for—they had no idea of how Corrupt The People in charge were. They were "Field Agents" following orders, some of these people were GOOD people, and no doubt Manipulated on some points along the way. But back to business.

The funny thing is, if people involved in this nonsense, this "Mobbing", knew how these Agents are they would be afraid of them—never mind that they would be afraid of their Government. Quite frankly the Public has no idea what it is Involved In when it gets involved with a Target, and this "Situation". We are talking about the Psychology used in Creating a Fascist-Movement, Programming used by Cults, we are talking about the Dangerous Misuse of Psychology, period.

Notice the similarity between Nazi Germany and the Jews, and this "Situation". There are "Targets", insensitivity towards other Human Beings is ENCOURAGED, Citizens turn on Citizens, Government Agents encourage the Negative-Behavior and ACTIONS towards the "Targets", in the end it "Feels Good" for the Public to follow someone else's orders and commands and to "Go with the Flow", no one wants to Think for Themselves, and all "Guilt is Dispersed Amongst the Collective". They inflict Traumas and Torture to the point of NO LONGER THINKING TWICE ABOUT IT. This is THE RECIPE for how Fascism is created. People are TOLD what TO BELIEVE, what TO THINK, and what TO DO. People simply feel "Entitled" to violate another person, simply because someone is TELLING THEM TO, and they think they are "Justified" based on what they THINK THEY KNOW. They are TOLD WHAT TO THINK, are TOLD what to DO, are TOLD what to "SEE AND EXPERIENCE". They stop THINKING FOR THEMSELVES and stop TAKING RESPONSIBILITY FOR THEIR OWN ACTIONS—any "Guilt for Actions Taken", for "Abnormal-Behavior" in regards to the Target, is Dispersed amongst the Collective.

What makes these people Idiots is that they do NOT know, they have been Lied to and Manipulated. But since these Agents have "Badges and Authority" no one questions these Agents, or their MOTIVES, the way they would normally question anyone else in life "Saying These Things" or asking them to "Do these Things". Many studies of Fascism have shown that people "Respond Blindly and Compliantly to Authority" even to the point of hurting others—this is what happens with the Public and these Agents. Normally ordinarily people, the Citizens, would be "Suspicious" and question the "Motives" of anyone "Saying the things they are Saying" and of anyone asking them "To do the things that they are asked to Do"; yet in this Mobbing no one questions a thing—of course these Agents are also Misusing Psychology though, and Programming.

It turns into Compliance to Government-Fascism. Here a "Badge and Authority" clearly make a Difference, without actually saying anything about the "Quality and Type" of people with the Badge—it gets down to a "Stereotype" a "Simplified Form of Thinking". In this case "A Badge" = "Telling the Truth", which it most certainly DOES NOT. The Agents engaged in this COINTELPRO and this "Situation" are in it because of their own "Warped-Personality-Defects", because they are Corrupt, they are not "In This" for "Truth". To top it off, no one would Normally "Act Out Against Another" simply based on what they have heard from someone else. So these Citizens get Used, and Manipulated for Agent's Motives which in my case and others, has nothing to do with TRUTH. In the end it is about Blind Compliance to Government, Authority, in the end it is about predisposing people to Fascism, and to create Fascism. These Agents are engaged in, and implementing, an Illegal-Program, put in place by Corrupt-People over them, with a Similar-Mentality.

These ILLEGAL-PROGRAMS, MOBBING AND COINTELPRO, marks Citizens, Targets, to be Destroyed through the "Legal-System".

This is I believe, about a Future Planned for New World Order. If you do not want to believe that last statement, knock yourself out, but it sure as hell is NOT about Democracy; and if you look on the Internet this "Mobbing" is prevalent across the World, and seems to be "Growing in

Volume". More people are being "Targeted", and "Terrorist-Attacks" are the "False-Pretense" for "Investigating" Harmless Citizens: good people who care about the "State of Things" with an "Opinion or Viewpoint" contrary to "Corrupt-Government". Or more to the point Corrupt-Politicians within various Governments who are responsible for the Corruption in the first place. And these Corrupt-Politicians have all the Power to begin with.

These are the ones who have written and "Passed" these ILLEGAL-LAWS, IMPLEMENTING ILLEGAL-PROGRAMS, to "PROTECT THEIR POWER AND SELF-INTERESTS". They are going after "Citizens who are in the Way". They are "Preparing the Way" for their "Next-Step", a World-Economic-Crash, and New World Order. They are "Designing and Creating" a Population which attacks others within it who are against what Government "Says or Does", in this way they are "Training the Population" to "Conform to Government" and to "Attack those who are Critical of It" and to attack those who are for "Positive-Change". In essence those Without a Brain, Idiot-Citizens, are being "Trained to Attack" those who are "Capable of Thinking for Themselves". There are "Those who Do as They are Told" attacking "Those who Do as they Think on Their Own". You have those "Who do not Think for Themselves", those WITHOUT SELF-EMPOWERMENT, attacking those "Who practice Independent-Thought, Who Think for Themselves", those WHO ARE SELF-EMPOWERED.

Only an Idiot does not see this as ILLEGAL IN AMERICA, and only an Idiot does not see this as Fascism. Only an Idiot does not see this as the Creation and Implementation of Fascism. This is about "Democratic Ideals" is it? This is about, and because of, "Terrorism" is it? When "Terrorists" have already CHANGED AND TRANSFORMED A DEMOCRACY INTO A NATION OF FEAR, BORROWING "TACTICS AND MEASURES" FROM A POLICE-STATE, THEN "TERRORISTS HAVE ALREADY WON". This is when the Ideals of America and Democracy SHOULD BE RAISED AS A STANDARD BEARER, THE CONSTITUTION RAISED AS OUR FLAG, AND "DEFIANCE BE GIVEN" TO ANY THAT WOULD SEEK TO DAMAGE OR MAIM OUR COUNTRY BY "CHANGING OUR COUNTRY FROM WITHIN", AND GETTING US TO "TRAMPLE OUR CONSTITUTION, FREEDOM, RIGHTS, AND IDEALS" OUT OF FEAR. This IS what being an American is about. This IS when being an American counts the most—those willing to give up our "Constitution, Rights, Ideals, and Freedom" for "Terrorists" can KISS MY ASS. Are you an American, or are you "More Afraid of Terrorists"? DON'T TREAD ON ME.

Just like any Documented-Fascist-Study, these Idiot-Citizens, based on a Badge and knowing absolutely NOTHING about the QUALITY AND TYPE OF PERSON WEARING THE BADGE, simply give in to their "Authority": a Job-Title, and a Badge, a god damn STEREOTYPE like this tells you anything about the Type of Person with the Badge. Stereotypes are SIMPLIFIED FORMS OF THINKING, "shortcuts" for people who do not want to think. The last time I checked Corruption, is the Misuse of ones Job, among other things, so we are talking about Extremely Stupid Citizens.

So in the BEGINNING, these Corrupt-Agents start spreading LIES AND RUMORS, and a False-Profile about the Target. My Fake-Profile was also built on previous Lies and Manipulations as well—which happened behind the scenes, and started "Secretly" as a direct result of this Pedophile-Cop. A lot of people begin enjoying talking "Gossip and Nasty-Innuendos" about the Target, they get "caught up in it" and encourage others to do it. I call this the "Rumor Mill" and it is actually a Tactic to get people "Desensitized in regards to the Target", to create "Ideas" for the Agents involved (Ideas for "Rumor-Mongering"), and to hopefully "Dredge Up" information about the Target which people were keeping to themselves. The Agents can and do then take any Information that comes up as a result of this "Rumor Mill" and make it "Fit" this Artificial-Profile, it is a "Given", it is Standard Operational Procedure. It is Government-Agent Text-Book-Procedure. And of course like any Gossip, and because of the Intent of the Agents, "Being Accurate" is NOT IMPORTANT here—I noticed this CONSISTENTLY. Further more this "Feeding Frenzy of Gossip" is "Built-Up" by Government-Agents, who through Lying and Manipulation, encourage people to take the Next-Step which is Actions. Like the "Gossip and Character-Assassination" this too "Builds-Up" and is again encouraged and Controlled by Agents. It usually starts off slowly, so that people get their hands "Wet", slowly Filthy, slowly Indoctrinated, gradually Desensitized in regards to the Target.

In the Beginning

All of this starts off with Heavy Lying and Manipulation of the Public, by the Government Agents. Unproven "Allegations" (In my case "things/and points" taken out of Context) are given to the Public which they believe, it all starts here. Facts, Details, Lies and Manipulations are made to "Convince People" that you are someone other than you are, and that you have done something you have not done: a False-Profile with a Set-Up/Framing, and/or Programmed-Event. Surprised by these False-Allegations, Set-Ups and Manipulations, the Citizens then "Begin to see you Differently" as the Feds have CONSTRUCTED an Artificial-Profile which centers around these False-Allegations, these Heavy-Lies, and at least in my case, Manipulated-Evidence. They take "Pieces of Your Life" and "Events from Your Life" and REMOVE CONTEXT to "Build a Fake-Picture of You" which they "Sell to the Public". Like a Messy-Divorce in which the two fighting for the House and Money, Deliberately "Take Pieces of each Others Lives" Out of Context to "Misrepresent and Mis-Portray" the other—to get what they want. It is all Character-Assassination, and in a Messy-Divorce what each person is doing is Creating a Fake-Profile of the other. These False-Allegations many times are built on "Large-Lies, Manipulations, and Deceits", extreme accusations in order for the Public to take part in this—normally nobody in the Public is going to take part in this nasty behavior simply because you smoke marijuana, or other "Piddly Offenses"—thus the Agents have to LIE BIG and Manipulate-Big. Unfortunately I made this easy for them, but I had never dealt with anyone so Corrupt before, or this "Situation" before, so they got away with it at the time.

This is where "Knowledge is Power" and why the Agents tell people NOT to speak to you about what is being said, about why this "Situation" started (initially they do not—it is beside the point in my case—they "had started" years before), and is why these people are told "Not to listen to You", and why anything you do say gets "Whitewashed and Re-Written" by the Feds when you do say what is True: never mind that they will "Re-Write" anything you say "To Fit" this Artificial-Profile anyhow. Remember the Agents are Lying, and have Constructed an Artificial-Profile, and therefore take what you have said that is True (or anything else for that matter), and "Make it Fit and Conform" to this Artificial-Profile. They do not want that the Public knows it is being Lied to and Manipulated.

The Public is "Surprised" by what they "Did not know all Along" and begin to "Cleverly Perceive You" and "Figure You Out" in this "New Light"—the "Light of Character Assassination". Especially over time as your "Actions and Words" get continuously "Whitewashed" and "Re-Written" by the Agents—as they keep Lying to Maintain the Manipulation of the Public. The Agents CONSTANTLY take what you SAY AND DO OUT OF CONTEXT, to fit Their Lies and so they can continue to get away with Manipulating The People—with Programming the Citizens.

The Artificial-Profile is built on these Deliberate-Lies and Manipulations, and everything you do and say, is "Made to Fit" this Artificial-Profile. People's Perceptions are "Deliberately Altered" to "Fit this Profile", they become, and are Programmed, around the Artificial-Profile so that "All they See and Experience" conforms to this Artificial-Profile. They "See and Experience" everything you "Say or Do" within the "Context that is explained to Them", as they are "Told how it Is", so they think that what they "See and Experience" with you, is True. They "See You Differently Than How You Really Are" and as a result "Experience You Differently" and "Interact with You Differently" as a result. They have changed in how they Perceive You and Think about You. Quite simply since they "Think you are Different" than how you really are, they "Begin to Re-Write Everything" that involves you, through these "New and False Perceptions". YOU THE TARGET have not changed, THEY HAVE. Thus how they act with you, changes as well. They Literally treat you as if you are someone other than you really are, and in fact it is THEY who have changed.

The Creation and Maintaining of an Artificial-Profile: What it is Like

If you are in this "Situation" I recommend keeping friendships with Intelligent, Questioning, people, not Ignorant Jackasses—as I have come to learn the hard way. Quality is what counts here.

Everything you do, and every INTERACTION you have with people, gets "REDEFINED" by these Agents, as they "Invent Causes and Excuses" for why you say and do, what you say and do. They literally "Re-Invent" what it is you are saying and doing to "CONFORM" with this "Artificial-Profile" they are "Selling". Thus it is "CONSTANT", or else people will stop seeing you in this "Light", because IT IS NOT TRUE. Therefore if they are going to MAINTAIN this Artificial-Profile, this CONSTANT-LIE, they have to persist in it Daily, otherwise "Normal-Life" and Normal-Perceptions of people, take over again and they begin to see the Reality and Truth of you again. Except for Major Idiots who swallow this Programming entirely, people will BECOME DE-PROGRAMMED. So this Lying and Manipulating of The People around you is Persistent and Constant as they do not want PEOPLE QUESTIONING THINGS, or "Looking on their Own" at things. They do not want The People involved "Thinking on Their Own" rather "Thinking what they are Told to Think".

As the Target, it is hard to believe, never mind Frustrating, as you watch a bunch of people you "Used to Know", turn into Chronic-Ignorant-Jackasses, never mind complete strangers, who somehow feel they "Know All" to the point of making "problems" with you: when in fact they haven't a Clue. You would expect that friends and Family could see the obvious DIFFERENCE IN THEMSELVES, but they do not. You remain the same, and THEY CHANGE completely. This is the same Psychology used for creating Fascist-Movements and Cult. The People you Used to Know have become Programmed, no longer "Seeing things as they Are" but as they have "Been Convinced and Told" to see things: their Perceptions have been Altered to Conform with the "Desired Viewpoint" of the person and people doing the Programming.

All of it Deceitful and Hostile, and the Citizens, friends, and Family are completely Clueless to this—yet somehow "See All" in "Your Clever Ploys to Deceive Them" as you speak out, and tell the Truth. Of course "You can not be telling the Truth" as it is Contrary to what they have been Told by Agents, therefore of course you the Target must be Lying. Never mind that "They Never Knew You" and that "You have been Fooling Them All Along": quite frankly they are Complete Idiots, Imbeciles, Dupes, and Fools. They are getting their hands dirty for Criminals, who Lie to these Citizens Daily, for the sake of Maintaining an Artificial-Profile.

I have given some of the most obvious points to Programmed People, I mean there was no way that they could not see the Truth of such basic obvious facts, and yet when the Agents were done with them, these people had their heads up the Agents Asses again.

To do this they Used "Staged-Events" which of course were Manipulated, and over here in Europe this has been done again, in September of 2010. Taking "Themes from my Past" which I was writing about, and Misusing "Tapes" based on Manipulations again. Things were Taken out of Context, in September, all over again, and what was "Given to People" is not all that was "There": naturally Context was Removed, and people were Programmed as a result. Just like in America these Jackasses get away with "Pulling the wool over the Citizens Eyes", except this time I am conscious of what is going on.

In short, people, Citizens, friends, Family, lose their heads and turn into Complete-Idiots. They end up "Making Weird Nasty Behavior" with you, and are so ignorant, that they actually believe "They are helping you with a Mental-Problem (old Soviet B.S.)", and actually believe that "You Deserve what you are going Through" because of "Things which you have done and are guilty For"—which if you have been paying attention, is a complete Lie and Farce. They are doing things based on Character Assassination, "Things and Events" taken Out of Context, "Hearsay and Gossip", Manipulated Evidence (How it "Looks" as opposed to how it IS) used to Manipulate and Program both Individuals and the Public, Manipulated/Programmed-Events, and in the end a Fake-Profile. Again this is what Agents have to do, and say, if they are going to get people to take part in this "Mobbing". And eventually these Idiots are Used to Create Psychological-Snaps: these "Snaps" being for "Manipulated-Dirt" which makes the Agents Artificial-Profile, "Airtight". I will get to these later. The following below, is from a letter to someone, and gives a good example of Programming the Public, Creating Fake-Witnesses, and what it is like.

An Example of Programming the Public and Individuals: How Programmed-People and "What they Experience" is not Reality, and is not "Correct": How Agents Create "Fake-Witnesses"

The Public began Targeting my kid at the age of two. These Spooks "Sold This" as "Helping me out with my Sensitivity" in regards to my kid and life and people. Which was a Lie and Farce. I did not have a "Sensitivity Problem", this was in fact what The Clown was trying to "Condition" by Misusing and Manipulating the Public. The Public, absolutely stupid, allowed itself to be Misused to "Try and Create the Very Problem" it was Supposedly "Helping Me Out With". Talk about Idiots.

The first time this happened was at the playground, I was with my kid, the two of us alone, except for a woman walking by which I did not see at the time. My kid was looking in her direction and began crying for no reason. I turned quickly and saw this woman stop what she was doing but not before I saw what was happening—she was pointing a finger accusingly at my kid and making all of these facial gestures like she was extremely angry at my kid. Of course my kid was scared and I was shocked—this woman quickly walked away while I was still in shock. Who does this, what type of person does this to a two year old? Well lets look at some points here.

In real life and Context, a "Disturbed Person" does this. A Hateful Person does this, and for no reason I might add.

Behind this "Surface Level", behind this "Cover-Story", it was being done for The Clown. It was Her idea. And again this says something about HER. Now again as we look at the "Real-Level" , the "Spook-Level", who does this because they were asked to? Either a "Nasty, Disgusting, Hateful, Person" in the first place who did it because they could care less, or someone who has been "Convinced To", under False-Pretense, because of Lies and Manipulating in regards to that person, and done by The Clown and Her Buffoons. Now these Spooks can also claim it was "One of our Spooks", and maybe it was—but Job-Title does not work here as an EXCUSE, it is STILL nasty and disgusting and says something about the Spook involved. In fact this whole thing does—it says something about The Clown. SHE decided to Target my kid, a two year old, and did so, to "Get to Me". It is SICK, not a matter of being Professional, and it is extremely "Personal" on Her end as well (She can misuse this, you figure it out).

Now lets look at this from "Normal Life". What is a normal reaction to this behavior? Minus my "Shock" most parents would be angry at the person who did this, this is more than an issue of "Violating Social Norms" it is disturbing. And it is sick.

Well after this, this began to escalate, and to happen more and more, until my kid and I were both constantly and continuously attacked when we went outside—this became constantly reoccurring. Like anyone when this is going on, you begin "Looking out for your Kid". And I might point out here we are no longer talking about "Normal Life" here, as the Context of "Normal Life" does not even exist anymore when it is happening all of the time. It reached a point of, "If we go out, we will have problems". The Public, Idiot-Citizens began doing this all of the time—each person only looking at "Their small Part" unaware of how much this was actually going on—that is because they were Compartmentalized, being Told only Enough to do what they were told to do—not Told the real reason for Doing So. If people had understood HOW MUCH this was going on they might not have taken part, as they would have understood a few things—they would have understood they were being Used, Lied to, and Manipulated to do something nasty and for very nasty reasons.

So I became watchful which is understandable given the circumstances. There is not a parent on the planet who would not have been. Now on this point the Spooks claimed I was "Paranoid and Sensitive"—it is an obvious Lie. Apparently they were already claiming this. This was the "Sales-Pitch" to the Public. They TOLD the Public that I was "Overly-Sensitive" in regards to my kid which again given CONTEXT was a Lie and Manipulation on the Spooks end. THEY JUST DID NOT GIVE THE PEOPLE CONTEXT, they did not give the Truth, Proper-Context so that people could make an informed decision. So the Public believed that they were "Helping Out" with a "Sensitivity Problem", that I was "Paranoid and Sensitive" when in Reality they were being Used and Misused to "Create a Problem". They in fact were being Used to "Create and CONDITION" the very problem they had been TOLD I had in the first place, which was a Lie and Manipulation. These Citizens are Idiots.

UNTIL THE CLOWN STARTED THIS BULLSHIT THERE NEVER WAS A "PROBLEM". And She made these Citizens look like Idiots by using them to "Create and Condition" a Problem that was not there in the first place, and Used and Misused them to create the problem She said I already had. They were Brainless-Idiots. None of these people even considered "How much this was happening to me and my kid".

And as time went on, of course I was "Looking" and as my kid was "Singled-Out and Attacked" of course I had a problem—no parent would not. But these Citizens were TOLD I had a "Sensitivity Problem" and so THEY SAW THINGS THIS WAY, NOT AS THINGS REALLY WERE, NOT REALITY. The "Context" had been REMOVED, HIDDEN. This is called Programming. The People were "Programmed to See" what they were TOLD TO SEE. Their Perceptions had been Altered so that they SAW AND EXPERIENCED THINGS differently than how they REALLY WERE. This is a way of "Controlling People", of Manipulating Them for Control of their Minds. It is a "Form of Subtle Brainwashing", as The People are "Programmed like a Computer Program", functioning as they have been "Programmed To", and it starts by Telling Lies, and "Explaining Actions and Experiences in Advance with Lies".

Thus they ask people to do something and based on knowing what they are doing to me and my kid, these Spooks of course can predict the Outcome and Reactions based on this. And since they know this, they have "Lies Prepared" which they give to the Citizens for our Reactions which the Spooks know we will have—and based on all of the Information (which they do not give to the Citizens) and based on "Normal-Behavior given the Circumstances" in regards to me and my kid. In other words they know what my kid and I will be like based on all of the facts and the Real-Level, the Spook-Level which is going on, which they have NOT shared with the Public. They know that how my kid and I are in this Weird-Situation is actually not "reacting differently" than anyone else given the circumstances. In other words given this Weird-Context, the psychology involved on my kids end and mine was normal and predictable. And based on this the Spooks "Prepare Citizens" with Lies, explaining "Our Behavior" before it has occurred, so that they do not see it as it REALLY IS. The People are Whitewashed with Lies so that they "See Things Differently" than they really are.

My "Reactions" and my kid's had already been "Explained" and in a way which was Deceitful, before these Citizens got involved for "Particular-Events and Circumstances": "Staged-Events".

And so many people became CONVINCED of something NOT TRUE, and became more Convinced, after Taking-Part in "Helping the Spooks" and "Seeing and Experiencing these Things": and they were not even "Seeing and Experiencing Things" CORRECTLY. They were not seeing and experiencing things as they REALLY WERE, because I and my kid were different than how they had been Told, they were merely experiencing things as they were TOLD TO, as they were TOLD THINGS WERE. In reality it was all a Lie, Deceit, and Manipulation. This is Programming: it is Modern Day Brainwashing.

HOW ELSE DO YOU EXPLAIN PEOPLE'S WARPED VIEWPOINT AND HOW THEY ARE ABLE TO "COMMIT SUCH BEHAVIOR"? THIS IS CULT-BULLSHIT AND BEHAVIOR, which is why Cult's are "Watched Closely" and is why "This Psychology is Illegal". But you will note that when GOVERNMENT DOES IT, it is "Fine and Legal". The real reason being that the Public, the Majority of People, do NOT KNOW this type of thing is going on.

The idea of "Normal Life" got thrown out the window from the beginning. When you know based on many experiences that you and your kid are going to be "Singled Out" and "Attacked" then this is Reality—it is not even a matter of "Being Conditioned", this happened every time we went outside. So like any parent, I was watchful and protective—again like ANY parent. And based on Life and Circumstances. The Context is not "Real-Life" as it "Normally Exists" at that point, but it is Reality. These Idiot-Citizens however were not TOLD THIS. Context was Hidden and Deliberately Removed so they did not know the Truth of the matter.

They were Told to "Do This" and Told "Real-Life" was that me and my kid were simply "Sensitive". In other words both the Citizens, and I and my kid, were on "TWO DIFFERENT LEVELS" and only I, my kid, and these Spooks knew the Real-Level/the Spook-Level. These Citizens were Told to "See Things", and THEREFORE EXPERIENCED THINGS, from a "Viewpoint" that was not accurate. They did not "See and Experience Things" as I, my kid, and these Spooks did. They were Ignorant-Dupes, "Telling Themselves" and "Seeing Things" from a "Real-Life-Perspective" in their Minds, when it was NOT. Two-Different-Levels, Two-Different-Perspectives, and the Citizens had the FALSE-PERSPECTIVE, and as a result of what these Spooks Told them, and as a result of these Spooks "Telling them WHAT TO SEE": their Perceptions and Feelings had been Manipulated—Programmed. And to top it off, once these Citizens "Experienced these Events" with these "False-Perceptions" it became "Burned in their Minds" this way, it became "Engrained this Way" in their Minds. And again FALSE. They became "False-Witnesses" in one sense, and the Spooks became "Airtight in any Argument to the Contrary (Truth)". THE CITIZENS DID NOT EXPERIENCE THINGS AS THEY ACTUALLY WERE—thus they were "False-Witnesses". CONVINCED of what they "SAW and EXPERIENCED" without actually knowing the Truth.

Now lets look at it from a "Normal-Perspective" and put one of these Citizens in this instead. Lets switch things. So I can take any parent here and this will work; and just to make a further point if I were to go out the door today and do this to some child and parent, the parent would have a problem with this and me, and the child as well.

So the parent knows their kid is being Targeted (and the parent as well) and every time they go outside, every time nasty behavior is being engaged in, and being engaged in by strangers and everywhere you go. Playgrounds being a favorite. Naturally they are going to be watching and this is not about "Being Sensitive". To top it off people are doing something they should not be doing, and have no business doing. That these Idiots "THINK THEY HAVE A GOOD REASON" is irrelevant to the Targets—the Targets know that this is not True. And they know this based on Information, Reality, that these Idiot-Citizens do not know—and because they have stopped asking questions. So naturally as time goes on both the parent and kid become "Wary"—and you can claim this is "Conditioning, a Conditioned-Response" but it is not. This is a matter of Reality at that point. In fact all would be the same way—including you. But the Citizens do not get this. And over time and many "Staged-Events" you begin to resent the Stupidity of these Idiots. If for no other reason, because if this was being done to them they would not like it, and they would understand how "Nasty this Is". Further more they would understand you DID NOT have a "Sensitivity Problem" in the first place, that this was merely a Lie, Farce, and Sales-Pitch to get Idiots to take part in this in the first place. Which parent would not be disgusted by this? To top it off, these Citizens are being Used to TRY and "Condition Sensitivity in the First-Place"—which makes them extremely Stupid and Gullible. Having said that, this is not about being "Conditioned" it is about what is REALLY GOING ON. It is about "Protecting your Kid" from nasty people, it is about what any parent does—and this applies to this Weird-Bullshit as well. Because no matter how you slice it, it is Ugly-Behavior, it is "Cult-Behavior", and it is Detrimental to you and your kid, no matter what.

So as you see, "Sensitivity" is not an issue here: REALITY IS, and theses Citizens do not have the Proper-Context for "Seeing things Correctly", because Agents gave them an IMPROPER-CONTEXT and IMPROPER-PERCEPTIONS to BEGIN WITH. This is because this is about what the AGENTS WANT, it is not about TRUTH. And in this case this was about ATTACKING ME AND MY KID, it most certainly was NOT ABOUT "Helping us out with Sensitivity". If the "Roles had been Reversed" for any Citizen they would have gotten this. But these Idiots were Programmed and could not think their way out of a paper bag at that point. THEY BECOME CONVINCED THAT WHAT THEY EXPERIENCED WAS REAL, WAS TRUE. And once they have/had "Experienced things Incorrectly", this "Incorrect-Experiencing" is how things "Remained in their Minds"—they became "Convinced that what they had been Told was True". THEY BECAME "FALSE-WITNESSES".

Quite obviously from the Target's Reality (THE REALITY) what the Citizens Experienced was False. Remember this is about the TRUTH, not about "Perspective and Viewpoint" in one sense. They were TOLD "What my Viewpoint and my Kid's Was" (Sensitivity), they were TOLD what our Reactions would be and WHY (Sensitivity), and these Citizens were TOLD so/given answers based on what was NOT TRUE. The Reality was, I was behaving like any parent who's Child is Targeted by strangers—and again the ONLY WAY this exists in REAL-LIFE, is in this "Situation". Normally NONE OF

THIS BULLSHIT would have been going on, BUT WITHIN THIS CONTEXT, which was Real-Life, OUR BEHAVIOR WAS NORMAL.

THEIR BEHAVIOR WAS NOT "NORMAL", they had been turned into Brainless-Jackasses.

Their own "Perceptions and Viewpoint" was based on being TOLD WHAT MINE AND MY KID'S WAS, and they were NOT TOLD THE TRUTH ABOUT OUR "VIEWPOINT AND PERSPECTIVE". Thus their "Viewpoint and Perceptions" as a result of THESE LIES AND DECEIT, were FALSE.

They are given the "Surface-Level" and given a "Normal-Life-Perspective" when what the Target is EXPERIENCING is not based "On Normal Life" it is BASED ON THIS "MOBBING". This "Mobbing" is what the Target is experiencing, and it is REAL-LIFE. If you do not explain this to the Citizens and try to "Sell-Things" as "Normal-Life" to the Citizens (in regards to the Target), then already CONTEXT AND THE TRUTH OF THE MATTER ARE BEING REMOVED. Duh! You can not claim "Normal-Life" in regards to the Target, can not Sell-It this way to Citizens and claim to be giving the Truth to the Citizens. Real-Life is what the Target is experiencing, and their "Thinkings, Actions, and Perspective" based on it, based on REALITY. When you are in the Target's Shoes, Real-Life DOES INCLUDE this Mobbing. When these "Agents Sell a Perspective" here, they "Sell a Version" in which the "Mobbing is NOT Included". This is Ridiculous.

If you "Omit Context" from "Real-Life", then you are not talking about "Real-Life", you are talking about a "Version that does NOT REALLY EXIST". THERE IS ONLY ONE VERSION OF THE TRUTH. Thus they Sell to Citizens a "Level that is not Real" to keep them in the dark. The Spooks get people to look at things from a "Normal-Life-Perspective" and if you are really going to understand the Truth and the Target you have to look from a MOBBING-LIFE-PERSPECTIVE, otherwise Context is being REMOVED, and it is being REMOVED TO HIDE THE TRUTH. This is Hidden so that people can be Manipulated and their Perceptions Altered—it is one of the first steps for Programming People into "Seeing and Experiencing Things" inaccurately. And these Idiots became "Convinced after these Staged-Events" and based on "What they Saw and Experienced", none of which was Real. They were "False-Witnesses" with a "False-Experience" and this is Government-Agent stuff, and it is Standard-Operational-Procedure in regards to this "Situation", COINTELPRO, Programming, "Mind-Control", because it is the same "Dirty-Tricks" with the same Psychology being Misused.

Look at it CLEARLY AND TRUTHFULLY.

The only ones NOT ACTING NORMAL WERE THE CITIZENS, and these Idiots furthermore were doing things to someone who in Real-Life they would normally NOT DO, and to top it off they would not do these things to anyone FOR THESE REASONS. NEVER MIND A KID. In other words in REAL LIFE, for the "Reasons Given" they would not do this Bullshit to anyone in the first place.

This is the TRUTH of the matter, but these Citizens can no longer see this. They are convinced and believe that they are "Justified and Entitled". How is that for Screwed-Up? THIS IS PROGRAMMING. People literally let someone else "Control their Thinking and Actions", based on their "Perceptions being Controlled". Once these Spooks start "Controlling your Perceptions" you can forget it. This is where Programming STARTS. And over TIME, once they have been in contact with someone long enough the Citizen simply takes things for granted, and believes everything they are TOLD, and believe in everything they are TOLD TO DO. Eventually the "Programming Process" becomes complete and Citizens accept things without question. Once they do, they are Manipulated Heavily, and in the case of The Clown who is extremely Corrupt, they are absolutely MISUSED. The Clown gets satisfaction out of doing this. And this says something about The Clown, about HER PERSONALITY. In fact this entire Scenario I have just described and explained DOES.

This Bullshit became so "Prevalent and Extreme" it reached a point where my kid at the age of TWO (2–3) "Could Tell" that we were about to have a problem with someone and would stick her tongue out at them (And the Citizens did do something—my kid was accurate). Think about this. And again this is part of The Clown's Personality—"Being Extreme". This is why you will see these points about Her come up again and again, and eventually it was Her undoing with me.

So if you are paying attention this is how it worked and WORKS:

They Plan-Events ahead of time. These are called by me Staged-Events.

They tell people how the Target will "Think, feel, React" to the Staged-Event.

The Target's reality dictates how the Target will "Think, Feel, React"; the Reality is this "Mobbing" and what is going on within it.

The Citizens from the beginning are not given all of the Information, CONTEXT is REMOVED and Information is HIDDEN. They are not given the Target's REAL-PERSPECTIVE or ALL OF THE FACTS.

The Citizens are given a "False-Viewpoint" of the Target, they are given a FALSE-PROFILE of the Target. And then they are Lied to, based on this.

From this False-Perspective, FALSE-PROFILE, of the Target that is given to the Citizens, the Citizens are "Prepared and Pre-Lied To" in regards to the Staged-Event. They are TOLD "The Target will react this way because of this"—"Because he is Sensitive" or whatever fits the Manipulations and

Lies they are Selling at the time. The possible and most likely reaction/s are "Explained to the Citizens" from a "Viewpoint of the Target" that is FALSE. And the funny thing is these Agents are able to "Predict your Reactions" based on knowing what all of the facts are, and based on knowing that your "Viewpoint and Reactions" are based on what they are doing to you and your "Mobbing-Life-Perspective". The Agents "Predict your Reactions" based on your Real-Profile.

Thus when the Citizens "See and Experience the Event" with the Target, they in fact do not have a "Clear Picture" of what is going on with the Target, and therefore do NOT have a clear idea even, of what they are "Seeing and Experiencing". They THINK they know what they are SEEING, they THINK they know what they are EXPERIENCING, they THINK they know why things are the way they are, they THINK they know what the deal is with the Target, they THINK they know the PERSPECTIVE of the Target, they THINK they know why the Target "Reacts or Acts" the way that they do within the Staged-Event, and the final irony is, they "Think, See, and Experience This" based on being Lied to and Manipulated. The "Reality is they are not Thinking on their Own", they are "Thinking, Seeing, Experiencing" based on how they were TOLD TO. Common-Sense, Reason,

and the word "NORMAL" go right out the window.

NO ONE IS "THINKING AT ALL", THEY ARE LETTING SOMEONE ELSE DO THEIR "THINKING AND ACTIONS" FOR THEM. A Programmer is "Doing the Thinking and Actions" the Citizens are TOTALLY-CONTROLLED, THEY HAVE BECOME BRAINWASHED.

They haven't a clue as to what "Normal" is at that point. And it is "Looking at the Bigger-Picture" that points to this, and that points to REALITY, yet they do not see this.

And "Normal" here is: They would "Normally" not be doing this to someone in REAL-LIFE.

What else do you call this? Take your choice, Programming or Brainwashing? You decide. But no matter what "Name you give It", "An Idiot by any other name is an Idiot".

They "Normally" would not be "Attacking a parent and Child" even if they were the most sensitive people in the world. In REAL-LIFE, they would not Think this way, Act this way, or even "Experience things this Way" if they were "Wearing their own Perceptions". They would in fact realize that in REAL-LIFE if they made this Type of Behavior with someone, they would "Be Attacked", there would be "Unpleasant-Consequence" as a result. And how they would "See and Experience Things" would be different as a result—their "Experience" would be different, BASED ON REALITY.

All Citizen-Behavior to the Contrary, points to Programming. Their Thinking, Actions, and how they "Experience Things" is not their Own, it is someone else's, and that is the point of Programming. To get people to do things, say things, believe things, they Normally would not, in order to Affect the Target, to Attack the Target, to "Get at the Target", to do what the AGENTS WANT them to Do. It is Control of their Minds, and it is in fact, a form of Brainwashing. The Citizens become Whitewashed and as a result everything you say or do is Whitewashed. It is also Standard Operational Procedure for Spooks and Agents, it is also ugly and disgusting if you are Me, my Kid, and at times my Girlfriend.

Quite simply these Idiots are WEARING ANOTHER PERSON'S "VIEWPOINT"—and it is not even the Agents REAL VIEWPOINT, merely one which THEY WANT THE PUBLIC TO WEAR in regards to the Target. That is what "Programming" is, and it is also known as Brainwashing.

The Public becomes "Mind-Controlled", their "Minds Controlled" by the Altering of Their Perceptions, they SEE AND EXPERIENCE things, the Target, in a way that has nothing to do with the Target. They are what I call "FAKE-WITNESSES" to many a Staged-Event/Programmed-Event as what they "See and Experience" is FALSE, and therefore they are "Witnesses" to something which has NOT occurred as they "Think they have Experienced It"—this is DELIBERATE ON THE AGENTS END. Thus they are FAKE-WITNESSES, capable of saying they "Saw and Experienced What the Agents Told Them" when in fact what they "Experienced" is NOT as they believe: and this is how the Agents want it—that is the point of the Programming. To turn everyone against the Target, by Manipulating the Public into "Seeing and Experiencing the Target" as different then they really are. These Agents would have no way of getting the Public to do so, otherwise. Thus for this Mobbing, it is "Standard Operational Procedure".

"Programming" is actually Modern-Day-Brainwashing, this is the "NEW-SECRET-FORM" being used and used OPENLY without most realizing it. They "Accept and Absorb" the Agent's "Thinkings" which have been put into their head: the Citizens "Accept and Absorb" the Programmers "Thinkings" and then DO as they are TOLD, REGARDLESS OF HOW HARMFULL IT IS to another. OF COURSE THIS IS BRAINWASHING, and of course this is what is used in Cults. It is Illegal.

Now just as this "Programming is Invisible" and is known of by another more common name, so it is with this Mobbing and Staged-Events. These "Staged-Events" and this "Mobbing" and all that these Citizens DO, is also actually TORTURE IN A FORM THEY DO NOT SEE. This is Modern-Day-Torture. They are literally inflicting Traumas on you DAILY, yet somehow "Think and Expect" "Nothing to Happen"—this is dumb beyond belief. What has happened to their minds, Programming, is invisible, and what they are Doing and Engaging in—Torture through Traumas, is invisible. NEITHER ARE INVISIBLE TO THE TARGET OR GOVERNMENT AGENTS however. And the Behavior of these Citizens, like any Cult, becomes Extreme once they are Totally-Programmed. Once they are Totally-Programmed, the Corrupt-Agents, begin going for Psychological-Snaps by Misuse of the Public, for Creating Manipulated-Dirt, which I will get to shortly.

So for my readers the piece above gives a very good look at these Idiot-Citizens, and shows how "Reality goes out the Window" for these Citizens because of Agents. It shows how Programmed-People "Experience Things Incorrectly", in a way that is FALSE. And because of WHO THEY TRUSTED. It shows the effect Programming has on the Public, how Staged-Events are USED FOR PROGRAMMING THE PUBLIC, how the Public has no Idea, and how Individuals are turned into Fake-Witnesses (That example was in Europe).

It is Illegal and Disgusting, never mind that the Citizens would not like it done to themselves, and that you are innocent. And all because they are Ignorant-Jackasses. Everyone around you thinks "They Know" what is going on, and why they are doing what they do, and think that YOU ARE CLUELESS, and in point of fact they haven't the slightest idea.

So then this "Mobbing" happens, and people who "Know All" begin Saying and Doing Things, they have no right doing and saying based on what they do not know. They have been totally fooled, and have become Total Dupes and Pawns, they are absolutely Clueless, and actually working for the "Bad Guys". They would not like these Agents based on what I know, would not have liked the Racist-Dispatcher or the Murder (try watching a kid bleed to death sometime—and in no small part due to "Race"), and would not have liked the Pedophile Cop, it is obvious. But as it is, these Citizens got Used and Misused for attacking my Mental-Health, created and Inflicted Traumas, Engaged in Torture as it is defined (and by the U.N.) and were used to Push-Me into Psychological-Snaps. They were Used Deliberately for this, and it is OBVIOUS. And this was to create "Manipulated-Dirt" by reducing my Mental-Health to absolute zero. What took place was Extremely-Illegal and Extremely-Dangerous, and if it happened to me, it no doubt is happening across America, as it is a Government-Program to do so.

Now. As the Citizens get more COMPLIANT, things become more Extreme. They are asked to do MORE, various Laws are Broken in even more different ways as more is done to you—it is all Illegal anyhow, and you suffer immensely. These Dirty-Deeds and Staged-Events are used to further

Program these Idiot-Citizens. They become "Extremely-Desensitized" and "More Convinced" as time goes on, "Doing Things" and "Thinking Things" (what they are TOLD) without question.

It is a long list of what these Idiot-Citizens do. People come and go from your house or flat at will, Citizens "Violate Social Norms" in the "Rudest and most Offensive and Unrealistic of Ways", take and do what they want, damage your property, even stand around AFTERWORDS laughing about it for you to see. This is COMMON, many times they have the "Perpetrators" stand around afterwords smirking or laughing and in a way that "Lets you know" that they are "In on the Crime"—which you will note is not "Normal or Common Behavior", especially for criminals. But since they work for the Feds, for Agents, then it is fine because they have been Told-To, and are "Protected by the Feds". IN OTHER WORDS AS LONG AS THE CRIMINAL BEHAVIOR IS ON BEHALF OF THE FEDS, THEY HAVE NOTHING TO WORRY ABOUT. This is American, and a Government Program at that?

This is "LEGALIZED-CRIME", Reversed-Legality, and is in regards to VIOLATING ANOTHER CITIZEN. In this case the Public is attacking an Individual, not the other way around. You can say what you want, but when I smoked marijuana, it was in my damn flat, and my decision to smoke, and the only person "Harmed by this Behavior" was me. The Feds and their Indoctrinated-Citizens are getting away with Everything, yet YOU are the "Bad Guy", and none of these people even know you. You are legal (I smoked pot, so what), and they, ALL of them, are ILLEGAL yet you suffer and the Laws only "Work for Them" as they are working for "The Law"—Corrupt-Agents. It is Cult, it involves "Programming People and their Perceptions", and getting people to act in ways they NEVER WOULD, and for Government at that. It is "REVERSED-LEGALITY".

And keep in mind what this was really about: a Pedophile Cop and Racist Dispatcher.

It is SURREAL BEYOND BELIEF, EVERYTHING HAS BEEN REVERSED AND IS BACKWARDS. Your mind is in turmoil and you are "Cooking" like a pressure cooker, over it all. It continues to go on DAILY, and WHEREVER you go, thus you are always under Intense Psychological Pressure, without any relief as it "Follows you Around"—see the movie "The Game". They literally have people follow you around, both Agents and Citizens, who are Deliberately-Conspicuous and who "Interact with You" in Public through a "Variety of Ways". In fact "Whistling" is one of their "Standard-Calling-Cards" to let you know that they are following you around and watching.

And again this involves the Agents who are conducting surveillance, and then the "Citizens they Telephone for Help". This involves a list of

Clueless-Idiots, Citizens who have agreed to "Help the Agents-Out" in regards to you. So the Agents who are watching you, have a bunch of Citizens come and "Help them Out", and again it is done in a way to let you know they are watching. A lot of these Citizens feel that they are "Big Time" as they get to play "Government Agent" the attitude of some of these people is Unreal. Their Heads become "Swollen" as they go about "Taking part in This" and for reasons they do not even know: they are Complete-Imbeciles. You as the Target experience Disgust among other things. Many times as you go about your life in Public, Staged-Events and Programmed-Events are Set-Up using these Brainless-Idiots, and these Staged-Events are simply used to further Program the Citizens—these are "Used as Proof", that the "Target Is" who the Agents say "The Target Is". It is a Lie and Farce, of course.

"Staged-Events" and the Build up to Psychological-Snaps

So as you begin to "React to this weird Bullshit" your "Reactions are Deliberately Misconstrued" to "Fit a Fake-Profile", "Actions and Words" get Re-Written, and the Public becomes more Programmed. The Agents "Stage-Events", using Citizens, in which each Citizen has a part to play, "Lines and Actions" like a "Script for a Play", which centers around the Target. The Target is in the "Center of this Play", and the "Words and Actions" are Nasty, Unpleasant, Surreal, and many times Aggravating, or Traumatic. In many ways what the Citizens are taking part in resembles a "Sting Operation" and they get to "Play Law Enforcement". It is the same kinds of Tactics Used by Law Enforcement for making a "Bust", a "Sting Operation", with the "End Goal" for Agents being to "Damage and Traumatize" the Targets Mind—the Citizens at that point could care less. Many of them in fact think they are "Cool" for taking part in this.

It is about attacking the "Mind and Psychology of the Target", it is to create Trauma by way of Torture, and is to Traumatize the Target, and for DESTROYING THE MENTAL HEALTH OF THE TARGET. And the more Corrupt the Agents are, the more Dangerous they are, Using and Misusing Citizens for the Agents own "Agenda": in my case they were being Extreme, and Pushing-For Psychological-Snaps, which are Dangerous to all involved. In my case they Needed-To as the whole thing they had constructed ran the Risk of being Exposed and coming down on their heads. For the Target every day is a constant barrage of this, some days worse than others, and since it is going on "At Work", you are forced into going. As this goes on the Citizens become more Desensitized, more "Whitewashed and Programmed", until they accept everything the Agents say WITHOUT QUESTION. At this point they are Totally-Programmed. When it reaches this point the Agents begin "Pushing for Psychological-Snaps"—keep this in mind.

These Staged-Events are also about making the Target "Look Bad". And there are many ways they have of Manipulating the Citizens, to do this. This involves Work, Individual Friends, Family, "One to one dealings with People", Strangers, People who you see consistently and or daily at various places. Such as people who work at the various stores you go to routinely—eventually they become involved as well. These Agents follow you around, do so through the "Citizens Helping" and make sure it is "Conspicuous", they make sure that you know you are being followed around, and everywhere you go consistently, The People become involved. Again it is Surreal, and "Staged-Events" are Arranged with Citizens that you know and don't know, which are many times Aggravating and Traumatic. Again the point is to Traumatize the Target and to "Attack the Mind and Mental-Health" of the Target. Does any of this sound Mentally-Healthy? And as these Staged-Events take place, your Reactions to a "Variety of Events and Set-Ups" is noted and "Re-Written" by the Agents, to fit this Artificial-Profile. The Agents are doing several things here at this point—they are Profiling you for a Real Profile kept to themselves, they are Traumatizing you the Target, they are Pushing-For Psychological-Snaps with the Target to "Make Dirt" (Manipulated Dirt), they are Selling an Artificial-Profile, and they are Programming the Citizens. The Citizens are acting as any Cult would. These Agents are Convincing Everyone that you are someone else entirely different, that you are "The Person They Say You Are", which of course is a Lie. They of course do this, to get the Public to Attack You.

Over enough time, and after enough "Staged-Events" everyone becomes "Convinced" that YOU ARE A MAJOR CRIMINAL and a MAJOR MENTAL CASE, everyone becomes Convinced that YOU ARE THE LIAR, and everyone becomes Convinced that you are the MANIPULATOR, and many times the "Aggressor", and they then BEGIN ACTING OUT AGAINST YOU in regards to what they "Believe". In Reality, you are "Wearing all the Accusations" that the Agents are guilty of themselves. None of which is True in regards to yourself, all of it True in regards to them—the Agents. So in the "Beginning of This" things "Escalate" and the Target begins to experience "Nasty Actions and Words" from the Citizens. As the Target you wonder what the hell is going on, and WHY. Very quickly, as people suddenly turn on you, and ACT LIKE TOTAL STRANGERS with you, and as a COLLECTIVE, you are at first "Confused and Shocked" and eventually begin to get Pissed Off, which is "Normal given the Circumstances". Who the hell wouldn't? You have NOT changed,

THEY HAVE, and they start ganging up on you, "Mobbing You". It is Nasty and Surreal as an Understatement.

Having said that since because these Idiots think you are "Someone else other than you Are", they think they are "Justified" in what they are Doing, which they are NOT. They are Idiots, acting out on False-Information, and have NO RIGHT doing what they are doing in the first place. Imagine how Pissed Off you would be, based on this alone—again to be Pissed Off about this is Normal, and it is no wonder why. Having said that the Agents involved get this, they get everything going on with everyone, as they are the only ones with all of the Information.

Later, over time, as you "Retaliate", as you Release-Hostilities, which is going to happen and these Agents know it, these Idiot-Citizens think you are the "Aggressor"; how god damned stupid can anyone be? This Releasing of Hostilities is going to happen because it "Keeps going on and Escalating", and reaches Extreme-Proportions, as you are constantly Violated and Traumatized. Never mind that in "Real Life" anyone would "React Angrily" to what they are doing in the first place. And nothing has changed with you, you just come into work one day, and it "Starts". You have not "Changed" these Idiot-Citizens have, and eventually you begin to "Cook" as it becomes Absurdly-Extreme. The Citizens "See You" as the "Aggressor" which is a pathetic joke.

They were the Aggressors, they were Illegal, they were Stupid and Non-Questioning, and they Committed Conspiracy, never mind Violated your Rights, engaged in Torture and Traumas, and did so for a very LONG TIME before there ever was a "Releasing of Hostilities" or a "Psychological-Snap". You are Responding to Nasty-Behavior initiated on their end first, yet they are Programmed to Believe that they are "Justified" and if you "Respond or React" that they are "Victims". If they are engaging in Behavior against someone, that they would normally Not Do, because they know it is wrong, yet do so and then think "You have a Problem" when you "React", then we are talking about people who are Stupid-A-holes, Not "Victims".

What had I ever done to any of these people, that I worked with as one example, that warranted this kind of Behavior from them? It was ABSOLUTELY UNJUSTIFIED and eventually led to several Psychological-Snaps, and one big one. Talk about ignorance on their end, it is no wonder. Over the years I had several Psychological-Snaps, and in some cases "narrowly avoided" several which would have been Fatal for the Citizens involved (Which the Feds knew—they obviously did Not like the Particular Citizens they were using). And in several of these cases, this was done with Total-Intent, and knowledge of the "Outcome" by the Agents involved. None of this is the

"Product of Normal Life" I might add, or was a result of "Normal Me". At this point "Life is not Life" and the Target no longer begins to resemble his or her self—which is of course the point. At this point, no one is their self, and it is being done to "Create Manipulated-Dirt". Quite frankly the only way this can happen to anyone, the Psychological-Snaps that I experienced, is by being in this "Mobbing"—there is no other way.

So after "Wearing the Agents Perceptions" long enough, the Public, becomes Totally-Brainwashed and does everything for these Agents without even questioning it. And eventually, once the Agents have this "Level and Degree of Control" over the Citizens-Minds, they begin Using and Misusing the Citizens, PUSHING-FOR PSYCHOLOGICAL-SNAPS with the Target, to "Acquire Manipulated-Dirt" in regards to the Target. This is about Destroying the Mental-Health of the Target through Traumas (and by Legal Definition—Torture), until their Mental-Health has DETERIORATED to the point that their "Mind Breaks or Snaps" in some form. These are "Preceded by Releases of Hostility" by the Target, but eventually as Traumas are "Heaped On", and as this "Continues Through Time", eventually these Psychological-Snaps happen. Which is what these Agents want in the first place. They want something "Damaging and Incriminating" in regards to the Target which "Proves the Agent's Points". Again the Target has been "Marked to be Destroyed" through the "Legal System", and all of this is incredibly Dangerous and Unhealthy. It is the Illegal-Use of Psychology, Psychology Used by Cults, and "Everyone but these Corrupt-Agents" gets Used and Misused.

What is a Psychological Snap and "Manipulated-Dirt": for what Purpose: Defining These

So when you have a "Situation" where Reality goes out the window, where Citizens throw Reality and Common-Sense out the window, you are left with a "Situation" that is Surreal and Unhealthy. As Citizens engage in Weird and Traumatic Behavior, Singling-You-Out, for these "Events and Behavior", your Mental-Health eventually Deteriorates. The Citizens are completely Ignorant about this, and Incredibly Stupid as well—they do not get this, and do not get that they are being Used and Misused for this very purpose. And in the beginning there are what I call "Hostility-Releases" by the Target, and then later Psychological-Snaps. It is these "Snaps" that are dangerous, the Hostility-Releases can be unpleasant for both the Target and the Citizens involved, but it is these "Psychological-Snaps" that result in Damaging-Behavior, and they are in fact a product of a Psychologically-Damaged-Mind. This is for Acquiring-Manipulated-Dirt.

Manipulated-Dirt is when the Target has a "Psychological Snap" and violates the Laws and Citizens, "Looks Bad" in some extreme way, and it is the product of this "Mobbing" not the Product of Real-Life. It is also the Product of a Deteriorated Mental-State, Deteriorated-Mental-Health. It is the "Product of Torture and Traumas", which are "Disguised" so the Citizens do not realize this. Many of these Staged-Events are Traumatic and Modern-Day-Torture in disguise, so that Idiots do not recognize it for what it is. And the only way anyone could ever get there is by being in this "Mobbing", we are not talking about "Normal" here. It is the "Cracking and Breaking" of one's Mind and it is the Product and End-Result of Trauma and Torture. Which is what the Feds and Agents have been doing to the Target. Only the Citizens do not know this, because they are IDIOTS. So a Psychological-Snap is when one's mind "Breaks" and the Target "Reacts Out" in regards to the Public, in some way that is Indicative of an Unhealthy Mental State, and in a way that is Illegal usually—that is because at this point the Target has been "Pushed Over the Edge", and no longer "Sane".

These "Staged-Events", Traumas, and this "Mobbing" and all that these Citizens DO, is all actually TORTURE IN A FORM THEY DO NOT SEE. This is Modern-Day-Torture. They are literally inflicting Traumas on you DAILY, yet somehow "Think and Expect" "Nothing to Happen"—this is dumb beyond belief. What has happened to their minds, Programming, is invisible, and what they are Doing and Engaging in—Torture through Traumas, is invisible. NEITHER ARE INVISIBLE TO THE TARGET OR GOVERNMENT AGENTS however.

In fact the Work-Place is a common "Trauma-Shop" for this "Situation" as one "Must go to work to Survive"—and the amount of "Dread" is Un-Real. You literally have to force yourself into going to work, knowing what will occur—the Mobbing, Traumas and Torture, and Surreal-Behavior. Also outside of work wherever you go people working for these Agents, Citizens show up, and interact with you in "Strange Ways", which the Normal-Public for the most part does not notice (whistling). This "Situation" follows you, as Agents have you under Constant Surveillance: and Conspicuous-Surveillance at that. This is to keep "Constant-Psychological-Pressure" on the Target, and even outside of work, you experience Staged-Events and Programmed-People, and very much Nasty-Behavior done by the Citizens. You are "Constantly Watchful" as you know you are being Followed Everywhere. You also know to expect Nasty-Behavior, Violation of Social-Norms, and at some point Staged-Events.

Now over time and with many Traumas, your Mental-Health Deteriorates. You begin to have "Hostility-Releases", and there are several versions; the ones I speak of here are very different than the ones I refer to as being a part of "Normal-Life" and a "Response to the Violation of Social-Norms", these are different than "Normal Behavior and Reactions" for Violation of Social-Norms. These are a Precursor to a Psychological-Snap and are a result of this "Mobbing". The Agents know this.

Things begin to spill, boil-over, as you are Constantly-Violated and Traumatized by people who have no idea of what you have been through, have absolutely no idea of Who you are or What the "Situation" is really about—I did not either at the time, but I did know I had not "Deliberately Interfered with an Investigation" which is what these Lying Feds said I did. In fact nothing I had done in my life "Justified" what I was going through. This is dangerous and it leads to "Psychological-Snaps", the result of no "Rights or Protection". For a LITERAL COMPARISON you are a "Jew at the height of Nazi Power"—you have absolutely NO RIGHTS and you are treated like Dirt, it is sick and disgusting. The Citizens have turned into a bunch of Brainless-Animals, and Feel-Justified in what they are Saying and Doing to you, and they expect you to "Accept It", based on everything they don't know. Which is STUPID, and TWICE AS STUPID AS THEY ARE BEING USED TO DESTROY YOUR MENTAL-HEALTH.

They have become Imbeciles, and over time, they BECOME CONDITIONED into simply accepting, what it is they are TOLD by these Lying Agents. They stop practicing Independent-Thinking, they stop practicing INDEPENDENT-THOUGHT, and simply let these Liars do their Thinking for them, never mind letting them "Pick their Actions" for them. And of course this is what the AGENTS WANT and is what they are TRYING TO ACHIEVE WITH INDIVIDUALS AROUND YOU, AND THE PUBLIC. Again it is a Sick-Misuse of Psychology, it is what is used for Fascism and Cult—and as we know people in Fascist-Movements and Cult become "Brain-Washed", Programmed, and Fanatical. Somehow this is all lost on these Citizens...And in the meantime while all of this is going on, you experience "Weird and Unpleasant Circumstances", Traumas Implemented for Control of YOU, and to Control Your Mind and your Actions and Reactions. Again this is MK-ULTRA Material, and it is ILLEGAL. It is by the United-Nations Legal Definition, Torture.

Again for those of you who do not know of this MK-ULTRA Program, it IS a Government Program, "Exploring the Use of Traumas" and Implementing Traumas on Selected-Victims, for the sake of "Mind-Control/Programming", for the sake of Brain-Washing, for Controlling People's Minds and Actions, for breaking their Will, to get "Desired Results and Responses" by those doing the "Programming", the Torture, the Brain-Washing and "Mind-Control". This MK-ULTRA Program was eventually officially exposed, and a number of victims received "Financial Compensation", no doubt a piddly amount, for their "participation". The American Government, people involved with this Program actually, destroyed the Files and Evidence of all of this MK-ULTRA Program however. No doubt to protect their Sick-Corrupt-"Careers". At any rate it is a matter of "Record", and in spite of what Lying-People's mouths say, is still going on, and in fact HAS REACHED A POINT OF "PUBLIC-IMPLEMENTATION". It is in fact being used on people like ME, people in this "Mobbing". And it is being done by very Sick-Corrupt-Agents: I think the reader can see at this point that this is Dangerous.

Agents Creating "Psychological Snaps" to Benefit themselves and their "Cause"—"Creating Manipulated-Dirt" to Make their Artificial-Profile "Air-Tight"

Now since I have had problems with Corrupt-Agents, I will be speaking mostly within this Context. In fact throughout this book, this is just about the only Type I speak of. Now, this entire Government-Program may be Illegal (It most certainly should be—it is NOT AMERICAN), but there are those within it who feel they are just "Doing their Job", and who have some "Professional-Conduct" within "This Context". They are "Following the Rules and Protocols", they are following a "Code of Conduct" of their Agency and for this "Mobbing". These people are more Fair, Realistic, and "Just" about things, all things being "Relative" here on this point—they are also LESS DANGEROUS. There are others who are Totally-Corrupt, like the Feds who got me in this "Mobbing", who will Lie, Falsify and Doctor Evidence, Manipulate people based on these Lies, put "Citizen-Dupes" in harms way knowing full well the Target is near a "Psychological-Snap", and who will construct an Artificial-Profile based on these "Snaps" and "Manipulated-Dirt", Staged-Events, Lying, and a Misuse of the Citizens. These are the real "Pigs" and my observations are based on these Types of Agents. With that in mind, back to business.

These Idiot-Citizens are Used to Create something that was not there in the first place, so that THEN they can be Totally-Convinced—they are Misused through Psychology to Traumatize and Torture the Target until the Target's Mind-Breaks. THEN the Agents have something "Provable" and which "All can See", except of course the Citizens do not really know anything in the first place—this merely "Proves All" and makes everything "Said and Done" "Airtight" for the Agents: which is why these Agents are going for these "Psychological-Snaps" in the first place. They want "The Dirt", the Illegal Actions, that happen as a result.

It is for the Agents, who now have "Witnesses and Victims" in regards to the Target, however it is not really needed at that point for "Convincing" the Citizens. BUT it does make it all "Look Beyond Dispute" and "Beyond Reconsideration" in regards to the Citizens who have no idea of what is going on. And that is what these "Snaps" are for—for "Creating Believable Manipulated-Events" with "Convinced Witnesses", it is "Manipulated-Dirt" and this resembles very much what the Agents do with "Programmed-Events" using "Programmed-People". I have experienced these in America and Europe. Without Programming People, there is no "Mobbing", and anyone involved with the Target who thinks they are Not Programmed is an Idiot. Without the Mobbing there would be no "Psychological Snap".

So as I was saying these Corrupt-Agents, are pushing you towards Psychological-Snaps, putting Citizens in harms way, to fit an Artificial-Profile, and to make "How they Look"—how "Correct they LOOK", "Air-Tight". They are doing this to "Create and Acquire Manipulated-Dirt", Artificial-Dirt. They are Corrupt, engaged in Corrupt-Behavior and Activity, and could care less about anyone they use—look at what they are doing to the Target to get an idea of the Type. Having said that, you the Target, should be able to figure this out, and know better than anyone. You have to start "Looking" and putting things together: and whatever you do, do not "Rely on the Perceptions of Others". These Agents will CONSTANTLY try to "Program Your Own Perceptions" through these Citizens. Theses Agents will try to get you to believe that they are "Good People" by way of "Others", when in fact they are Lowlifes who are Bombing You with Traumas. They are in fact trying to "Traumatically Bond with You". This is Sick Shit.

And in this case they are going for Hard-Core-Brainwashing with you the Target. This is what they will try to do to you through the Traumas and Torture, and Weird-Traumatic-Events, and if you begin to experience "False-Objectivity" which I cover in the "Tactics Chapter", then you should be WORRIED. I went through this at one point (1998–99), after they acquired this Exotic-Techno and they found they were "Safe" to "Push some Issues" (they needed to Create more Dirt because they knew I was close to figuring everything out), and they got a lot of FALSE-STATEMENTS as a result of this—this is Brainwashing and you are not even seeing your own experiences as you experienced them, but as you are Told They Happened (Forced Programming—"Old Style Brainwashing"). This is the result of Brainwashing and Traumas and Torture. Strengthen yourself, and tell them to "Stick It", as you sure as hell do not Deserve what you are Going Through, and these Agents are Sick-Bastards. Remember these Psychological-Snaps are for "Creating and Acquiring Manipulated-Dirt" by way of Traumas and Torture, and all of this is Illegal and Dangerous. None of which could happen in Real-Life, or should be for that matter.

Thus these Agents know what they are doing is Dangerous, and they are "Using People" REGARDLESS. As they are Creating a "Mentally-Unstable Person" they KNOW that they are putting Citizens in Harms-Way, who the hell could go through this "Weirdness" and "Remain Stable"? Does this "Look Like" it is about "Mental-Stability", Legality, and a "Good use of Psychology"? Everyone from the Target to the Citizens is being Misused, and when these Agents start Pushing-For "Snaps", everyone from the Target to the "Citizens Specifically Used", are put "At Risk". And remember this is so the Agents can have "Manipulated-Dirt" to "Neutralize the Target" and to "Have something to Use for Court". Again see a movie called "The Game". They don't care about anyone but themselves. And the funny thing is, when these "Snaps" happen, and they have Manipulated Dirt, no one even stops to think of "How or Why you would break the Law", when Government Agents are watching.

If things were to go into a courtroom based on these "Snaps", what you have done under Extreme Psychological Stress, and what the Public and Agents have done to you FIRST, then Their Actions, the Real Circumstances would have to be looked at (Minus everyone Lying), and in fact Things would have to be put into CONTEXT—normally. In fact Normally it would be "They who are in Trouble" as what they are doing is Extremely-Illegal and Dangerous, and Conspiracy is among other things what "Terrorists Engage In".

Having said that things will not go "Justly in Court" because these Idiot-Citizens will do anything for these Corrupt-Agents and if you think about it, ALREADY ARE. And again that is the point of the "Manipulated-Dirt" in the first place. And keep in mind that what these Idiot-Citizens "Think they have Experienced and Seen" is not as it has happened—but they will "Give the Idiot-Version/Programmed-Version" in court: they are FAKE-WITNESSES, Programmed-Imbeciles, and have their head up some Agent's/Programmer's Ass. These Idiots do not get that they have been Used and are being Used. But to make a point, under fair and objective circumstances these Agents would be Exposed and in Trouble—they are committing a long string of Felonies along the way, they are in fact Extremely-Criminal in what they are doing, and how they are "Using People", and people they have Programmed at that. Cults are looked at very closely because of "Programming" and these Agents are engaged in the same Psychology, and also Misusing and Manipulating People with it. They are literally "Controlling People's Minds" so that they do not in fact Experience Reality as it is.

So if things could "NOT HAVE HAPPENED OTHERWISE" under any other circumstances, and you the Target have a "Snap", then do not wear the Blame. I have had several "Psychological Snaps" and regardless of how nasty they were in regards to another person, I have gotten over any "Guilt". If it could not, and would not happen, under any other circumstances otherwise, never mind these Weird-Extreme-Circumstances, then do not "Beat yourself Up" over it. This is due to a Misuse of Traumas and Torture, and unlike the Citizens, there is "No Choice" involved.

These people had "Clear Brains and Minds" when they started, they allowed this Programming to occur, and did so by Trusting Lying-Manipulating-Corrupt-Jackasses, who may "Represent Government" but they sure as hell do not "Represent America, Democracy, or the Constitution". And anyone you have a "Snap" with, and certainly in my case, is no doubt going to be "In up to their Ears" in this Conspiracy and in terms of Actions and Words. It is a good thing they did not ask one of these Citizens to shoot me, just to make a point—everyone is still Responsible for themselves and what they do, regardless of Agents. This is where most people lose their heads, and let Corrupt-People, engaged in Corrupt-Activity, "Do their Thinking" and "Do their Actions" for them. This is where all of the Citizens screw-up. And they did, not just with me, but with Bryan as well. No doubt with Kaye for that matter.

Again, people are choosing to get Involved in Conspiracy and making their own Decisions to do this, and what you end up doing if you have a "Psychological Snap" would not have happened otherwise—it is that plain and simple. Also they have not asked themselves some very basic questions, and simply chosen "To go with the Flow", and allowed themselves to be Manipulated and Lied to, and Broken the Laws in regards to you FIRST, Regardless.

Everyone has to take some Responsibility—so if this happens and you are the Target, do not beat yourself up. Look at what you are going through, realize the Intent and Motivations of the Agents Involved, and realize that this CAN and WILL happen to ANYONE that this is done to: to Anyone in the same Situation, Circumstances, and "Environment". This is what the Agents were going for in the first place. Who among these Brainless-Idiots, the Citizens would be DIFFERENT? This is what these Agents are going for, only the Citizens do not know this, because they are too busy being "Know-It-Alls". Which obviously they are not; so Idiot-Citizens of my past "Tell me what you don't Know"—think about this and let it sink in.

I can easily accuse any of you of being Pedophiles and Racists at this point—that is what you were "Helping out For". Idiots. But as long as you are "White" and an "Adult" what have you got to worry about? At least you were "Well Informed" and on the side of "Justice", and to think I had no idea I was working with a bunch of Racist-Kid-Fuckers all along. Doesn't take much to be a Know-It-All, but you sure as hell "Fit my Profile", know what I mean?

And if any of you with "Personal Scars" from taking part in this twenty years ago, come across this book, I am sorry but you will wear some Responsibility here—I most certainly will NOT "Wear it All". I "Wore Plenty" which is "What, Why, and How" people got scars in the first place, and if you can not "Sort it Out", and want to "Play the Victim" then I suggest you take a good look at what is in these pages, take a good look at what you were involved in, take a good look at your "Own Legality", and ask yourself "Why"? And most importantly of all, take a good look at these Agents, and ask yourself, "How Clueless were They?". Some think they were victims when they were not, they were merely Manipulated and Programmed, and really only one can claim anything, and quite frankly this person was "Used and Misused" by Agents, "Reckless Endangerment". This person was a "friend of a friend" and in the end, this person was Used by these Feds like everyone else, and although I obviously was not proud of myself, I still have to say that what this person was involved in was Illegal, and what they thought they knew was a Farce. As an added point, the Feds knew exactly what they were doing.

Having said that this WAS the one major thing I did in all of this Mobbing and it got "Added to my Artificial-Profile" and "Proved It" all. Which is a sick joke in light of all they had to do to "Achieve It", to Create-It. All the result of a Psychological-Snap.

Later they would use "Citizens" in Staged-Events in which they clearly did not care what happened at all to their Citizens, and in the worst of ways. They clearly were going for getting me in prison, and clearly considered their Citizens "Disposable". And again no doubt based on their desire to "Get To Me" and based on their not "Liking of the Citizen" in the first place. I can safely say this based on what I know now—there is no doubt.

I am not proud of some of my "Hostility-Releases", and even less proud of some of my Psychological-Snaps, but there was indeed a "Method to the Madness" with these Agents, and whatever you want to say these "Snaps", or otherwise, were not the product of Normal-Mental-Health or Normal-Circumstances—these Agents, in case you are not paying attention, knew a lot more than they were telling people. Minus one event, which I "beat myself up for" and for a long time, everything else was a FARCE, LIE, MANIPULATION.

Including this one thing I "beat myself up" for, which was a Direct-Result of my "Work Place Environment" and this was a "Farce" in one sense as this was Deliberately Created and could NOT HAVE HAPPENED OTHERWISE.

Think about this. And this "Nasty B.S. at Work" was a direct result of a Lie and Manipulation of someone else in my life, to Cover-Up a Pedophile-Cop, who got his "Girlfriend" beaten and Raped due to his own Negligence. And it is Lies and Manipulations which got me in this "Situation", never mind "Bryan" and "Kaye". And these further Lies and Manipulations were over a Racist-911-Dispatcher and a Murder-Cover-Up. Think this is all "Coincidence"? Think any Detective with half a brain would think so? Think any Detective with half a brain, would believe these people simply because they "Said Otherwise", and write it all off to merely "Looking that Way"?

What these Citizens do to you on behalf of Corrupt-Agents who are not telling the Truth, is Cruel and Unusual to say the least, it is Torture by Legal Definition, and Conspiracy, and without your Rights and without the Laws Protecting you from these people and this Behavior, you will have a Psychological-Snap, it is not possible otherwise, and you will take it out on some of these people. This is the result of "Traumatizing the Target" and Torture by definition of the word. This is how normal life works—given the circumstances, which is "Why these Circumstances were Created" in the first place.

Anybody thinking they could "Get through this Otherwise" is Lying to their self, or Others. The Agents know this and remember this is what these Agents are going for, in order to have something "Incriminating" which it isn't given the Circumstances. It "Looks Incriminating" but it is all based on ILLEGAL CIRCUMSTANCES AND BEHAVIOR and could NOT have happened otherwise. CONTEXT is removed for them to look good. If you get them into a court these Agents will have to Lie, to avoid a "Disaster", and they will have to Lie anyhow as the Public would be very angry if they knew the Truth. This is why COINTELPRO "Supposedly Ended". Especially people who realize how much "Risk was Taken" with themselves and others. This is certainly how I felt about a lot of this as I "Uncovered It" (in regards to my Past), and certainly how I feel about a lot of the "Stuff I have Uncovered Here" in Europe, as The Clown is worse than most. She is also "Caught more than Most" and is "Waging a Campaign" to "Cover Her Tracks", both with people around me, and with Her Own Agency—which is why She has delayed my Book two years.

As an example of the kinds of things they do which lead to "Psychological-Snaps", you can look on the Internet, there are a lot of people, experiencing a lot of Cruel and Unusual Behavior by the Public and Agents. If you look you will see that it is all Illegal and Extreme. This Extreme Behavior leads to "Extreme Results", over a period of time. This is not hard to Figure Out, this is merely a matter of the Public feeling "Entitled and Justified" in regards to the Target, based on Lies, Deceit, Manipulation, Programming, and in the end Character Assassination.

Ironically the Public believes, that this is all "Normal" and part of your "Criminal Profile", as the Public has been Programmed and Told to "Believe This"—this is in fact an Artificial-Profile, and your "Snaps" get added to this. These "Psychological-Snaps" are to "Totally-Convince-Them" as many times they are "Around" to "See These".

They do not know that the Agents are Pushing-For Psychological-Snaps, and that these are to be "Used to Convince the Public" that you are what they "Say you Are". These Citizens are Complete-Idiots, they are being Used to Create a Criminal-Profile which DID NOT EXIST until they started to "Help Make One", they are being Used to create the very thing that did not exist in the first place—so it can be Misused to Convince the Citizens: it does not get more Sick or Stupid than this. So it can be "Proven" that it was "Always So". These people are Idiots being Used to Create something that was NOT THERE. Further more when a "Snap" happens and a Citizen is the "Victim", this only "Further Separates" the Citizens and Target, which is what these Agents WANT. They have "Something Fresh" and in regards to one of the "Actual Citizens". It is a Farce, a Sick-Misuse of Psychology and People, Manipulations for more Manipulations. Layer upon Layer of Deceit.

These Agents knowingly sent their Dupes, into Bad and Potentially Dangerous Circumstances HOPING that something would happen, not giving a rats ass about their Dupes, merely wanting to "Acquire Manipulated-Dirt" on me—so they would have something for Scaring me, Intimidating me, and for keeping my Mouth Shut, in case I decided to "Speak Out" about what I knew in regards to the Police Department. You will note, that minus this "Mobbing" what they got, would NOT have happened—THEREFORE THEY HAVE TO LIE IN COURT AND THEY HAVE TO LIE TO THE PEOPLE THEY USED. Having said that what they have is piddly in comparison, considering what they had to do.

What it was Like and About: "Shutting Me Up" to Maintain the Lie and Farce

Extreme-Situations are arranged to get a "Desired Response" to fit this Artificial-Profile. So they are Adding-On to it and Building-On it, Constructing-It. The only person who knows what is going on is you, and these Agents. It is extremely frustrating to say the least. You know what they are getting away with and they know you know (this is here in Europe after I sorted many things out). To make it worse they rub it in your face that they are getting away with it, it's Sick. You look around at The People getting involved and realize they are Clueless-Idiots, you experience extreme disgust as these people Violate You in a variety of ways, and in ways that they would not like either. Having said that, at the beginning of this, I did not understand a host of "Things and Violations" going on, could not understand why people felt "Justified", and I thought that people "Were Mistaken", including the Feds. What was going on "Did not Fit" and "Was not Warranted" in regards to ANYTHING in my life that I had done "Wrong"—which is actually an extreme understatement. Talk about none of it making any sense.

And at some point, when it was finally brought up by these Idiots that I had "Interfered with an Investigation", I spoke out about this, but it was clearly too late. They (Citizens) all "Cleverly Perceived" I was Lying about this, and there was no "Fooling Them". This is because it was a Set-Up by the Feds in the first place, and they were "Hoping" to get something they could use "To Put Me in this Situation"—this is why the van with the Telephone Number parked outside my window. So the Feds had no "Intentions of telling the Truth", it was all about Framing-Me in the first place, and Inventing "Excuses for putting me in Mobbing". So when I did Speak-Out to the contrary, the Feds Told these Idiots that I was Lying. And again at that time, to be clear here for my readers, I thought the Feds were "Mistaken" in regards to this, that they did not get the Truth here in regards to anything I had said to Idiot-Tim: a lot of the "Confusion" at the time on my end, and my "Inability to put Things Together", was due to the "Weirdness, Trauma, and Shock" of the beginning of this Situation. It all affects your mind, and your "Mental-Health Plummets". And this is a Direct-Result of Trauma, your brain literally does not function correctly, you can not even begin to think straight, and that is just the beginning.

Keep in mind that I had made myself quite clear about this guy Orlin and these Drug-Dealers to many people outside of work, who worked for the Police and Feds, and in fact when I did start to bring this up, and point this out to people at work, I was "Bombed with Traumas" to "Shut Me Up" and these Idiot-Citizens were TOLD TO NOT LISTEN AND TO NOT BELIEVE: these Idiots actually Took-Part in Covering-Up a Racist-911-Dispatcher and Murder, and behind that, a Pedophile-Cop as well. Until that point I had never had reason to speak of this to people at work. They were Dupes/ Idiots.

If this "Situation" had not been going on, I would have "Sorted this Out" at the beginning, and these people would have had a much "Harder Time" of it all. And keep in mind, this was after "Making myself Clear" from the beginning, in regards to Orlin, many, many times, and having said so to a variety of people, outside of "Work", who worked for the Police—many times. These people had nothing to do with, or any relationship with, anyone from my job at the restaurant, and so these Feds were easily able to "Keep these Facts Away" from the Idiots at my Job. So when I began bringing this up with everyone at work, these Agents actually had these Idiots Shut-Me-Up, in regards to what "They needed to Know". They had a these Idiots do a Cover-Up, in regards to themselves, in regards to Information they "Should know About". It does not get dumber than this.

The only person I had spoken to about Orlin who did have any contact with my "Job" and some people there, was my girlfriend Penny. Penny, who I had repeated this to that I had no "Eternal Grudge against Orlin" at least several times, probably more. And it was to make myself clear so that there would be no problems.

Yet when this all started she "Seemed to have forgotten This" obviously, and Conveniently. I have no idea what Pile of Deceit the Feds Sold Penny, but she was obviously Told something different than the Truth; and she sure as hell didn't tell anyone that I knew, that I had told her several times before, that "I had no problem with Orlin". She certainly did not "Communicate This" to anyone at my Job who was Mobbing-Me, and no doubt she was Programmed: although it is possible that she was "Promised Rewards" by the Feds as well—such as the Feds "Pulling Strings" for her to help her get a good Job Later (I picked up on this later). Having said that I did not like this guy Orlin at all, and for good reason, he was a Lowlife, but I could care less about him as he was not a "Part of my Life" and I was actually glad to see that he was going after real criminals—Penny would not have liked him either by the way for the same reasons I did not. All of this "Swept under the Mat", and Penny strangely enough never brought this up. I still have no idea what they "Told her and Sold her": as to "Rewards" the Feds obviously did "Reward Her" when all was done.

So I initially "Spoke-Out" and said "Hey this is Bullshit I didn't Interfere with a damn Thing", because I had known my "Own Intentions" in regards to calling the number on the van, and I knew my "Own Intentions" in regards to speaking to Idiot-Tim, and I certainly knew where my head was at in regards to Orlin. That said, no one listened, they had been Told I was "Lying" and this was Used as an "Excuse" by the Feds for "Attacking Me More" with the Citizens—the Citizens believed that "I was Lying" and in Reality, the Feds were "Using the Citizens" TO SHUT ME UP.

The Feds were Lying, Manipulating, and Programming these Idiots and Used them to ATTACK ME, to MOB ME, to Shut Me Up, and for the FED'S

SAKE, to PROTECT THESE LIARS AND THEIR MANIPULATIONS. Which I eventually did (shut up), but not before also bringing up Orlin with a few people, and in relationship to these "Plumbing-Ninjas", as I did begin to put a "Few things Together". And again I was "Attacked and Mobbed" with Weirdness and Heavy-Extreme-Traumas, to the point of being "DISTRACTED"—this was actually the Psychological Affect on my Mental-Health of these Traumas. Once the Traumas start you begin "Forgetting Things" (this is common) and this is a result of Hard-Core-Traumas; my "Forgetting" was also a Direct-Result of my having to "Deal with Things and People Daily", I was distracted by the Idea of more as well. So eventually I Shut Up and it all looked like a Moot-Point anyhow; at that point these Idiots were Totally-Programmed and they could care less about "Truth" or what they were doing. Quite a few were "Enjoying Themselves".

The fact of the matter was THE CITIZENS ALREADY BELIEVED WHAT WAS NOT REAL, Believed the Agents Lies, to the point of "Acting Out Against Me" in the first place. They were already "Convinced" and when I spoke the Truth they "Already Knew" that "I was Lying", geniuses that they were. This was the result of PROGRAMMING, and when I spoke the Truth, they only "Saw and Heard Lies and a Liar" therefore they "Attacked Me More" at request of the Agents, because of it. They thought they were "Experiencing Lies and a Liar" as well I might point out, and this IS HOW THEY EXPERIENCED IT, even though THE TRUTH was something else ENTIRELY DIFFERENT.

These people were and are "FAKE-WITNESSES" to "Me Lying" about all of this, and if you were to ask one of these Idiots this today, they would "Swear It Was This Way" when this is not even close to the Truth: THE "EXPERIENCE" BURNED INTO THEIR MINDS THIS WAY, FALSELY AND INCORRECTLY, BECAUSE THEY WERE PROGRAMMED. Complete Dupes and Imbeciles and because they Trusted Corrupt People, and Trusted a Corrupt-Programmer: someone who Misused Psychology to get them this way in the first place—this is an "Abuse of Trust" issue. They "Trusted" the absolute Liars, Manipulators, and Lowlifes, who they would not have liked if they knew the Truth.

IN REALITY I was telling the TRUTH and these Agents were using these Idiot-Citizens, to Shut-Me-Up, so that these Citizens would not know the Truth. THIS IS THE KIND OF STUPIDITY I DEALT WITH, and you will note that the Citizens started it all, based on what they Did Not Know. And the Traumas were Extreme, and based on other Lies and Manipulations which had taken place in my life before this "Mobbing" officially started: the ones which had started because I knew about this Under-Cover Pedophile-Cop. This is all COINTELPRO.

I had had no idea of this as well (until 1997 I believe), or what had been said behind my back, never mind what had been done behind my back. It was not an Investigation, it was all a Dirt-Creating-Mission, one big Smear-Campaign. An "Illegal Misuse of the Legal System" to "Discredit Me" and to engage in "Character Assassination" behind my back so that they could engage in more of the same. It was "To get at Me" because of what I knew in regards to the Police Department. The entire thing was one long drawn out "Dirt Creating Mission": THIS IS COINTELPRO. Your "Tax Dollars" at work.

They had committed previous Lies, Deceits, and Manipulations behind my back, before I was even in this "Mobbing Officially", and had done so, by Manipulating people in my life and around it, and my girlfriend at that time. In her regards they Programmed her for a Programmed-Event, and I did not get to the bottom of this until here in Europe. Which is a good thing. And in this case again, I had reasons of my own for "Talking to Biff", I had reasons of my own for "Saying what I Said", and just like with my Telephoning the number on the van, and my speaking to Idiot-Tim, what I said was not about what was Real. It was about "How I did Things" to "Find things Out", my Tactics and Techniques, which I had in no small part developed by watching Informants, Cops, and Feds, never mind my own Ideas on how to do things. And the Feds knew this—but since this was not an "Investigation", since it was about "Discrediting Me" they "Made this work for Them". Which is what it is "All About" when they are trying to "Discredit Someone" and Engaged in Character-Assassination, IT IS ALL ABOUT HOW THEY CAN MAKE THINGS LOOK—when they are doing this it is NOT ABOUT REALITY OR TRUTH.

And this all comes down to the FIVE (Earlier I have said it is 3 and 4 but have added on) BASIC RULES OF THUMB FOR AGENTS: We'll make it work for us. If it works its good. First "Us" then the Target—this is about the Agents "Looking Good", about Manipulating Public-Perceptions. And Attack what the Target does not like. The last is, IF IT IS EASY, THEN IT IS GOOD. The Clown as an example has gotten away Manipulating People in the dumbest and easiest of ways. Many times it is Reverse-Psychology, The People can not believe they are being Lied To because it is so obvious that they are being Lied To. Brainless.

And again the Feds knew what was Real and True, they knew "What and Why", and they Lied and Manipulated People Regardless—just like they did at my Job at the Restaurant. Between me talking to Biff, "Tapes", a few people I had contact with, and my girlfriend, they had done a lot of

Lying, Manipulating, and Programming behind my back. And just like these Citizens at work, they Used the same Tactics and Techniques for Manipulating and Programming People at this point in my life. So when my "Mobbing at Work" started, they used this Previous-Bullshit, these previous Lies, Deceits, and Manipulations (a Fake-Profile from earlier), to convince these Idiot-Citizens to engage in the worst of Traumas (it is a safe guess—again at the time I was clueless), and again based on WHAT THEY DID NOT KNOW. It was Extreme, and I had several Psychological-Snaps because of it. And I did not even understand at the time why a lot of this was so Heavy, Extreme, and Traumatic. To say that my head was in Turmoil is an understatement, I had all the classic signs of someone suffering from Trauma and Torture, it was extreme. To claim that my head and mind "Were Normal" is absurd, and these Feds got this—this was the point.

I put this together here in Europe, thanks to The Clown, and I realized that the "Themes of these Traumas" had to do with previous Lies, Deceit, Manipulations, and a Programmed-Event(s) which occurred before. If I had actually been "Guilty of their Character-Assassination" I would have figured this out at the time. As it is the Feds were walking on thin ice, by doing this. If I had known what was going on and why, they would have been "Caught" as an understatement: which is why they "Bombed Me with Traumas" when I began Speaking-Out about Orlin, and Not Screwing-Up an Investigation. Quite frankly what they were getting away with behind my back could have been proven for the Lying and Manipulating it was, never mind "Their Motives".

Again all COINTELPRO Bullshit, again all of it involving Standard Operational Procedure: Character Assassination, Misusing Information, Misusing "Evidence" (Taking Illegal-Evidence-Out-of-Context), Lying, Deceit, Misusing Citizens, Manipulations, Programming People's Perceptions, Staged-Events, Programmed-Events, Hiding Information, and Framing. These Citizens were Absolute-Idiots, both the ones at work and the ones before them, before my "Mobbing" started, and everyone GOT USED AND MISUSED. And I got to suffer as a result of these Brainless-Jackasses. And twenty years later, and millions of dollars later, so do my Kid and Girlfriend as well. Does it make sense? How much sense does any of this make? And now to top it off, I now have "Someone in Charge", The Clown, who is even "Worse" than these Corrupt-Jerks from America, in many a way, shape, and form. The Citizens here in Europe are Idiots, and the Citizens in America were Idiots.

They "Knew All" and "Did All" based on "What They Knew". The Pedophile-Cop did not go through a damn thing, and I guarantee you neither did this Racist-911-Dispatcher. I DID. And simply for being in a "Position to Do the Right Thing". Now my Family goes through this Bullshit as well.

And based on what a bunch of Brainless-Idiots were "TOLD TO SEE", based on what these Idiot-Citizens were Programmed To See, Hear, and Experience. And this is why these Idiots are TOLD NOT TO SPEAK WITH YOU, about what is going on behind your back. So the Agents can get away with it, with Manipulating Friends, Girlfriends, Family, Co-Workers, the Citizens and Public. All of it Standard-Text-Book-Tactics.

Needless to say this is Dangerous, and Government has no business doing this kind of thing. It is a "Movement", Localized-Fascism, it is Cult, and most people look at Fascism and Cult as "Kooky and Undesirable" for a reason. This is what we have Laws for, and you can't create a Government Program that exists in "Secret" enabling Agents and Citizens alike, to go "Breaking Laws" in regards to another Citizen. You can not go running around Committing Conspiracy, and Terrorizing and Traumatizing a Fellow Citizen, because Government tells you it is okay. You can not "Single a Person Out" and do this to them—Sick-Idiots do this to another Human Being. This is Ridiculous, and this is what Nazi Germany did to the Jews.

A Final Look at "Psychological-Snaps" and this "Mobbing" and COINTELPRO: The "Bigger Picture", a Summary of this Chapter

This is also MY LIFE and what I have experienced for 18 years (April 2011), all of it under "Humanitarian Intentions" and "Vigilante Justice", total old Soviet-Style Deceit in regards to "Dissidents", where the Target is labeled with a "Mental-Condition" or as a "Problem", and in Reality the Government is opposed to them due to THEIR MENTALITY. This is COINTELPRO, the Target is "Marked for Destruction", through the Misuse Of the "Legal-System" and it is anything but Legal. In my case I had information which could get the Police Department in trouble, on several points, and Corrupt-Agents decided they would "Fix Things", confident in their ability to fool everyone. The "Final Gist of It" is that this was all about a "Pedophile Cop" and a "Racist 911-Dispatcher" and a "Murder Cover-Up". These Citizens were absolute Dupes and Idiots, who got their Hands Filthy, and engaged in all sorts of Corrupt-Illegal-Bullshit for Corrupt-Agents "Covering these things Up."

But the Bigger Picture is that this Program is Targeting Citizens, and Lying and Manipulating the Public, because of people who are a threat to a Corrupt-System. It IS the Ground-Work, the Foundation for, Fascism, and Fascism in America—these Citizens are being "Trained and Conditioned" to blindly follow Government. These Citizens are Idiots, "The road to hell is paved with good intentions". The Agents deliberately Mislabel you as something you are not, be it CRIMINAL OR PSYCHOLOGICAL, or as in my case no doubt BOTH, and then through a Misuse of Psychology and "Staged-Events", REVERSE-LEGALITY, FAKE-PROFILING, they MISREPRESENT who and what you are to the Public (Character-Assassination), to gain support of The People, the Public, the Citizens, the Idiots.

ANYONE can be pushed into having "Psychological-Snaps", anyone can be Pushed-Into-Snapping and in the case of the Agents involved and "Controlling this Mobbing" this is for "Creating and Acquiring Manipulated-Dirt". Look at what this is "Used For".

Again it no doubt depends on how big a "Threat" the Target is. In my case I could have "Caused problems for the Police Department", but so what? Look at what was done, look at what I knew. These people were and are Filth. Yet I PAY for their Severe-Crimes. And no one "Pushed-Them" into doing a damn thing, their "Mental-Health" was how it always is—they were Lowlifes. They had spent, and have spent more time with me, a marijuana smoker (past tense), than taking care of a Pedophile Cop and a Racist Dispatcher. It is disgusting. And for this I was Targeted, had COINTELPRO running behind my back, was made out to be someone I was NOT, so that friends and Family would stab me in the back, I was Set-Up and Framed, and eventually Attacked and Mobbed by a bunch of people who did not know the Truth. I was Traumatized and Tortured for what they did not know. To top it off I was Targeted for Psychological-Snaps, for the sake of Manipulated-Dirt, so these Feds could "Look Good" and so "Citizens would be Absolutely Convinced". Brilliant.

Keep in mind that there are supposed to be Laws protecting people from this type of Deceit, and at some point, after enough ABUSE, people are going to have a Psychological-Snap. None of this is Psychologically-Healthy, it is not intended to be, but this is actually "Normal Behavior" (I am not saying "Healthy-Behavior") given the Weird-Circumstances—most would, again it comes down to THE CONTEXT OF REAL LIFE and the Context of this "Mobbing". Again anyone put in this "Mobbing" can be pushed into these "Releases and Psychological-Snaps", and if one thinks of what is being done to the Target it is easy to see. Usually these "Hostile-Releases" are a sign of the "Pressure-Building" and a DANGER SIGN of an impending Psychological-Snap. In fact these "Hostile-Releases" are to "Relieve-Pressure" that would otherwise lead to a Psychological-Snap, and if more "Traumas" are "Heaped On" this may in fact happen. These Hostile-Releases are what "Qualify as Healthy Psychology" given the circumstances, they are a "Release" and are a clear "Warning Sign". Having said that, when these "Hostility Releases" happen the Agents may, and in my case did, have these Idiot-Citizens "Attack Me with more Traumas" which is stupid. When they know what is going on, and this is what they do, then the Agents are clearly going for a Psychological-Snap: for Manipulated-Dirt.

If you work somewhere with somebody who is a Target, and this happens again and again (Hostility-Releases), it means that these Agents are going for a Psychological-Snap, and it means the Target is "Nearing One". It also means that the Agents you have contact with are Extremely-Corrupt and Extremely-Dangerous. They are concerned about nothing, only concerned with Destroying the Target, and willing to Use and Misuse Citizens to do so. Like anyone who is Corrupt they only care about themselves and what they can get away with, not the Target and not Citizens in spite of what these Citizens believe. Does it look like the Citizens have a clue? Having said that this is what a lot of these stories in the News are about, in regards to "People Snapping and Killing Co-Workers": it is actually this "Situation/Mobbing", COINTELPRO on the "Next Level", and very Corrupt and Dangerous Agents behind it all. Not all the News Stories are this way, but I have picked up on "Tells/Details" which have indicated that more than a few were.

Not one of these citizens could go through this "Situation" and with Agents Pushing-For Psychological-Snaps, without having one themselves—it is impossible due to this "Situation", that is the point. And to top it off, regardless of the last "slight detail", there are people, Citizens, Programmed-People, who will "Attest to what they have Seen": it is a Farce, Manipulation, and Lie, and these Citizens are what I call "Fake-Witnesses" to many a thing. They are Brainless-Dupes. What they have done to the Target is Extremely-Criminal, and only an Idiot can claim that the Target was of "Sound Mind" and that what they the Citizens did, was of no "Affect" on the Target. And at that point claiming "It is still the Target's Responsibility" is a Farce: they merely "Feel Justified" in what they have done, and do not actually know "What or Why". Like many of the Nazis were with the Jews.

These Psychological-Snaps are a product of these Citizens, and when the Agents get "What they Want", it of course involves one of their Citizen-Dupes being "Targeted Themselves". Of course the Agents knew it would come to this, this is how Blind and Stupid these Citizens are. They are Used for this. They are Lied to, Create Traumas, Reduce the Targets Mental-Health to nothing, believe they are Justified (Their Feelings are Manipulated), could care less, do not think about things "From a Real-Life Perspective", expect "nothing to happen" (many are even amused or think it is funny), and are Manipulated to Push the Target into a Psychological-Snap(s).

When this happens they are "Further Convinced" of Lies about the Target and the Fake-Profile Built around the Target, are so stupid they believe that this is a part of the "Real-Profile" of the Target, and then believe they are "Justified to Attack and Traumatize" the Target further which leads to more Psychological-Snaps in which they become "Further-Convinced and Justified". And to top it all off, they believe they are "Witnesses" to "Who and What" you "Always Were". They are "Convinced" and Fake-Witnesses, and you will note that "They took part in Nothing" and what they "Did" was of "Little Consequence". They have ABSOLUTELY NO SENSE OF REALITY.

And people in Cults, and people who are Brainwashed are not known for their SENSE OF REALITY, are they? Does it look like these Citizens are any different?

I could do a fraction of what was done to me, to one of these Programmed-Idiots and get results. And at times I did do this, and got "Hostility-Releases" for a variety of reasons from these Programmed-Idiots, and did indeed "Get Results" as the Citizens feel "They are Above You", are Arrogant, and get angry quickly in response. They can not walk MINUTES IN YOUR SHOES, never mind the Bigger-Picture.

Quite frankly they were angry quick, and you will note here, that what I did, was actually IN RESPONSE to what they were doing to me—and again this shows how "Out of Touch" they are Due to Programming. There were also times I did this Consciously simply to "Make a Point" with the Citizens. Like my example though, they think they are the "Master Race" and you are a "Jew at the height of Nazi Power". As far as they are concerned you the Target have "No business Responding", which quite frankly is a Pile of Deceit. Many of my "Responses" were well within the Context of Real-Life. Their own anger in response WAS NOT. And there were almost fights, several times as a result, and as a result of not being able to "Wear Real-Life"—again a Product of Programming. I also called the police in regards to some of these Brainless-Idiots: there are "Consequences" for these Idiots if they are not careful. And the funny thing is, if one of them screws up "Big Enough" they get Mobbed as well—I have seen this happen to several people. And just to be clear here, these were "Normal Hostility Releases", "Violations of Social Norms", not a build up to a Psychological-Snap.

Now as the Target, and as one who has looked into things, you are at a total disadvantage, as the Agents know the Laws, and as they go running around after you "Flashing a Badge" and telling everyone they have "Legal Justification" for doing what they do—it is a Farce, but since this Secret-Program is on paper they get away with it for this reason combined with

Lying, Manipulating, and False/Staged-Events: and of course Programming. To top it off, you as the Target have "To Prove" what is going on, a Conspiracy, with you at the center, which is near impossible. Basically you have to be rich, and then some, as no one you know even believes that you have "Reason for taking anyone to Court". They all believe the Agents, all believe in what they are doing is "To Help or Punish the Target", and are too damn Stupid to Look or Listen—unless you have a Badge of course. Thus between these Corrupt-Pigs having "Legal-Justification", Lying, False/Staged-Events, having a Badge, a "Job-Title of Authority", and Programming, they get away with what they do.

You lose all of your Rights and Freedom, and to top it off, people do things to you which are Extremely-Unpleasant, and many times Illegal. You experience Torture and Traumas and this goes on Constantly—who the hell would not have Psychological-Snaps? And you can not claim that any of this is "Normal" and that these Psychological-Snaps are a result of a "Psychological-Profile" that already existed. As a point why would the Target, if they are "Being Themselves", do what they do when they have a Psychological-Snap, knowing there are Government Agents "Watching Them Constantly"? So.

The Agents know how to "Cheat the Laws" and so come up with ideas and ways for Harassing you, in which they can get away with it. They actually have a Hand-Book for this, obviously. And as they are watching you, 24/7 and know your whereabouts constantly, doing things like breaking into your flat, is easy—again see the movie "The Game". They like to do things like this routinely. Stealing and Damaging Property is another big item they like to do. "Gas-Lighting", stealing objects and returning them later is another "Common Thing". And better yet, and a favorite of Spooks and Agents, is "Using Famous People" on television and radio, to "Deliver Personal Messages" by way of "Subtle Delivery" to the Target—they "Specifically are Told to Use" key words, phrases, sentences, and Themes which are Relevant to your life. Specific-Things currently going on in your life and they many times say these things in a way to "Convey" they have been "Contacted" and are "Speaking to you Personally": many times they are "Quite Obvious" in their Delivery. This is to be "Surreal and Affect your Mental-Health" it is to destroy your SENSE OF REALITY. Sports Shows are a favorite—although I have had a host of "Famous People" from Talk-Radio (even a few times from T.V.) get involved on behalf of these Feds and Spooks, and these people are Idiots as the Agents Investigate ANYONE THEY HAVE CONTACT WITH. You would be surprised by the list of people, and this is why in the movie "The Game" you see Michael Douglas's character

"Spoken To" by someone famous from television, on television—this Movie is actually based on, and written by someone with Insider-Knowledge of this "Mobbing". It is a "Feel-Good-Film" for Idiot-Citizens engaging in this "Situation", complete with "Fantasy-Ending", and it was not "Written by Accident". I have never felt like Michael Douglas's character at the end of this movie and neither did any of the Citizens who "Experienced Hostility Releases" under Normal-Context, never mind those who were Used by these Agents for Psychological-Snaps.

Now while all of this "Unpleasantness" is going on, they have colleagues, friends, Family, act as if you "Deserve It", and literally laugh about your circumstances right in front of you, it is Surreal, and make jokes and innuendos as well. As I say, all Social-Norms and Mores, go right out the window. Respect, Friendship, and "Normalness" get totally thrown out the window, and annihilated. NOTHING IN YOUR LIFE IS NORMAL and it goes on Non-Stop. You are Violated Routinely and Daily. Of course you are "Psychologically-Stressed", Angry, and "Cooking". It is Surreal. And keep in mind that this does not even involve Heavy-Traumas.

People betray you, including people you used to be friends with, the Laws stop working, you are Violated Repeatedly, Experience Traumas Daily, you are followed constantly (and these people let you know they are—usually through whistling), your belongings are stolen or damaged, where you live is constantly and obviously entered, and there is no end in sight. There is no "Safe-Zone" and that is the point—to keep you Psychologically-Stressed, and wired. I knew five people in America this happened to, and I saw one guy on French Television (an American) from the DoD, who also had this happen to him. You LITERALLY are turned into a "Jew" at the height of the Nazi Power in Germany—there is simply no other comparison I can think of to give. No offense to anyone of Jewish Ancestry who suffered—but this is what it resembles very strongly, and shows that this is a Dangerous-Precedent. This is EXACTLY what it resembles, in terms of your Rights, the Betrayal of "Friends", HOW YOU ARE TREATED, and your ability to seek "Legal-Recourse": never mind the Behavior of Citizens. The only way it is different is that you do not "Have to worry about Gas-Chambers", or live in a "Designated-Ghetto", but other than that it is all the same. Having said that if you are me, you also have to worry about "Reckless Endangerment", and Electromagnetic-Technology.

Life becomes Oppressive and Repressive, and loses all of it's joy. To get an idea of the "Feeling", see the movie "In the Name of The Father". This will give you an idea of the "Feeling" and the "Justice" although the subject matter is a bit different.

You feel powerless, there is nothing you can do, there is no Legal-Recourse, as the "Law" is involved. And many times they will actually have Police take part in the "Events" as well, just to "Demoralize You" and let you know, that the Police Know about your "Mobbing" yet won't do anything about it. Having said that, as I get to later, if you catch these people, the Police still have to do their "Job"—so this is a "Head-Game" not Reality. Document everything, keep a Journal, as this is actually important—it is still "Documentation". I got this tip from a lawyer who used to deal with taking Corrupt-Agents to Court—you never know what may happen in the future and it is best that you have such "Documentation". Having said that, until you get on your feet, you are "Powerless" and you have NO RIGHTS AS A CITIZEN, it is Crazy and Dangerous, and should not be happening in America—what a Farce. To top it off, if you try to speak to ANYONE INVOLVED in this "Situation" they will act as if they "Do not Know" what you are talking about and Deny It. Which is Ridiculous, or would be, if it were not a Government-Tactic.

Folks I once read that people who explode in a Destructive-Spree, like a Killing-Spree, do it because they FEEL POWERLESS and try to claim it/own it, their Powerlessness all at once. They try to "Take Back" their "Power" in one Big-Act. Which is why there are a lot of "Work-Place-Shootings", these are actually a result of this "Situation" and the Target having the worst of Psychological-Snaps. In THIS "Situation" you actually are Powerless, nothing is "In your Head" on this point, and you experience the worst of people and life. Folks this "WAVE", THIS LOCALIZED-FASCISM, is a practice of Dangerous-Psychology, and is a Dangerous-Government-Program, and no one on the planet has the right to be doing it, and certainly not in a "Democracy". This is for the "Benefit of Government" NOT PEOPLE, Society, Citizens. These Agents were "Protecting" Corrupt-Law-Enforcement-Employees, CRIMINALS WITHIN LAW-ENFORCEMENT AND CORRUPT-PEOPLE WITHIN HEALTH-SERVICES WITH CONNECTIONS TO LAW-ENFORCEMENT, they were not "Protecting" People, Citizens, Society.

Since when did Hitler and Nazi Germany become good god damned role models? Most don't trust the Government, and will even less when they loose all of their money and are directly affected by Government, yet when the Government comes knocking on their door and begins asking them to "Believe This" and says "Start Breaking the Laws for Us", in regards to a particular Target, THEY DO. When people, Citizens like myself, begin going to "Summer-Camps" people will feel differently. What is wrong with people, where are their brains? Folks this is Fascism, and due to Hitler and Ignorance, over here in Europe they call it "Mobbing", just to hide from the

Reality of what it really is. They DELIBERATELY MISLABEL IT HERE IN EUROPE just to further remove it one layer from what it actually is, to Hide It, they call it "Mobbing": which is Localized-Fascism.

I even saw a guy from the Department of Defense interviewed in English on French-Television with French under-titles, and he went through the same thing as me, for trying to expose Corruption within the DoD. He was trying to expose Corruption within the System. In his case they accused him of being an Axe-Murderer and having killed a friend or neighbor, although "Authorities Could Not Prove It". Of course since it was not True, he was NEVER CHARGED. Having said that if you are a good person, and they have no Real-Dirt, or have nothing "Heavy to talk behind your back About" then they have to INVENT SOMETHING which is what they did. His "Neighbor" or whoever it was "Disappeared" so it looked like it fit the ACCUSATIONS, but without a doubt, because his neighbor was "Properly Asked To", he took part in Framing Him, and simply "Disappeared". No doubt as a RESULT OF MONEY, OR THE THREAT AND "A TASTE OF THIS MOBBING", OR THE AGENTS THREATENED HIM WITH PRISON, FOR SOMETHING ILLEGAL HE HAD DONE, IF HE REFUSED TO GO ALONG.

Having said that this guy from the DoD had money and Legal-Options and "Recourse" which no doubt, most do not. This "Mobbing" is Illegal yet "Legal" for a reason, so GOVERNMENT CAN GET AWAY WITH IT. IT IS TOTALLY-NON-AMERICAN, and the IGNORANT-JACKASSES WHO TOOK PART IN THIS, WHO LET THEMSELVES BE USED, AND WHO "BELIEVED WHAT THEY WERE TOLD" CAN KISS MY AMERICAN ASS. So back to this guy from the DoD, and his "Disappearing Neighbor" who was "Payed-Off" or Threatened with Charges and Imprisonment, never mind "Mobbing".

I have seen this here in Europe with one person they "Used", I believe he "Had no Choice"—nonetheless he should not have done, what he did in regards to me and my Family. In his case the Spooks had him pull out his penis, and start masturbating in front of me with my child at the Public-Library from behind a Bookcase—courtesy of The Clown. Like I said some people will do anything for these Agents and get absolutely Filthy, but in his case I believe he was "Threatened with Prison". Thus these Agents have ways of getting people to do anything, including Frame You, if need be. As things now stand, these Spooks are also threatening me with this, with Framing Me, in addition to Death-Threats, as if I have not had enough of both. This is what Sick-Lowlifes with Authority do, when they Screw-Up and Screw-Up Royally.

In regards to this guy from the DoD, the Allegations, Character-Assassination, Accusation(s) had to be EXTREME in order to get people, the

Public to participate, as he had a good reputation no doubt, and was known for being a good person no doubt—which one can gather from his Actions of trying to Expose Corruption. No one loves an "Axe-Murderer" and so everyone no doubt took part in his "Mobbing" based on this. Without a doubt this "Being an Axe Murder" was the "Center Piece" of the Artificial-Profile, a False-Profile Built Around It. This is how this kind of thing works if you are a decent person. If you are a "Lowlife" are a major Criminal, then they DO NOT HAVE TO LIE, it is simple. It is called Court and Jail.

Unless you work for "Law-Enforcement" in which case "Your Crimes are Covered Up" and any people knowing of it are Framed and Destroyed.

So unless The People are already Previously-Indoctrinated from past experiences of "Helping Agents" with "Other Targets" (White-Washed, Programmed, Indoctrinated, and Predisposed to this kind of thing) or unless the person involved is really bad to begin with, then they have to Create and Invent Circumstances: which is what COINTELPRO and this "Mobbing" is for. This "Mobbing" is COINTELPRO on another level, it is "More out in the open at this Level". Like I said the Public, Citizens, Society, are being Indoctrinated to Blindly Serve Government: Fascism. The Theme is Compliance to Government, the theme is Fascism, Government and Cult. This is by GOVERNMENT-DESIGN, not Coincidence, and is to serve a purpose for Government, the Corrupt-Elite, on down the road here. Fascism IS when Government becomes Cult. As an added note, you'll note that, when you are not guilty of what they are accusing you of, and they know this because they have "Set This Up" and Framed-You, then the actual Criminals are the ones "Wearing the Badges".

All of this "Situation/Mobbing" is for shortly down the road here, when the Government CREATES AN EXCUSE for becoming "We the Government", as opposed to "We The People"; and this will all begin, Layers of Manipulations and Deceits, in 2012, giving all power to a Government controlled by Corrupt-Politicians and their "Friends", the Elite. This Mobbing, and other Illegal-Laws and Programs, Gray-Laws, are all for Preparation, the Groundwork for, New World Order. It is all about Creating a Population that is Compliant and Obedient to Government, and a Corrupt-Government at that.

CHAPTER 31

Corruption: What to do: Citizens, The People, Agents and Politicians: "Uniting in Goodness" and Fighting Back: A Matter of Energy: "Human Physics"

When the American-Economy-Collapses, there will be a lot of angry people, a good many of them angry because they were not informed and ignorant. This has to do with people choosing to be blind about the World they live in as there are a lot of signs of what is to come. As is typical, many people want to believe that they live in a World that does not actually exist. They believe in what they are "Sold" in the Media (which is Corrupt and Controlled) and by Politicians. They "Feel more Comfortable" believing in this Imaginary World. They choose to believe in a World that THEY WANT TO EXIST, yet does not. Until people begin LOOKING AND SEEING, begin seeing clearly and start dealing with obvious problems like Corruption, this will always be a World which is less than they believe. If you are going to change things for the better, you first have to Identify the problem—just like with an alcohol problem for example. If you want a better World then you have to work for it and put in time and Positive-Energy, for Positive-Change. Until then this IS the World you live in, NOT the World you WANT to live in. There will have to be a "War Against Corruption" if things are ever going to get better and improve in this World.

When the American-Economy-Collapses, and then the World-Economy, people should not "Loose their Heads" because of it, it is obvious that this will occur, rather people should remain CALM. The "People Reacting Angrily" will be an excuse for Implementing Martial Law, it will be an excuse for the Government taking over The People. When the Economy Crashes, people should LOOK AT THE FACTS FIRST as this is what has allowed the Economy to Crash at that point in time anyhow—people not looking at Facts, and people not being Active for Positive-Change. People will have to "Wake Up" and take Responsibility at that point, not simply follow the "Same Path" as before which is called NOT BEING CONSCIOUS, the Path of Experience. Simply running out in the streets and breaking some windows or anything else for that matter, will not do a Damn-Thing, and in fact the Government is prepared for this—in fact EXPECTS THIS and WANTS THIS. Anyone doing so, is falling into a TRAP, and will PROVIDE GOVERNMENT AN EXCUSE FOR DECLARING MARTIAL LAW—once it does it will never return to a "Normal Democracy". These people are Idiots and part of the reason for such an Economic-Crash in the first place, never mind that they are Typically Predictable. No, what will need to be done, is that people will have to start asking questions as to why and how this happened, and start PURSUING JUSTICE.

"Pursuing Justice" meaning that people will have to RE-ELECT OFFICIALS, call for Instant and New-Elections, and go after those in Government who allowed such an Economic-Crash to occur in the first place, as at the least they were INCOMPETENT at performing their Jobs, and in Reality Corrupt. And the theme will have to be "Corruption" as America is supposed to be one of the Richest most Powerful Countries in the World, yet somehow will be broke—it isn't being spent on The People, I guarantee you it is going to Corrupt-Politicians and this Elite. The proof of this is in the "Silence" in regards to "Trillions of Dollars Missing Yearly"—thus a "War Against Corruption" IS NEEDED. These Corrupt-Politicians and Their Friends will have "Drained America Dry" of her financial resources. There is no other alternative that will work, anything else will simply result in a "Government Band-Aid": an appearance of "Doing Things, Fixing Things" without any real intention of fixing anything. In fact that is part of the Plan—they will offer "Socialism" and say that "Capitalism" does not, and will not work. Which is a Total-Lie; of course it works it just does not work with Corrupt-Politicians stealing all of the Countries Money, and Deliberately sabotaging the System and Country. THE FACTS ADD UP TO THIS. So when the time comes, I recommend CALM, and I recommend NEW AND INSTANT ELECTIONS. A Calm

Logical Approach, not a Blind-Emotional one. IF THIS OCCURS THE ELITE WILL BECOME EXTREMELY WORRIED. They are expecting an Emotional-Reaction, not a Calm and Logical one. A calm reaction will automatically throw a monkey wrench into things—they are no doubt hoping for an EMOTIONAL-REACTION so they can declare Martial-Law. Thus "Custodial Facilities" all over the country for Millions of People, while in the meantime it is said there are "Not enough Prisons"; while the Government has said that these "Prisons are Reserved" for the "Great Mexican Invasion", which of course was a Lie so stupid that it was an insult.

Now having said all of that, I imagine that as we approach this Economic-Collapse the Government will begin to show signs of "PREPARING FOR THIS". There will be a Mobilization and Deployment of Military across the Country. No doubt the "Signs" will be obvious and "When and As" this is occurring take a good look, because this is your Future. I suspect that eventually they will "Quarantine Cities" and go from house to house confiscating weapons on some "Pretense" that will merely be a Lie and Excuse for doing so. The Real Reason will be that they do not want a population capable of standing up for itself and fighting back. The People are going to have to decide what to do at that point and make a Decision. IF things happen this way, and as they may.

Keep in mind that the Elite have all of this Planned, and the American People's response will no doubt be predictable. Based on this they may take extra measures, and take away fire-arms before this Economic-Crash, if they do, this is obviously a sign of trouble. No doubt "Martial Law" will be a part of this as well. At that point, I think "Pursuing Justice" will be a problem, as the Government will already have taken all Power and Control or will obviously be in the process of doing so, based on an Economic-Collapse they will have Created. It will all get down to, and be, "Excuse Making", and taking Power and Control from The People, as it becomes "We the Government" not "We The People". And the Elite within Government will have Created it All.

Any talk of things "Being Temporary" will be just that, "Talk", not Reality. This is what Creating this Economic-Collapse will be about in the first place, a pretense for taking Money and Power from the Masses. For Centralizing Government. If this occurs first, "Pursuing Justice" after an Economic-Collapse will no doubt be impossible, as the American Government will have Total-Control, and therefore no worries about "The People". Again this is a possibility. With that in mind, I provide the following piece I wrote a year or two ago. This brings up the issue, of Energy and Balance, and doing what is right. This brings up the issue of Action and Non-Action within a certain context—again the best and first line of defense should be a Calm, Logical Approach, and a call for Justice in this manner. Again when the time comes people should be calling for New and Instant Elections, and should be Calm. Having said that, what if The People are not given a choice?

Action and Non-Action: Good and Evil: Doing what is Right

You are a good person and the New World Order comes rolling around. They tell you you have to go with them. Maybe they give you an excuse, maybe they just take you to a "Camp". Or maybe it is your neighbors who are "Taken".

What will you do, what will your neighbors do? Be enslaved at a camp, a "Detention Camp or Center", a "Custodial Facility", where they may put you to work or kill you? What is the purpose of these camps? Obviously the purpose is to lock up those who are "resistant" to the Government's "overthrow of The People". All of this "Looking the Other Way" at that point, all of this "Ignoring Things" at that point, any talk at that point, will not work as there will finally be Action being taken by these Elite—and against The People, out in the open for all to see. Passivity at that point will not work. America will have reached a point where the Government has decided it is "We the Government" not "We The People". And I might point out, this would not even be possible if the Constitution had been adhered to. This is what the "Terrorism Scare" the "Red Scare of Our Times" was Created for. Again the Politicians are fallible, the Constitution not

Now, everyone has a right to question things. If you are good and against Corruption, Fascism, Abuse of Power, or simply believe in America as it was founded and meant to be, then you will have to fight. At that point if you want your Freedom you will have to fight, there will not be much choice—hoping for the better at that point will be stupid, if the Government has decided to take Action. Think about that word freedom and what it really means to you. Think about how much you take it for granted: I couldn't take it for granted if I tried, I have lived without it for 18 years.

If things get as far as I have just described, Violence that no-no word will be necessary at that point. Things are already near that point because too many people have been Blind, Lazy, and Fearful. Lets toss in Stupid as well. And the longer people sit back on their haunches, their asses, now, the more extreme it will have to be later. This is a matter of Energy, "Physics on a Human Level".

When Negative-Energy builds, a Counter-Active-Energy takes place to Balance it. This is a Law of the Universe. For every Action there is an equal opposite Reaction. In the beginning Lesser Energies are required to Balance things out. If Positive-Energies do nothing, in this case PEOPLE AND POSITIVE-ACTIONS, then the Negative-Energy slowly builds like snow on a mountain side, an avalanche waiting to happen. Give an inch take a mile. In the end if there is too much "Inactivity and Cowardice" then a "FINAL POINT" will be reached. All of the small accumulative "Non-Action" will result in one big "Mass of Action"—an avalanche.

History speaks for itself.

In terms of Past History people have ended up fighting their Governments because they ignored the Corruption within them until they had to take Action. The People tried living in Denial, being Lazy-Minded, Non-Conscious, Non-Active, and Apathetic. I can not think of an example where this is not the case. The Masses ignored the Corruption and Abuse of Government, because of it, until they were forced into taking Action. In all cases the Masses wanted to Ignore things, the Path of Experience, until they had no choice but to deal with the problem. They wanted to "Look the Other Way" and not deal with things Consciously or Intelligently. Thus the Negative-Energy, and Negative-Actions, of Corruption slowly built up until people were forced to take Action—and usually at a point when the Corruption began to totally Infringe on their Freedom and Existence. In this case, in order to compensate for all of the Negative-Energy, which built up slowly over time, a large amount of Positive-Energy and Action, all things being relative here, was needed to "Re-Balance" the problem—an "Avalanche of Action" occurred. Thus throughout History there have been Revolutions and "People fighting their Governments"—this is Human-Physics. The Law of Cause and Effect and "Balance of Energy" as all things seek a "Balance and Harmony", a natural state of being. This is "Physics on a Human Level", a part of what I call "Human Physics".

Now.

If you want to maintain your Lifestyle and Freedom then you can not live in Denial. You must occasionally take Responsibility for the Democracy you live in, and Speak-Out and/or be Active. Especially in this day and age, as we can not expect Corrupt-Politicians to be Responsible for cleaning up Corruption within Government—this is stupid and a "Recipe for Disaster", as many are about to find out. In this day and age for Good to survive it must get ACTIVE and be VOCAL for change. Otherwise right around the corner you will have to fight to survive, Good will have to fight to survive. Your fighting will be the "Lesser of two Evils". If Good waits too long then fighting for

Principle, Morals, Ethics, Freedom, will take place. The very thing a lot of "Good People" don't want to do, will take place. Because they don't know that being "Good" means standing up for Good and whats right, and from the BEGINNING. This means taking Personal Responsibility and not waiting for others to "Fix Things", like say Corrupt-Politicians.

It is Stupid and Lazy to wait until things are too late, to wait until problems are so large that they have to be dealt with in an extreme manner. How can Good Flourish if you don't stand up for it? This is why "Corruption is Flourishing". This is Cowardice and Inactivity. Lack of belief in Goodness itself. Avoidance will ensure that which you try to avoid. Corruption and these growing Orchestrated Problems exist for us to take on as a challenge and fight—a "Test from Above" on a "Collective Level", in one sense. Being INACTIVE causes the problems to grow until we are left with no choice but to deal with them. If you believe in Goodness then you have to stand up for it or else "Evil will Flourish". Being "Soft" does not mean being "Weak". If you are not willing to stand up for what is Right and Good, then this is a Lack of Belief in Goodness itself. Inaction is Action in this case, and the "Greater of two Evils". Thus Negative-Energy will build.

YOU CAN LEARN THROUGH CONSCIOUSNESS OR EXPERIENCE. Learning through Experience is the hard way, and in the end, will result in a change in your Consciousness so that you don't want or need the Experience again. In the end it is better to be Conscious in the first place and avoid the Experience altogether. It is best to recognize the problem from the Beginning and deal with it before it becomes too large, before it grows worse. Again this is a matter of Physics on a Human Level, what I call Human-Physics. Having said that, America and the World, are no where near the "Beginning" in terms of this "Build-Up" of Negative Energy.

You can't ignore a leaking pipe in your house, or a hole in the roof for very long. I am sure there are slackers who do, but eventually the problem deteriorates more and more, getting messier and messier, until we have no choice. The Negative-Energy continues to build, as a result of no Positive-Action, which will Re-Balance the Negative-Energy. Thus as the Negative-Energy builds, the Problem grows Larger, and a Larger Amount of Positive-Energy is needed to Re-Balance the Problem. This is the Law of Cause and Effect. When things reach a point where your entire first floor is wet and everything ruined from a busted pipe, or the upstairs is drenched and leaking through to the kitchen floor because the hole in the roof got bigger during an immense thunderstorm, THEN you have to take Action. But it didn't have to be that way did it? That was a choice made by anyone too slack, Lazy, Irresponsible, and Non-Thinking (not Conscious) to deal with the problem, head on, in the first place. It comes down to "YOU CAN LEARN BY CONSCIOUSNESS AND INTELLIGENCE OR BY EXPERIENCE". The funny thing is, after you learn the hard way, through Experience, you learn to do it the Conscious and Intelligent way anyhow. How hard is it to figure out?

So getting back to earlier, Energy. An extreme protective Balancing-Measure will have to take place. You will have no choice. Again this is a Law of the Universe, Cause and Effect in regards to people, and History proves this.

The "Nature of Evil" and Corruption will leave you with no choice.

You will have to make a choice. The Evil of others who want to take your Freedom, Control You and/or kill you will leave you with no choice. YOU WILL HAVE TO MAKE A CHOICE, IT IS BETTER TO MAKE IT NOW. It is better to learn through Consciousness while there is still a CHOICE AND OPTION, to start Repairing Things before they reach a Final-Point. The Final-Point being when all of the "Inactivity", Laziness, results in one big Mass of Action—an "Avalanche of Action", as Corrupt-People who have been given "Total-Leeway" in regards to "Doing what they Want" feel they have reached a point where they can simply "Take All". In this case Citizens fighting their own Governments for Freedom and Survival, the Right for Self-Control, will no doubt take place.

When this time comes, I assure you there will be other affects as well. Certain Countries will begin invading and conquering other lands because the so called "Moral and Free Western Countries" will be busy fighting their own people. There is a strong probability for this, although there seem to be scenarios in place to avoid this kind of thing, such as the "Anti-Christ" (an "American Version") and a "Staged Alien Invasion". I imagine a combination of all of the above just to put the World into total Shock and Chaos, so that these Demented-Nit-Wits can get what they want. No matter what, all of this will be Preceded by Total Economic-Chaos, and by Total Design. This Economic-Collapse will be THE BEGINNING and is guaranteed. Everything else, afterwords will be "Obvious in Retrospect" and admittedly I can only make "estimated guesses" based on "What is out There" and is "Visible".

But no doubt the future is going to suck and all because of the Corrupt-Few and a whole lot of people too Lazy, too Lazy-Minded, too Afraid to stand up for Good and Goodness. AGAIN THIS IS LACK OF BELIEF IN GOODNESS ITSELF.

When people come to your door it will be too late for changing these people, or the situation itself. You will have to make a choice about your "Worth and Value" in the world and what you can do to better it, and what to do to live. And the value of The Peoples lives who threaten your own. It is harsh, but that simple. Again at that point you will not have a choice. You won't be given one. Take a look, these Elite are already making choices that affect us all, for Themselves, based on What They Want.

Do I advocate violence? Not at all, I advocate ACTION and being VOCAL so violence can be avoided. I advocate Responsibility before the water pipe explodes or the hole in the roof gets bigger. I advocate doing something before it is too late. I advocate the PATH OF CONSCIOUSNESS before we are all forced into THE PATH OF EXPERIENCE. Much more people need to get involved and need to start caring. MORE PEOPLE NEED TO GET SELF-EMPOWERED, VOCAL, AND ACTIVE. More people need to stand up for What Is Right. More Positive-Energy has to be "Put Out". It is simple.

Think of your own kid, or kids. Think of your Family, think of your own kind. Think of their future and your own. I and my kid have suffered plenty due to Exotic-Techno (and still do: 2011), Government Agents, and "Various Things". I have two fingers up for these Corrupt Spooks and their Ignorant Pawns—the Truly Brainless of Society. It is a combination of Non-Thinking-Jackasses and Corrupt-Jackasses, just like in my life, that have made the World into the mess it is, and who have made the World into the dangerous place it is.

If you are against Corruption and against Misuse of Power, they are against you. These Elite, and the Illegal-Programs they have "Passed" are Targeting INDEPENDENT-THINKERS: people who are Capable of Thinking on their own, as opposed to "People who Think what they are Told to Think". These Elite and those working for them—whether they are Conscious of this Fact or not, these Demented Nit-Wits ARE fighting for the World they want, and they see Good People as an obstacle. They want people who "Listen and do as they are Told" not people who "Question their Authority" and who "Question what they are Told". Think these "Custodial Facilities" are for Corrupt Government Employees? No, they will be for people who have a problem with Corruption and Corrupt Employees.

Take a good look at that No-Fly-List, this is NOT about Terrorists, and any claims about this "List being much Smaller" are no doubt a Lie—it seems like a lot more than thousands are on it, doesn't it? Take a good look at Guantanamo Bay, and the Fact that you do not even get a Fair-Trial yet sit in prison and are TORTURED (Never mind Electromagnetic-Techno which makes Torture a Moot Point). Put the two of these together to get a better picture—this is not the way America is supposed to be, and American's should be Alarmed and Concerned. THE CONSTITUTION SHOULD COME FIRST, NOT POLITICIANS AND THEIR "MOUTHS". The Constitution is Designed to protect The People from Corrupt-Politicians and Government—so if it is being walked all over this is clearly a Warning Sign. You have got to be an absolute Idiot to believe otherwise, and must be "Hanging Onto" an America that clearly no longer exists AS A FORM OF DENIAL. If America was the way it used to be this kind of thing would not be going on. So where is the "Argument"?

THE WORLD THIS ELITE WANT, THAT THESE CORRUPT-POLITICIANS WANT, is not a Good World, and most people would not agree with the World they want to Create, thus they are doing what they can in Secret until it will be too late for the Masses. These people don't care about the Masses, they are so out of touch, Megalomaniacal, Egotistical, that they have set out to change THE ENTIRE WORLD IN THEIR IMAGE. They have decided to change the Entire World, so that they can "Thrive and Survive". The Future they want is going to be much more "Controlled" and less enjoyable as a result. The "Form of Socialism" they want to put into place is going to be Very Restrictive to say the least: it will actually be a Police-State, and all of the "Comfort" that comes with one. Think about it, there will be no Capitalism so that people outside of Government can not acquire Wealth OR POWER, so that "Equality, Power, Wealth" will remain ONLY with Government. This is about "Taking ALL Power and Control and Wealth" to the point of it being Non-Existent for others—it won't even exist as a possibility for anyone outside of Government. Again this is all a Plan for Centralizing Power at the top in Government, which is what any Dictatorship is, and making sure it is NOT DISTRIBUTED AMONGST THE MASSES. That should give you an idea of how "Bright the Future" will be. It will suck.

Life "Without Freedom" at that point will be "Known to Suck" and Idiots who stopped being Americans, who began throwing away their "Freedom" and "Everyone Else's Freedom" because of a Large World Trauma, 9/11, the Twin Towers, and "Terrorists" will finally get what the Constitution means and why: they will finally get of what I write here. FREEDOM IS EVERYTHING and life without it sucks, and will suck.

This Elite and their "Team", are and have, set out to change the World into a place where what they do is acceptable and no longer really has to be Hidden. How hard is it to figure out? You don't have to do anything against these people and you will never understand these people. They are against you for being you. They will search you out for your MENTALITY. You are a threat to them, to THEIR Freedom to do as they like and want, to do as they please. Until they have the World they want they will always have to worry about Consequences. Really think about that:

You are a threat to them and their Freedom, to do as THEY like and want, a Criminal Lifestyle.

They are about their own Self-Gratification, Self-Fulfillment, and their Egos—they are Megalomaniacal, Psychopaths. If you want to understand them try comprehending that. I have had no choice but to understand this Type of Corrupt-Person, my Government Mobbing, and The Clown, left me with no other choice. It is unpleasant to "Stick one's Mind" into their "Mentality and View-Point" which is why most people can not "See" this Type of Person or "Understand Them"—having said that one can simply look at ACTIONS. Now I share with you based on my own Experiences and Government-Agent-Education, and you can learn Without the Experience. Believe me Consciousness rather than Experience is preferable on this point, and if you have someone to help you out on this point, then there is no reason to "Reinvent the Wheel" here.

These people who want this New World Order, the Elite, The People they have bought out, this Fascist-Cult under them—this Team, have a Power-Monger Rapist-Mentality (and a Total Sell-Out Mentality) and you will in all likely-hood never understand them: it is uncomfortable to stick one's mind into "Their View-Point" and "Psychology". Most people, the Masses, can not relate to these people, it is difficult and disgusting to wear their View-Point long enough to understand them, and therefore they are "Difficult to Comprehend". The other half of this "Equation" is the Idiot-Citizens who believe what they "Hear and See" on television, who believe what they are TOLD TO THINK (Which is what Television is Used For by the Corrupt), as opposed to being Independent-Thinkers. Until people become SELF-THINKING, things will continue to spiral Downward. Between these Two-Groups we have a "Recipe for Disaster". We have "Corrupt People", and "Idiots going Along".

These Elite and others with them are Power-Mongers. You will never be able to reason with them, in their warped way of thinking "Join Them, or Obey Them", there are no other options. Given the large amount of people needed to "Convert the Masses", the Masses are thus turned on themselves. The Idiots of Society attacking the Thinkers of Society. Those "Who Do and Believe What They are Told" against those "Who Question What They are Told". The Blind attacking those with Sight and Vision. Slowly bit by bit Society is dirtying it's hands more and more for these Elite and the Governments, slowly Society is "Defiling and Debasing" itself more and more and "Making Itself in the Elite's Image".

The World-Populations are being INDOCTRINATED for Fascism, for a New World Order. Take a good look and contemplate. I have already heard Obama speak of a "Citizen Police Force" (Which is what my "Mobbing" is about), and much of what he speaks of is Socialism. The Elite WANT SOCIALISM to be the new form of World-Government, and they want a "Society that Polices Itself". In reality it will be TOTAL FASCISM and the Masses are being prepared for this. That will be the new "Next Level" that the Blind-Masses, Sheep do not see. If and when the Government steps out into the open, this will be the beginning of it all. At that point, Government will become Cult. THAT IS WHAT THESE "ILLEGAL PROGRAMS" ARE FOR. This is what this "Mobbing" is for. This is what "Grey Laws and Programs" are for. It is "Preparation" for what the Elite want.

Reality and Experience will eventually wake everyone up though. As I currently see it, Mankind has chosen to learn through Experience rather than through Consciousness and Intelligence, which means people "Will Learn the Hard Way". Thus things will eventually get "Messy", and the Elite will loose Control in a major way. It will be some unforeseeable mess that no one wants or has predicted. Both groups of Idiots will pay, both the Idiot-Citizens and the Power-Monger Types, and anyone else for that matter who is not prepared. No doubt as the Elite make their "Bid for Power" and "Final Conquest for the World" something will go wrong.

These Elite and their Team enjoy the suffering of others, could care less about others, especially people they consider to be "Lesser Beings", the Majority of the Masses falling into this category, and they do as they like. Think about this. The World is a big Play-Stage of Monopoly for them, and YOU are merely a game piece for these Whacko-Nit-Wits, you are an inanimate object. They like what they do: The Clown and Her Buffoons are this way. Causing injury to me and my Family, nearly destroying my relationship with my girlfriend, and an assortment of other things has been done without even blinking on their part. Hard to believe right? This is the Mentality of people like this, Power-Mongers, they are Power-Mongers and set on their own "Self-Fulfillment" and nothing else: they have "Power-Issues", a "Rapist Mentality". Anyone that gets in their way of this, is an Enemy, and is attacked. Yet I assure you these Power-Hungry-Jerks do exist, are real, and in my case hold powerful Government Jobs. People like this are Attracted to Jobs of Power and Money, and Government is a "Honey Pot" for these people, especially as the Job gives them the ability to act with Virtual-Impunity. Think about it. All because of Power, "Illegal Laws", and "Blind Compliance" of The People, they are able to get away with almost anything. And now they Plan to "Take All" by Deceiving the Masses, for what looks like the last time. This Economy-Collapse will be the Beginning of their Plans for Total-Control and Total-Power. You can forget about your Freedom. And I can tell you from personal experience, life without Freedom sucks.

So what to do? You must become "Solidified in Goodness" and stand up for it, and Fight for it. You have to become "Firm and Strong in Goodness". You must become Self-Empowered, Active, and Out-Spoken.

For Good to survive, for Good to exist it must Fight. WHO SAID YOU HAVE A CHOICE? IT IS THE NATURE OF EVIL TO MAKE CHOICES FOR YOU, especially with this Elite and the World we live in Now. If You and your Freedom are attacked you will have to perish or fight, or you will end up at a "Detention Center" where I doubt the experience will be "Pleasant", "Temporary", and no doubt for many even "Survivable". They have yet to properly take care of Citizens who have survived "Weather-Disasters" never mind "Prisons full of people disgruntled with Government".

Look what happened to the Jews in Nazi Germany, many who were "Rounded up and taken to Camps"—this IS the Mentality of people in Power who are "Deranged Power-Mongers", and the FACT OF THE MATTER IS, that the Government DOES HAVE PRISON CAMPS, has acknowledged these and calls them "Custodial Facilities". Notice the name "Sounds Temporary" and "More Pleasant" than Prison. Yet why would they let any of these people go later, why would anyone be released, especially given the reasons they would be locked up for in the first place? THEY WON'T BE. And Prisons are not for "Temporarily Holding People", and these Prisons all across America were certainly not constructed for holding Millions of People for merely a matter of days or weeks—usually people are locked up for a long time. Why would they "Let people Go" who are "Mad at the Government" a few weeks later? In fact if you "Look" at the "Patriot Act" these people "Fall into the Category of Terrorists". I imagine people would be locked up indefinitely, given the circumstances (look at Guantanamo Bay), all of it of course against the Constitution, which will be a "Moot Point" at that point. Like I say, if they start going around "Collecting Firearms" and "Quarantining Cities" you can pretty much forget about any "Pursuit of Justice"—having said that if this does not occur people should remain Calm and call for New-Elections. People should take the best approach for getting the Job done.

Having said that YOU WILL HAVE TO MAKE UP FOR ALL OF THE "PASSIVITY AND FALSE GOODNESS" LATER IF YOU DO NOTHING NOW. Give an inch, take a mile, and the Elite are preparing to TAKE ALL as they have been allowed to do anything they want, and practically right out in the open—with Total-Impunity. The Public has not been holding these Corrupt-People ACCOUNT-

ABLE, and so these Unaccountable-Politicians believe that they can "Take All" as they have already gotten away with everything else.

A "Re-Balancing of Energy" will have to occur, and no doubt will occur. Eventually the Elite with their Out-of-Touch, Out-of Reality-Ideas, will fail and fall, but at what price? That depends on The People. If you look at "Prophecies", Religious or otherwise, this has been "Predicted" and is said it will last about three years. I think the question is, whether you believe in Prophecies or not, or are Religious or not, does anyone want any of this in the first place? These Prophecies are a "Warning" of what can occur if enough people do not take Action and Speak-Out, to divert the worst case scenarios with Positive-Energy. If you look, regardless of Prophecies, you can see that everything is being "Set-Up" for Total-Control by Government. This is not a matter of "Coincidence": Time, Action, Money, and Energy have been put into this, for Creating this. In my opinion "Coincidence Theorists" (I borrowed this term from John Judge) have "Avoidance of Reality Issues": "Coincidence Theorists" are Coaches for Idiot-Citizens who get to look smart and sound smart until Reality catches up. Once Reality catches up, their "Reputation" will be a "Moot Point" anyhow.

My question is, IF YOU CAN'T STAND UP FOR GOOD, FOR GOODNESS, THEN DO YOU BELIEVE IN IT REALLY?

And others have to bear the weight until you realize this.

It may very well be YOU fighting in the streets in time, if you don't react now (while there is still time). When Reality sinks in. When Reality comes knocking on your door, when you wake up one day and find out some "EVENT" has happened, some Orchestrated Pretense, and you have lost all of your FREEDOM. It all depends on your own ignorance, it all depends on whether you want to learn through Consciousness or Experience. The time to act is NOW, as there is LITTLE TIME.

BEING GOOD DOESN'T MEAN BEING WEAK, IT MEANS BEING ASSERTIVE AND BEING PREPARED TO "TAKE AND MAKE ACTION" FOR THE GREATER GOOD OF ALL.

BEING GOOD DOES NOT MEAN LOOKING THE OTHER WAY. BEING GOOD MEANS "FIGHTING FOR GOOD".

Thinking this, that "Being Good" means being a Passive-Jackass, is dishonesty with oneself. If you ignore it long enough YOU will learn what it means to Fight and be Good. To Fight for the survival of Goodness. You will learn the hard way the Truth of the matter. THE TIME FOR BEING INACTIVE IS OVER whether you realize it or not. And you will realize very soon the Truth of this.

Again if Goodness does not Fight it will perish. A lot of poor bastards will learn this the hard way. All of this is based on Info at hand and projections of the future. Too many people have looked, are looking, the other way because they are confused about what being Good is about. BEING SOFT, BEING GOOD, DOES NOT MEAN BEING WEAK. Period. There is no argument to the contrary.

How about when your own Family is threatened? I bet you don't even think twice then, it is No Issue. At that point most begin contemplating other possibilities, violence becomes an issue then, A VIABLE OPTION. So what World would you have your kids inherit; a World of Total Control and no Freedom? Not standing up for what is right, not doing the right thing, being Blind, being Afraid, not Looking, being Lazy, none of these things are good excuses for watching the World deteriorate, none of these things are a good reason for being INACTIVE when the future of our loved ones is at stake: when the future of Civilization and the World are at stake. Other Good people are your Family. People need to TAKE ACTION NOW while there are Peaceful-Options.

People have to become SELF-EMPOWERED realizing they CAN DO something and then get Active in whatever way shape or form comes to them. I used to speak to people and try to make them Aware of the possibilities, before my "Mobbing" started. This is one good way, it increases Awareness of subjects and places "Book Markers" in peoples minds. Later if they encounter evidence on their own they will think of what you have said and think about the possible validity of it. Another thing is demonstrations. Putting out your own single page fliers and posting them across town is another good idea. You never know who is going to be curious and take a look; depending on how you present the Info you may make a few people reconsider the facts. The point is realize YOU CAN DO SOMETHING other than panic and feel like a victim, you can indeed make a Positive-Difference. Just the act of doing something Positive will make you feel different, give you an outlet, and you will feel less hopeless, afraid, and "Blocked". Doing things increases AWARENESS and CONSCIOUSNESS, brings more light on these Corrupt Politicians and Elite, and gives them less room to maneuver. If the Elite were not afraid of the Masses they wouldn't be Hiding so much in the first place, BEING ACTIVE AND SPEAKING OUT MAKES A DIFFERENCE. People have got to Push Back against the Elite, The People they own (Sell-Outs); and this Team. Good people have got to start putting out Positive-Energy with Positive-Action: if you think of this in terms of Human-Physics this is easier to see and understand.

People have to stand up for what is right and the time is NOW. Later there is no time.

I suffer now and so does my kid, thanks to Illegal Government Programs and Total-Idiots. Later it will be you and your kid, your Future, your kids

Future. Believe it, time for Delusion is over. I AM DOING WHAT I CAN. I have been in a "Situation" for 17 years involving all sorts of Nastiness and Oppression, and yet I am writing this book and getting the Info out. How easy do you think it is? It is very difficult. I am assaulted daily with Exotic-Techno, go through weird daily events involving people, and receive death threats (as this book nears completion) on a near daily basis. And to top it off, the person in charge is Mentally-Unstable and Dangerous to say the least—this is not merely a "Tactic to Scare Me" as I and my kid have had "Close Calls" already, and as I have come to realize this person in charge is actually a Psychopath. This is a matter of Actions not Words, and it is a long list of Proof on my end.

BEING PASSIVE IS NOT AN OPTION, neither is waiting for other people to solve Our Problems. IF EVERYONE HAS THE SAME ATTITUDE, OF WAITING, NOTHING WILL HAPPEN FOR THE BETTER. We each have to do our own Individual Part and take Responsibility in changing things for the better. People have got to become Self-Empowered. People have got to realize that they are not alone. People, the Masses, have got to determine the real Source of Problems in this World, and Unify. Whether you are a Citizen, Police Officer, Government Agent, or Politician. There is too much Fear, Inactivity, and Laziness going on in this World today, and people, regardless of their walk of life, have got to be willing to fight these school yard bullies, these Elite. This school yard reference is a fair analogy of the Bigger Picture. People have got to STOP "Going Along with The Flow".

Everyone is too afraid to do anything, instead standing around doing nothing but watching, and wishing someone else would do something. This is ridiculous, this is the Sheep Mentality. NOTHING CAN BE SOLVED OR FIXED THIS WAY. So be SELF-EMPOWERED, get Active, be Involved, Speak-Out, and increase Awareness of Problems that exist and that need to be taken care of. You can't fail if you give it your all; if enough people don't get involved at least you can live with yourself knowing that You have done the best You can do. Watching other people do nothing is no excuse for being Inactive. Each person must do their part, each person has to take Responsibility for their own Life, and in the end each person has to live with themselves. What will you do when you get to the "Other-Side", when you meet God in whatever form you imagine? Tell him or her you didn't have "time" for doing the right thing? That your "Job" required you to have no Ethics or Morals? That you were "Tricked into Ignorance"? Most of you have got it easier than I do, so what is stopping you?

So what has gone wrong with this world? The main thing is that people are not standing up for What is Right. They have become complacent because Democracy and the System was working, or the Land in which they live, with its System was working. But this is no longer the case. These Corrupt-People in Governments realized they could keep getting away with "More and More", and so have (give an inch take a mile). As usual people don't want to do a damn thing unless they have to, so again History is near to Repeating Itself. The stakes are far higher this time though. The Corrupt-Politicians of many lands have bonded together to take Total-Control of the World this time, and they have a lot more Technology at their disposal. Much of it so far ahead of the Public-Sector as to resemble something from a Science-Fiction movie. And this time, I believe, the Government(s) will take "Preemptive Measures" first, "Disarming the Population" before actually overthrowing it's people(s): and keep in mind that the Government already has "Custodial Facilities", prisons, and this again indicates "Preparation" and is in and of itself "Preemptive" and a Warning Sign. The Corrupt-Leaders have learned from History, the Masses have Not.

Now the System is in a serious State-of-Decline and everyone wants to pretend, in spite of many Growing Signs, that the System is the same. That Democracy (lets use America as the example) is what we still live in. That we are Free. Like an onion the outside looks fine but the inside is rotten and decaying. Corruption is eating up America from within and the only thing left to it is Appearances.

But if people don't speak out to maintain the Democracy they live in, they will lose it. If people do not practice Democracy they will lose it.

Slowly at first because Corrupt Elements in a Democracy know that that is the only way to really do it, if you are totally Power-Oriented, a Power-Monger. Like a frog in a pot of water on a stove. The frog cooks slowly, the temperature slowly rising, until it is too late. It doesn't know it is being cooked, that its life is in danger. Like in a Bugs Bunny cartoon, people, the Masses are asking for a sponge and soap.

Understanding the Mentality of Corrupt Power-Hungry People, Psychopathic in regards to other peoples lives, is not required. How can anyone in their normal lives understand this Type of Person? We are talking about Powerful-Psychopathic-Criminals and their Team, that is all. We are talking about Organized-Crime within Government and Industry. They are simply criminals in powerful positions, think about it. It is less difficult to believe than many Folk imagine. It is simply Organized-Crime in Politics, how hard is this to Accept or Believe? Take a look around you the Signs are quite obvious and are everywhere. Everyone has Eyes and

Ears and should be able to hear and see the Signs. Everyone is capable of looking at Facts and Actions. Or are they?

Unfortunately many still can't see what is going on because they are living with their "Democracy and Freedom glasses" on. They are viewing the Present-System through the glasses of the Past, which colors and distorts Present Day Reality. People are merely believing in an Image in regards to America these days, not looking at Reality. Again the only thing left is "Appearances" and quite frankly even this is "Shrinking". This is Dysfunctional and Detrimental to all.

Our Founding-Fathers were not so blind.

The price of looking the other way now, is a Massive Shock to ones psyche and mind later. Obviously Freedom and Peace will be totally gone. It is better if people start seeing clearly now and start speaking up. If you don't start preserving your peace of mind now you'll lose it later. If you do not work to maintain and preserve Democracy, then you will lose it. Believing in an "Image and Stereotype of America" does nothing for it, it is actual Actions and actual Caring that counts.

If we don't look and do something now the Negative-Energy will continue to grow and build until it reaches it's limits: a toppling point, a Turning-Point, critical mass, A FINAL POINT. At that point there will be a "Corrective-Procedure", an "Avalanche of Energy", by which the Masses take back their "EMPOWERMENT". Having said that the Elite are prepared for this and have all sorts of "Plans and Toys" ready for Conquering the Masses.

Usually this is called Revolution, fighting in the streets.

Again if people don't take back and own their own EMPOWERMENT now, their right by birth, and realize their own self-worth now, most will end up doing it later. Or perish. Blindness, Cowardice, and Inactivity will finally dissolve. "When and How" we choose to have EMPOWERMENT is a choice of the Masses. Not "If". These Criminals in Political Positions, and their "Wave of Idiots" will leave you with no choice. WHO SAID THEY ARE GIVING ONE?

Again "Being Good" does not mean being Passive and Inactive.

Being "Good" requires that you stand up for it and fight for it. IF YOU REALLY BELIEVE IN IT. If Good does not fight, than Evil will win. How hard is that to understand? New World Order is another name for Fascism, Fascism is a term for when "Government becomes Cult". Is this what we want? Is that Democracy? For now most people are sitting in the middle, Inactive, not wanting to make any choices, wanting to Pretend, so their "Seeing and Hearing", their Choices and Learning will come later. But it doesn't have to be that way. The Path of Experience is a dumb Path to take. People can not keep waiting for others to "Fix Problems" this is NOT ENOUGH. YOU HAVE GOT TO START GETTING ACTIVE AND VOCAL.

So what will happen?

THERE WILL BE FIGHTING IN MANY COUNTRIES, revolution. Countries that enjoyed the greatest freedoms, and/or lived in the greatest denial, will have the bloodiest fighting. The World will be Politically and Economically Unstable, and no doubt other Countries will see this as an opportunity to do things that they've been thinking about in the first place. War will break out. It is almost a sure thing. The chance for grabbing other Countries, their Land, and their Resources, will be opportune, there couldn't be a better time for doing so.

I wouldn't be surprised if more than a few countries are waiting for this N.W.O. , to try and take control, for this very reason. Some of the most powerful countries on Earth will be busy fighting their own people. Who will there be to stop other Countries from invading other Lands? No one. So think about it, I offer it as food for thought. As an extra point these other Countries sure as hell don't want the N.W.O. to work. It's a threat to everyone and every Land. It is going to be a "Power Grab" by many Nations, no doubt, when the time comes. The Possibility certainly exists, and from a Military View-Point, this would be the time to do so. All of this because people were too Inactive, too afraid, on a personal level. Again think about it. Things will "Snow-Ball" out of control, as World-Leaders see other World-Leaders making a "Grab for Power and Control". "World-Stability" will be at an all time Low, and by Design. The question is, how predictable will things be for the Elite, once they begin "Shaking and Stirring things Up"? This World functions on a basic-degree of Stability, Predictability, and "Normalness", which very soon will not exist.

Things like an "Alliance of the Middle-East" led by one figure, and other Powerful Countries of The East, have to be looked at, and considered here.

Personally, in my opinion, there is no doubt that N.W.O. will happen, but there will be many scenarios occurring more or less at once, and there is no telling what scenarios will happen, which will be successful, and what the outcomes will be. For instance, a staged U.F.O. attack or landing can happen, thus negating a lot of the "War-Like" scenarios, yet fulfilling the N.W.O. agenda of "World Unity"—which it won't be, it will be World Fascism. These Elite have many Plans and Options, and are considering all. In my opinion it will be a combination of things, possibly many at once, or at least so close together it will seem that way. The obvious things are that there will be a World-Wide Economic-Collapse, and an attempt at New World Order. In a dream I had one time, I saw Countries rioting against their Governments, due to a major world wide disturbance—there was the feeling of Economic-Collapse being the cause, people were in the streets and the military was out, and then people started going inside to watch television. A Staged-U.F.O.-Event was happening, everyone forgot about their Discontent and went to watch television instead. And these U.F.O.s were Military-Made.

As an added note here the water supply had been tampered with for at least a week in advance, having had Drugs and State-of-the-Art Micro-Chips put in by the Government to make people more Docile, Responsive, and Compliant to commands. To top it off the Exotic-Techno involved in my life, Electromagnetic-Technology, was being used via Satellite to make The People more Compliant and Docile, in an instant everyone went from Rioting to Totally-Passive and Self-Absorbed in the television as if they were Hypnotized and on Drugs in a matter of minutes. I could feel the "Electromagnetic-Technology" and this "Passiveness being Beamed". I was the only one left out on the streets, everyone else was basically Hypnotized, I watched as the Military moved into position on every street corner, owning the city street by street. At that point no one could even get out of their front door without being shot. Any chance of rioting or fighting was finished. It was that easy...This shows how EASY taking over can be for these Elite, regardless of this being a Dream. And keep in mind that if the Military is on every street corner with orders to "Shoot to Kill" anyone that steps out from their house, forming in any kind of groups will be impossible. At that point Government will have Total-Control, do you think they do not know how to do this? Do you think they do not have the means?

So I've spoken here about taking control of ones life, SELF-EMPOWERMENT, and speaking up and standing up for what's right. To fight Corruption and the Deterioration of Democracy, which is what Corruption is. Both on a Personal Level and a Group Level—people have got to work together at this and make it clear that Corruption is not acceptable. But what about people in Government? Government Agencies? People in various Intelligence Agencies? What are they doing? Well what I've already stated in the last few pages applies to All People, regardless of one's walk in life. THE DIFFERENCE BETWEEN A SHEEP AND A DOG, IS POWER AND A LITTLE BIT OF KNOWLEDGE, NOT BLINDNESS.

However the subject of people in Government Agencies is a whole different can of worms and the reasons for Agents varies. There are Good people and Evil people in Agencies, for simple terms. There are Power Struggles. There are secret splinter groups whose entire purpose is Illegal and Evil.

There are others who look like they are not so Good on the surface, yet are against the very Policies and Secret-Laws they are following. Good People posing as Bad People because they are doing their part to combat what is going on. Just like the Harry Potter books, where everyone is doing their part to combat Voldemort, maintaining their jobs and positions and using them to help combat Voldemort. Again it gets down to the "Quality and Type" of person, not Job-Title, not a Stereo-Type. Job-Title tells us what people do for a living, it says nothing about the Type of Person doing the Job—and I have had nothing but Corrupt-Jackasses get away with destroying my life because of "Job-Title". Things like "Corrupt-Agents" exist only in movies until you have them in your life. Currently I have to deal with The Clown, who in addition to being Corrupt, is a Psychopath.

Everyone has to do their part Collectively. There are others who are Fed-Up with looking at the all of the Illegality and Evil, and who anonymously get Info out to the Public. Some Agents may not have a choice but to stay in, once they are in—just like joining a Mob Family. No doubt many are "High Minded" when they join, and slowly Corrupted and Jaded as time and the dirtiness of their own actions sinks into their soul. With time they become Indoctrinated and Jaded—Corrupt. They start taking things for granted. I think this applies to the majority who are Corrupt: I've seen it happen with a lot of Citizens getting involved for these Spooks in my life, in regards to my "Mobbing". A lot of the Citizens start enjoying what they do. Just like the German Citizens of Nazi Germany dumping on the Jews. It comes down to no Accountability, Impunity, and No Guilt because any guilt that should be felt is dispersed amongst the Demented Collective. As a general rule when people get together in a Group/Mob, it results in an Intelligence Down-Grade, not an Upgrade.

Of course some Government Agents are just Evil to begin with, and enjoy inflicting pain and misery. These are the Psychopaths and of course they are Corrupt, having chosen their Jobs for the "Perks and Self-Fulfillment"—a "smart choice" in a Psychopathic-Way. They are attracted to the Power and the idea of no Accountability in the first place and are able to Abuse their Power and able to Violate others with it. They Hide behind their Job, and the "Protection" it provides, just like the Elite. They take no Responsibility for their own Actions, and they do not have to, and will attack you instead for Speaking-Out against them. They Violate and Attack others with Impunity because they can get away with it, because of their Job-Title. That's their idea of Responsibility. These people Violate, Abuse and Misuse their power for their own Self-Fulfillment. These are the Sickest of the Sick, and the most Corrupt of Agents. The Clown is one of these.

A typical comment from one might be, a common response to criticism, "Well you're not perfect.". Yeah but most people are not Psychopathic and "Rapist" in their Mentality. Like your imperfections even begin to compare to theirs. Everyone makes mistakes, but not everyone uses other peoples mistakes as an excuse for making more, for making Evil. For all intents and purposes these people are Rapists (Major Power and Self-Control issues here Folks), and unfortunately it is this last type that I have been dealing with—The Clown, who got this book started. Rape is a Power Issue, not a sex issue, and The Clown has nothing but "Issues"—this is the Type of Person I refer to as a "Power-Monger". Just like the Elite, these people are totally Power-Hungry, Power-Mongers, and could care less about Human Life. They have "Power Issues" and "Abuse Issues", and at the bottom of it all they have "Inferiority Issues". It is because of The Clown that I have a much better understanding of the Manipulations which occurred in my life in America, and the Elite.

Having said that I don't care who you are or where you are in life, your Responsibility for all is to do the Right-Thing first. Then your Job. Morals and Ethics come first or else Society falls apart. This is not simply about doing a "Job", this is about the Person Doing the Job. We don't simply pick a boyfriend or girlfriend based on "Hair Color" we look at who they are, the "Type of Person"—again it gets down to the Type of Person, not just a Job-Stereotype. Without a person to fill the Role, the Job is non-existent. The person, and Quality and Type of Person is what should come first, and people turning their heads away from Corruption, out of fear, or "Pretending to be only there for the Job", are Lying to themselves. Ignoring Corruption, is only going to make things worse, and we obviously live in a time when it is "Accepted and Encouraged". If this is to change, people have got to start exposing it, and make it clear that it is not acceptable—otherwise it will continue to spread like a DISEASE. As it is, it has reached a point where people are worried about "Repercussions and Retaliations" for "Whistle-Blowing", for exposing Corruption within Government. As I have mentioned already in this book and as an example, one man from the DoD exposed Corruption, and ended up in my "Mobbing", accused of being an "Axe Murderer", and he eventually moved to Canada. To expose Corruption has become dangerous, and this is Un-Acceptable. This is also a statement of HOW CORRUPT the System has become. Corrupt-People want the System filled with Corrupt-People underneath them, as a "Buffer-Zone", and this is where all of the Power and Control is at—in the System. If it is Corrupt then they have all of the Power and Control—this is, in a very real sense a battle between Good and Evil.

CORRUPTION IS A "SOCIETAL MENTAL DISEASE" THAT EATS SOCIETIES, THE PUBLIC, COUNTRIES, FROM THE INSIDE OUT. ANY ONE WHO HAS STUDIED HISTORY SHOULD KNOW THIS. CORRUPTION IS ULTIMATELY WHAT BRINGS COUNTRIES DOWN IN THE END. IT IS A SURE SIGN THAT A COUNTRY IS ON IT'S "LAST LEGS", THAT IT IS ON IT'S WAY DOWN. JUST LIKE ANY PHYSICAL DISEASE IT THEN SPREADS FROM PERSON TO PERSON, SLOWLY AT FIRST, THEN FASTER AND FASTER AS CORRUPTION AND "SELLING OUT" BECOMES MORE AND MORE ACCEPTABLE. Again it turns into a "Group Mentality Thing", a "Collective-Movement", and any sense of personal right or wrong, any guilt, any personal intelligence gets dispersed amongst the Masses (like with Nazi Germany and the Jews). IF YOU ARE NOT RESPONSIBLE FOR YOURSELF AND YOUR ACTIONS AND WHAT YOU BELIEVE IN, THEN WHO IS? Corruption in one very real sense is a "Movement of it's Own", and these Elite and their Team get this.

When you get to the "Other Side" and meet God in whatever form you believe, are you going to say, "I didn't have time for morals and ethics, I was doing my Job.". Do you honestly believe that this is going to make a lot of sense, does it make any sense now in this life? YOU are responsible for the keeping of your own body and mind, no one else. You are the keeper of your own soul. There are no "logical arguments" for being weak or passive, for being Inactive, for "looking the other way".

Just like any "Crime Family" The People at the top, the Elite, are hiring fellow criminals to do their work. As an example, how effective would the Mob be if they hired priests and nuns to do their work? People of Similar-Energy are needed to carry out these jobs and tasks. It couldn't and wouldn't work out otherwise. This is what has happened to Government. The Corrupt-Elite have united, bringing in more and more people within Government, to "Their Side". More and more people within Government have been Corrupted and it has gradually filtered down to Government Agencies, so that more and more Government Agents have been Corrupted. More and more Corrupt People are being Hired. Now it is being passed on to the "People", the Masses, Society, by SECRET-LAWS, GREY-LAWS, that are allowing "Mobbing", Localized-Fascism to occur. My "Mobbing" speaks of this. My "Situation" has followed me across three Continents, and three Countries, INDICATING INTERNATIONAL AGREEMENT AND ACCEPTANCE OF THESE SECRET-LAWS AND POLICIES. And it is Disgusting, Illegal, and "Misused" to say the least, and Misused by Corrupt-Agents for Corrupting and Indoctrinating the Masses.

In short Fascism is being laid down now for the FOUNDATION AND GROUNDWORK OF NEW WORLD ORDER LATER. Citizens of many Lands are being taught to Attack and Control themselves through very ugly behavior, and the ones being attacked are The People speaking out in the first place—Independent Thinkers, a Mentality. People who see what is going on, people WHO ARE AGAINST CORRUPTION, people who believe in Freedom and Democracy are being attacked, Targeted, Mobbed. In the end it is about those who are Corrupt attacking those who are against them, and using the Public to do it. In short the Citizens who take part in this behavior, this Mobbing, are Total-Idiots attacking the very people protecting them and their rights. If your fellow Citizen is being Mobbed, what makes you think you can't be? Your fellow Citizens rights are YOUR rights, how hard is that to see?

Already President Obama has talked of a "CITIZEN POLICE FORCE"; just what exactly do you think is being spoken of here? The Citizens, through these SECRET LAWS are getting their hands dirty and becoming Indoctrinated for a Government, Governments, that do not care about them or their Freedom, but instead want TOTAL-CONTROL of their Populations, and none of it for the greater good. This is about laying down the Groundwork for Fascism, the Citizens Controlling one another, blindly Complying with Government—this is about Government becoming Cult, and Citizens doing anything they are asked. How hard is it to see that the Groundwork for Socialism is also being laid down in America? This will be the "Surface" of this Fascism, the Deliberate-Mislabeling. Really that will be on the surface, what we are really talking about, just beneath this layer of Socialism being created, is Fascism. A FASCIST STATE is what it is all about. A FASCIST WORLD: New World Order.

Now, getting back to earlier, with a few tips, ideas, and suggestions. Good Government Agents need to start Secretly Documenting and Releasing Information. PEOPLE IN A POSITION TO DO SO NEED TO START EXPOSING CORRUPTION—PEOPLE NEED TO START WORKING FOR THIS ON AN INDIVIDUAL-LEVEL. And this is really what all of this talk about "Wikileaks" is about, and why they are going after the "Founder" of it, never mind them Framing Him which I am sure they are doing. No one likes an "Axe Murderer" or a "Rapist", and this is the quickest and easiest way of destroying a good persons reputation as the "Stigma, Accusations, and Insinuations Alone" are enough to distance people.

This is about Corrupt-People in Government trying to go after people exposing them, "Whistle-Blowers" and Informants, which is what this World really NEEDS. And these Corrupt-People exist High-Within-Government and are using their Power to try and Silence it all. THEY DO NOT WANT THE MASSES TO KNOW WHO AND WHAT THEY REALLY ARE—they do not want the Masses to know how Corrupt they are, and they are worried that eventually something "Extremely Damning" will be released in regards to themselves: like me with "The Clown and Her Buffoons" actually. We are talking about Criminals here, going after Good People trying to expose their business, which is ALL OF OUR BUSINESS.

People have got to do THEIR PART, and this includes all people in all walks of life—especially within Government. People like me are in no position to do so. I am doing what I can by writing this book, and I am swamped daily with nasty Manipulations, Death Threats, and Experiences...How easy do you think it is on my end? It is incredibly difficult, my writing and thinking are interfered with constantly via this Exotic-Techno. Others in Government Agencies need to start doing what they can. Simply "Following Orders" is not a good enough EXCUSE anymore for looking the other way—presumably people within Agencies took an oath to "Protect their Country", they need to start DOING IT.

CORRUPTION IS AMERICA'S GREATEST THREAT AND ENEMY AND DESTROYING THE COUNTRY FROM "WITHIN" (It IS "America's Terrorists")—anyone disagreeing with this can stick it. I have had twenty years of Corrupt-Bullshit and I am absolutely finished with it—I have lost twenty years of my life because of it, and my Family has suffered, and still does because of it.

Anyone from "My Past" who thinks they got the Truth from these Jackasses, never mind who believes they were not Manipulated, can Kiss My American Ass. You do not even begin to know. But back to business.

I recommend using the same Tactics as the Elite and their Team of Sell-Outs. These Tactics work on everyone, and for everyone. Investigation Techniques, and "Cloak and Dagger" Techniques work for all. These Techniques are based on observation of Human-Behavior and trial and error, thus there is no need to "Re-invent the Wheel" here. This is how these Government Agencies have become so good at Lying, Manipulating, and Hiding the Truth—it has all been "Perfected" over a very long period of time. This is also how Government Agencies have become good at finding out Information and Exposing it in regards to other Governments and Agencies, and/or then Misusing it. These Tactics work and will work against exposing Corruption and the Elite. People have got to do their part.

Find a few people you can trust and work together at exposing Corruption and Abuse of People, Abuse of the System. Document it, videotape it, wiretap it, do what you have to do. Then Secretly release the Info, to get the Info out. It is best to do it Secretly, anonymously, when releasing the Info or else you will become a Target, and "Watched and Investigated" to say the least: like me. It is a guaranteed thing these days, this is the World we live in. Furthermore, once you are out in the open, you are for all intents and purposes NEUTRALIZED. You will not be able to do much of anything without it being known. Eventually "Events" may be arranged for you to make you look bad and so on and so forth. But once you are "Known" it is best to pursue your path and be as loud and as visible as possible. The more people are watching the less likely it is that anything heavy and negative will happen. So no matter what the situation is, people need to get busy.

This Secret Organization of Crime, of Corrupt People, this Elite, are making a major push to hurry things up and enslave the World with N.W.O. There isn't much time, 2012 is the Chosen Date of this Elite. They are going to use Prophecies and Religion as a SMOKESCREEN for a lot of their Planned-Events. Many already believe in these Prophecies, and few know of this Exotic-Techno, so it will be easy to do. They are going to make it look like a lot of "Prophecies are coming True" to make their Manipulations believable, and this will all start in 2012. Again this is why things like a "Reformatting of the Internet" is Planned for 2012—this is their "Starting Date".

Really it is a trick to convince a lot of Religious People, into believing without questioning, that a lot of these "Staged-Events" are real. So a lot of people in America will be buying into these SCENARIOS from the beginning: hook, line, and sinker. It is a way for this Elite and Fascist-Team to pull the wool over everyone's eyes, (which they will like) seeing as how they don't like ANYONE WHO IS GOOD AND DOES GOOD; so they will be laughing all the way as they Manipulate a lot of Religious People with their own Belief-Systems, and Prophecies, to get what they want. But I digress.

So back to what people can do. Obviously Citizens can start working together to "Go After" those who are not so good, those who are Corrupt and who Abuse and Misuse the System. As an example, got that dirty cop in your city or town who is Abusing people, and getting away with it? Well how hard is it to set up "juicy circumstances" that they will jump on? All you need is a couple of people and a camera. Video tape it and release it. Of course make copies first, you never know how close the Media is with a particular police officer or department. It is this simple, and when the "Legal-System" no longer works, when it is "Impaired by Corruption", then it is up to the Citizens to take Control and do what is RIGHT. The same for Government Agents and others. Again it does not matter where you are—but when you do it, make sure the information gets out and is "Loud". Look at my "Mobbing"...Trying to be "Diplomatic" with Corrupt-Jackasses is a waste of time.

If something is very Incriminating the police can put a lot of pressure on any town newspaper or Media Outlet. I have seen this in my own home town, a few times, and I know because I saw the real circumstances occur.

They can shut the media up, and totally. So make copies before you release anything and keep the original. This IS how a lot of abusive people, police, are being caught. IT IS THE ONLY WAY THESE PEOPLE WILL BE HELD ACCOUNTABLE. The Police Departments will always protect their employees. Always. But if you have something on video tape they have to do something in regards to disciplining their employees—the Police Department wants to look good, not doing anything looks bad for business. Even then it most likely will not be a fitting punishment, but it will be more than it would be otherwise. This is the World we live in, not the one we want. That won't change though unless people start getting SELF-EMPOWERED and start doing something. PEOPLE HAVE TO START TAKING RESPONSIBILTY IN MAKING POSITIVE-CHANGES, WAITING FOR SOME ONE ELSE TO DO IT WILL NOT WORK. SELF-EMPOWERMENT AND UNITY ARE KEY.

As more and more people become Self-Empowered, Active, United, more and more people will become encouraged to do the same. As more people do it, the safer it becomes to do. Just like with Corruption, as Corruption Flourishes more and more people become encouraged to be the same way. So the same principles exist here for doing the Right-Thing, and the more over-burdened the Corrupt System becomes at keeping people "Quiet", "Silenced", the further back Corruption is pushed. Self-Empowerment and Unity spreads from Mind-to-Mind the way Corruption does except Corruption is like a "Fatal Disease", and Self-Empowerment is like an "Antibody" that Empowers the Individual and Collective. They are flip sides of the same coin, it is all a matter of Energy, Human-Physics. Corruption "Weakens the Collective", Self-Empowerment "Strengthens the Collective". Again the Elite have to be mirrored, their own Techniques and Tactics used against them, "Flipped" on them.

Now the same can be done by Government Agents and Politicians.

What people have got to realize is that N.W.O. will fail but the mess will be devastating and affect the World. There will be partial success, segments, mostly the large cities will be taken, but there will be resistance and fighting. Also other countries will become hostile and aggressive and "Make their Moves" while countries like America are fighting their own people. Such as a "Union of Middle Eastern Countries", led by one man. It will be a huge mess. So EVERYONE from all walks of life have got to do their part, just like in a Harry Potter novel. From Citizens, to Government Agents, to Politicians.

Most of what is going on these days is Illegal, or would bring about Outrage from the Public, the Entire Country, if they only knew. This is why Wikileaks is "Being Silenced" and "Neutralized". So if you are a Government

Agent, let the Public do the rest. Whatever you do, do it in secret as this guy from the DoD ended up in "Mobbing" (most Spooks are not this naive). Your end of business is simple. Government Agents and Politicians need to put MORALS AND ETHICS first in their lives again, say to hell with "Secrecy" and release those documents exposing Corrupt Government and Sinister Government plans. STOP BEING AFRAID ON A PERSONAL LEVEL, THERE IS TOO MUCH AT STAKE. People need to start doing the right thing and fighting for it. And this guy at Wikileaks is being "Framed and Stomped" to make an example of him, a High-Profile Target and Example, to try and discourage anything similar in the future—IGNORE IT and DO YOUR PART. Screw these Corrupt-Idiots, THEY BELONG IN PRISON, NOT GOOD PEOPLE.

Since when did exposing Corruption, become a Crime? And you will note, that it is not being stated for what it is in the Media, in regards to Wikileaks. They are "Stating it as a Security Risk" in the Media, NOT as the Exposure of Corruption, WHICH IT IS. These Elite-Jerks are MISUSING THE SYSTEM for their own "Protection and Benefit", to "Cover-Up" damaging information about THEMSELVES. It is all a Farce, and the Media is being Deliberately Misused to SELL this Farce.

Long term, although only a matter of a year or two now, Agents and their FAMILIES will end up paying just like everyone else. All sorts of people are going to have "Regrets" for "Going Along with the System", and ignoring Corruption, when the time comes. The Elite only care about Themselves, and many within this Corrupt-System are going to find out the Hard-Way. Or do all of you have some special planet you will go to right before everything hits the fan? I don't think so, and by the time you have regrets, it will be too late. Contemplate the Future. Contemplate the Motives of those over you with "Greater Authority". Because quite frankly there is nothing "Great" about it. Everyone is looking the other way, because there is still "Another Direction" to look at, yet very shortly Reality will be here, and a lot of people will have regrets. Once "The Future the Elite Wants" gets here, you will not be able to look the other way: there will only be "One Direction to Look At". Until that time gets here people think there is "No Harm", or simply want to believe that things will remain "Fine". This is a matter of "Wanting to Feel Comfortable", and "Being too Afraid" to Speak-Out. This is not good enough anymore.

For Politicians who are Good, or who Want to Do Good

Now lets go one step higher and address Politicians. In a Democracy there should NEVER be "SECRECY" in regards to Laws. Secrecy in regards to building Military-Technologies is one thing, Secret-Laws in regards to The People is another...Openness is a part of Democracy. A Government with something to Hide, is a Government to be worried about, and one which can not be Trusted.

But apparently Politicians are required at times to sign a "Pledge of Confidentiality".

For what purpose? Obviously to keep things Secret. From Who, and Why, and for What Purpose? In this case we have to ask if these Secrets are for the betterment of The People or not? That being the case, why do they have to be Hidden? It already stinks of Government doing things which The People are not going to agree to, or be happy with, which means that Government in these cases is representing itself, not "We The People".

But what is being said? I doubt anything with good intent. We are talking about Politicians who are supposedly carrying out the "Will of The People"—there is no need for Secrets, unlike the Military with it's need to have Secrets. Although the Military seems to have gone nuts on this point as well, developing and using Exotic-Technologies, Electromagnetic-Technologies, against the American People for the purpose of Manipulation and Control. Corrupt Politicians are creating ILLEGAL-LAWS, and the Military will enforce these ILLEGAL-LAWS. Both the Politicians and Military have been infiltrated by these Corrupt Elite and their Team, these Fascist-Cultists.

So what to do, and what about these Pledges of Confidentiality? Well if Lying, Manipulations, and Deceit are a part of these "Secret Meetings" then it needs to be Exposed—the American People should not be betrayed. Morals and Ethics must come first. With out Morals and Ethics, without taking Personal Responsibility, without fulfilling one's oath to one's country, The Country, Society, The People, will fall. Now this applies to all people, in all walks of life. Corruption will DESTROY AMERICA—so it is UN-AMERICAN. If it is Un-American and within Government, then it most certainly needs to be Exposed. What about ones Oath to Protect the Country?

I have been told by the Corrupt Government Agents I'm dealing with, The Clown and the Buffoons, that the Constitution is just a piece of paper with ink and words on it.

Well so are "Pledges of Confidentiality". So are ILLEGAL-LAWS written with Deliberate-Double-Meaning. The question is, when this is the View-Point, what should our choice be in Regard to such things? The choice should be based on Doing the Right Thing, otherwise how much sense does it make to "Follow a Piece of Paper" that says to betray your country, and your people? It makes no sense, in fact it is a piece of paper that is Illegal never mind the issues involved. Like anyone you are put into a Situation of having to do the right thing. Do not remain Loyal to Criminals who ask you to go along in their Criminal-Enterprises against the Country and People but rather of course, Expose It. This is not "Deceit and Backstabbing" this is doing the right thing. THIS IS what being an American is about.

Who says you have to be quiet and secret? People who in the first place are lying, cheating, and stealing in regards to "their Country", American Citizens, Government Agencies, and other Politicians. Skip peoples Job-Titles, we are talking about Psychopathic Criminals, and Powerful-Organized ones at that. There is no need to be "Silent" in regards to "Pledges" and "Oaths". YOU DECIDE THAT. Is this your idea of "Doing the right Thing", of being "Honest", by keeping your mouth shut about other peoples Crimes, Lies, Manipulations, and Deceits? This crap, this Corruption goes on, because The People do not know about it.

If you are a Politician or Government Agent who has taken an Oath or Pledge it does not mean you have to hold to it. Put your "Pledge and Oath to Country" first. Morality and Ethics also come into play, to put it bluntly—the laws were created for man, by man, NOT man created for the laws. The Truth is what matters most in the end, not some Pledge made to Hide "Illegal Secrets and Laws"—to Hide Manipulations and Deceit. The Country comes first, not a "Pledge for the sake of a Pledge", and one which Hides Deceit of Country and People. This is a contradiction and all mixed up. If it is wrong it is wrong, and needs to be exposed. Politicians and Government Agents need to speak out, and preferably anonymously, and NOW.

It is absurd under these circumstances to keep your mouth shut in the first place. Too much is at risk. We are all in this TOGETHER, these things affect us all, we all have friends and Family. This is in one sense, a real battle of Good and Evil. Use the same Tactics as these Evil People, but use these Tactics for Good. The Tactics work for everyone and can be used for Good; the Tactics themselves are neutral, it is how they are used that matters. Just like in the Harry Potter books, the Good need to work together to survive—Individually and Collectively. And it is because of a group of Negative-People doing these Tactics and Techniques, within Government, that the Country is going downhill.

This is not about "Following Orders, Following a chain of Command", this is not about "Following Laws", this is not about Following other Elected Officials with some authority, this is about Criminals in Government, using Job and Title and Laws to Hide behind—to Hide what they do. It is Illegal Activity, it is Corrupt Activity and if The People knew of it they would have these Crooks and Criminals in Prison. So do the Right Thing, it is not as if you do not have an ENTIRE COUNTRY who will STAND UP FOR YOU, for doing the Right-Thing. Everyone has got to stop thinking that they are alone, and stop being afraid because these Criminals are so Organized and Powerful. They too must worry which is why they Hide Things, and those Doing Good are not alone, they are NEVER ALONE.

So if you are a Politician go to these meetings and Secretly and Safely Release the Information: this Inside Information. You think these Criminals aren't doing the same? They are using Insider Information, or rather Misusing it for their own gain. They are Misusing the System, and Inside Information for Money, Power, and their own PERSONAL GAIN. They are "Hooking up Friends with Contracts", and massively overpaying them with the Tax-Payers money—"Money Laundering" in one sense. Just as it was once upon a time, 1000 dollars for a screwdriver—it is all Money-Laundering, Corrupt, and Illegal. Of course we have Trillions of dollars missing yearly, this is why. Friends of Politicians with Companies are being "Awarded Contracts" and being Massively Over-Paid, it is simple. Later these Corrupt-Politicians get some of the money back, Pay-Offs, a "piece of the pie", by coming to the company and receiving some Job-Title in which they get paid a Fortune. At least a Trillion Dollars a year is missing this way Folks. Where is our money? We could have the best health care system on the planet for this amount.

Good Politicians need to Counter-Act this Corruption with Positive Energy by Secretly Releasing inside Info about the Corrupt Going-Ons and Plans of these Political-Criminals. Good Politicians need to release info about these Political-Criminals and their Plans, for the GOOD OF ALL. The actions are the same but the intent is different. Again it comes down to Good and Evil, Morals and Ethics. Good and Evil are measured by intentions. Speaking out for the "Good of All" makes it in and of itself good.

We all need to Unite-Collectively. Citizens, Government Agents, and Politicians. This is what these Elite and their Team are doing, and their own Techniques will work against them. By people Uniting and Pushing Back against this Corruption, it will create an Atmosphere of "Good", and an Atmosphere and Acceptance of Speaking-Out against Corruption. It will become more Acceptable to Find, Seek Out, and Attack Corruption—once this happens a lot of very Real Powerful Criminals can be put away. The Government will become more clean, and no doubt over time, the World. This is a matter of Energy and Balance. Positive-People with Positive Action have got to start Pushing Back. This is not some impossible dream, it is easy to do, people just need to start doing it, and across all walks of life.

UNIFY PEOPLE, UNIFY. BECOME SELF-EMPOWERED AND UNITE IN GOODNESS.

Find your EMPOWERMENT AND SELF-WORTH in this World, become EMPOWERED and TAKE ACTION. Power is not Evil, Power can be used for Good. It's how Power is used that determines if it is for Good or Evil. Power is Neutral. Take back your own EMPOWERMENT and use it for Good, for the Good of all. Be Active, Action, not simply Words.

Folks there is a real battle of Evil versus Good going on right now; let's take back OUR BIRTHRIGHT OF EMPOWERMENT and make it a battle of Good vs. Evil. Shouldn't WE? From Citizens, to Government Agents, to Politicians, we all must start doing the right thing. Like a Harry Potter novel where all Unite against Evil, doing what they can for the Greater Good. There is no Honor, Courage, or Strength in keeping our mouths shut when WE see things that are wrong that need to be spoken out about. We need to become SELF-EMPOWERED and ACTIVE, WE need to use the same Techniques that this Elite and it's Team are using because they work against anyone, including the Elite and their Team. We do not need to "Reinvent the Wheel" on this point. People must start caring enough to take ACTION, waiting for someone else to do it for us will not work. WE MUST UNIFY, UNITE IN GOODNESS, FIGHT BACK, BE SELF-EMPOWERED, SPEAK-OUT, AND BE ACTIVE. The stakes are high and "The End" is near because the Elite have a 2012 Deadline for Their Plans. UNIFY, BE GOOD, AND FIGHT FOR GOODNESS: BEING "GOOD" DOES NOT MEAN BEING WEAK, BEING GOOD MEANS STANDING UP FOR WHAT IS RIGHT.

CHAPTER 32

Tips and Pointers for the Two Agencies Involved, Who are Being Manipulated and Used, by "The Clown"

For my Readers: the Following Below is for Different Government Agencies, and gives a good idea of "My Past History" as well as "Current-History" with The Clown, although it is far from "Complete".

My "Clown Mess" as of 2011:
A List of Problems for the Two Agencies Currently Involved

Here is a List of the Problems which is "This Mess". This "Mess" also happens to be My Life.

First of all the Agency in America Framed Me several times. As these were Corrupt-Employees, and as I have come to see, these Employees are not "Monitored or Supervised" carefully: thus this Agency no doubt is unaware of this.

Second of all the Agency of the Country I live in is Unaware of this.

Third of all the American Agency in Charge of The Clown may be Unaware of this as well.

As it is, The Clown's Agency is Unaware of a good many things, especially in regards to Whorbat (For readers a Fake Name), The Clown.

Now there are "More Problems" as well. All "Complications" due to The Clown.

1. This Clown has been "Keeping Me In" this Mobbing, even though I "Passed" the first time, many years ago. It is now going on year number eight since the "First Time I Passed".

2. This Clown "Kept Me In" the first time, and many others, based on Personal Reasons, not Professional ones. She in fact has NOT followed the "Rules, Protocols, or Standards" of this Mobbing, or even Her Own Agency. For this alone, she should be in Trouble. My Book is the only reliable way I have found, for communicating with The Clown's Agency, and is the only "Solution" I could come up with. As time passed, She began "Keeping Me In" because of this Book, again Personal NOT Professional.

3. As I have come to realize, She is "Personal not Professional" based on her "Personality and Profile". Which I have come to realize is about her being a Psychopath. Psychopaths by definition being ABSOLUTE MANIPULATORS. They are in fact "Predators". I will get to things which point to this later, having said that an "Interview" by The Clown's Agency would go quicker—which would also be to "Their Benefit". As it is I am in a hurry to finish this Book, because of what your Clown is "Doing to me with this Techno", never mind She and her Buffoons are "Spiking my Writing" in a variety of ways.

4. I doubt The Clown's Agency, or anyone else for that matter, knows that she is a Psychopath.

5. This American Agency has given The Clown Electromagnetic-Technology.

6. The Agency of this Country, has obviously not informed it's Field-Agents who are "Involved in this Mobbing" of this Electromagnetic-Technology, for that matter, neither has The Clown. This is an "Assumption", but if they are simply believing Whorbat because of "Her Rank", then they are dumber than they know.

7. IF these Field-Agents of this Country's Agency, DO KNOW of this Technology then they are Absolute-Idiots, in light of the FACTS AND INFORMATION I have given them. I will get to this later, but this involves several points, not the least of which is that they DO NOT KNOW what happened in America, and they have been Programmed by The Clown. No matter what, the Field-Agents of this Country that I have approached are a couple of Dupes.

8. What do you get when you give a Psychopath this Exotic-Techno? Picture "Hannibal" from the movies with this Techno, and about to be

Exposed. Think it has been pleasant around here? You can't imagine what I and my Family have been through. What we have been through has been absolutely sick and criminal.

9. This Whorbat has been Misusing this Techno, on a wide variety of people, for a wide variety of reasons. Having said that, the one common theme of Misuse of this Techno, is that it is "Misused for Personal Reasons"—many of which have nothing to do with "Being Caught", but merely being able to do it and get away with it. She has also been "Showing Me" how She can "Use it on Anyone" and "Get Away with It", including Two People who work for Law Enforcement, and who are Involved in this Mobbing—no doubt with this Countries Agency. No doubt all of this being part of Her Brilliant Plan for "Demoralizing Me". Brilliant.

10. At this point Whorbat is Misusing this Techno, to Spy on Others, and for "Protecting Her Ass". She is in fact Misusing it to "Keep a Lid" on what is really going on. The chances of her not "Spying on these Two Field-Agents" is zero. For either Agency let me know when this sinks in.

11. This Exotic-Techno is currently being Misused to "Create Fake Tapes", which have nothing to do with my "Personality, Feelings, or Thinkings". That would be because I have "Passed this Mobbing" already, many times, Whorbat is "Keeping Me In" at this point, as a way of "Controlling the Information about Her". Whorbat is Misusing both Her Job, and this Mobbing at this point, for Hiding Damaging Information about Her, which Exposes Her as the Corrupt, Sloppy, and Dangerous Agent She is.

12. If she had not "Held Me In", the first time, never mind many others, there would be no Book, and I would not be taking this approach of Informing Agencies of what is, and has been, going on.

13. Whorbat in 2010 Misused this Techno to "Create a Violent Attack", between Two Citizens of this Country, "Letting Me Know" all along the way it was Her. This was to try and scare me, to "Shut Me Up", it is in this Book instead. Very effective I might add.

14. What Whorbat did to this couple, getting the young man to attack his girlfriend, is in fact what she tried with me and my girlfriend many years ago. For The Clown's Agency there is in fact a record of a telephone call I made that night, to my parents, in which I spoke to my Mom, about "Catching a Thought being Pushed-Into my Head" followed by a "Push of Emotion", Intense Anger. In this phone conversation I clearly state that I recognized it was an "External-Influence" and NOT MINE. As a point to All Agencies Involved here, who do you think it was? The Russians?

15. Whorbat is currently threatening to KILL ME WITH FATAL ILLNESS, by way of this Technology. And when "She Leaves" at that—like "Cancer".

Take a good god damn look, does it look like I need to take this seriously? As an added point, neither Agency involved here even begins to know of what has been done to me with this Technology, never mind my Girlfriend and Kid. As it is we have two dead pets.

16. As a result of Whorbat's Heavy Misuse of this Techno, it became more and more clear that there was an "External Influence Involved" in my "Failing this Mobbing". If you think about it, Techno was Relied On because I could "Pass Normally". This has been "To Create Looks" for Holding-Me-In, not Reality.

17. As a result of my knowing this, and as a result of Whorbat's Showing-Off, never mind "Cheap-Excuse-Making", she came out of the closet with this Electromagnetic-Technology, revealing it to me, and others as well.

18. For the Agencies Involved, there should be MANY "Tapes" of me and my girlfriend "Speaking of this Exterior-Influence" which "Kept Affecting Me" at the "End-Tests of this Mobbing", and "Last-Day of this Mobbing". There should be at least about two years worth of me talking with my girlfriend about "Something Holding Me In" on the "Last Day". Obviously this Techno, was being Misused to "Hold Me In". Who else do you think it was?

19. As a result of Death Threats, and Clueless Agencies and Clueless Agents, I have "Taken this Book Approach". As you may have noted, this Book "Raises serious Issues" and Exposes this Clown, because this was the ONLY WAY I COULD GET IDIOTS TO LOOK AT THE FACTS. Never mind it was the ONLY WAY TO GET ALL RELEVANT INFORMATION OUT. As it is, it is not all included here. Neither Agency has all of the Facts, or knows what is going on.

20. The Clown has Delayed my Book, by OVER TWO YEARS NOW, and very much so with this Exotic-Technology, in fact almost entirely with it.

21. You two Clueless Agencies, need to ask yourself WHY.

22. I will give a copy of this Book to BOTH AGENCIES, and state this here so that The Clown's Agency has no reason to feel compelled to do a Cover-Up, by way of Hard-Core-Framing, or by Murder. AND FOR THEIR OWN CORRUPT-EMPLOYEES, LIKE THE FEDS DID FOR A PEDOPHILE-COP, AND RACIST-MURDERING-REDNECK. Thanks but I have had enough of "Your Justice".

23. For the Clueless Agency of this Country, YOU FIGURE IT OUT. I'll help you out: IT IS CALLED A COVER-UP.

24. There is no doubt that the Two Field-Agents of this Country, that I have approached, have been Programmed by This Clown. Based in no small part on a "File" that is FULL OF LIES, DECEIT, AND MANIPULATIONS. They have been Programmed, REGARDLESS, of if they know of this Exotic-Techno, or not, and regardless of this "File". They remain "Clueless and Programmed", because of what they DON'T KNOW, and because they are Unaware of the

Tactics being Used on Them. At the very least, like a couple of Idiots, they obviously do not believe this Clown is "Holding Me In". On the bright side, if you two Agencies ever get your shit straight and do an Investigation of this "Mess", Your Shared Clown, these Two Agents will no doubt become more "Educated", having "Become familiar with the Tactics" the hard way.

25. The Two Field-Agents of this Country, IF THEY DO KNOW of this Technology, can absolutely Kiss My Ass for not believing me. If you Do Not Know about this Techno, then "Your Viewpoint is at least Understandable". However if you do know about it then I have this to say: Everything is in this Book, because of Ignorant Backstabbing Idiots like yourselves. YOU LET YOURSELF BE USED FOR A CORRUPT AND DANGEROUS CRIMINAL WITH A "BADGE". Having said that "A CRIMINAL IS STILL A CRIMINAL" NO MATTER WHAT DAMN JOB THEY HAVE. Simply "Following Orders" in spite of the Evidence, Information, and Facts, does Not make you "Professional". Using Your Brain to get to the Truth, to the bottom of it all does: it is in this manner that you do not get Used and end up "Looking Cheap". Having said that if YOU DO NOT KNOW of this Exotic-Techno, aside from being "Sour with You", consider yourself "Forgiven".

26. The Clown has spent the Last Two Years, Re-Editing Her Files, AND NO DOUBT NOT JUST MY OWN, for Her Eventual-Exposure. You Two Agencies are now Informed. What exactly were you expecting Her to do? Why exactly do you think The Clown has Delayed my Book for TWO YEARS?

27. For The Clown's Agency you will need to contact The Last Agent in Charge, as he was familiar with me, and no doubt from the beginning, will recognize that something is wrong with The Clown's Re-Edited-Files, never mind other "Tells" as well. As an example he will tell you, I DID NOT KNOW of this Technology.

28. All of you Agencies are Incompetent, and this Book and my Life, are a RESULT OF THIS FACT. You can ALL KISS MY ASS, UNTIL YOU ALL GET YOUR FACTS STRAIGHT, AND TAKE RESPONSIBILITY, AND ACT PROFESSIONALLY. NONE OF YOU HAVE ALL OF THE FACTS, AND OBVIOUSLY NONE OF YOU KNOW WHAT THE WORD "JUSTICE" MEANS. Your "Agency Title" is not worth a Rat's Ass.

29. I will not "Wear this Bullshit Anymore", and someone somewhere had better get their head on straight. You can be Intelligent, Responsible, Professional, and "Clean this Up" by BEING SO, or you can escalate things. If that is the case, seeing as how I AM INNOCENT, AND NOT RESPONSIBLE FOR THE CLOWN AMONG OTHER THINGS, YOU HAD BETTER START PRAYING I DO NOT HAVE MONEY. IF YOU DO NOT TAKE CARE OF YOUR MESS AND YOUR SCREW-UPS, WHEN I HAVE MONEY, I WILL PUT YOU IN A COURTROOM, AND I WILL START WITH THE CLOWN'S AGENCY. Your "Files", and certainly the ones The Clown has,

has ALL OF THE FACTS, and you have NOT shared these "Slight Details" with the Agency of this Country. Why? And for what reason? And for Who? Either you have done this as an Agency, or Corrupt-Agents have done this. What kind of Bullshit is this? This isn't Professional, it isn't even Legal. Furthermore this started because of a Pedophile Cop and Racist, how good do You look here?

30. The Clown is quite aware that I was Framed in America. She in fact has been Framing Me as well, doing a Cover-Up of Her Mistakes, and FRAMING-ME with this Techno, so that I am "Kept in this Mobbing", and therefore She can Use it as an Excuse for "Monitoring Me" and "Controlling all People" I have Contact with, and therefore "All Information about Her" as well. She is Hiding Her Corruption and Mistakes from everyone, by DELIBERATELY "Keeping Me In" this Mobbing. Like everything else, this is NOT ABOUT AGENCY, it is about HER.

31. This Clown has been doing a "Wide Range of Things" for a Deliberately Inaccurate Profile of me, for "The Next Person in Charge". This includes "Techno Recordings" of "Thoughts Pushed into my Head", and other Tactics as well. Like "Emotion Pushing" set on an "Extremely High Level" for a "Very Slanted Version of Me" in regards to "Thoughts", and "Scripted Thoughts" while this is done as well. A "Collection of Deliberately False Recordings" made with this Techno no doubt has been made, for tricking anyone else down the road here. A wide range of things have been done with this Techno, for "Recording with this Techno" and no doubt for "The Next Person in Charge". What I have been "Noting for Months" as "Weird Tactics", which I can only figure are for this purpose of Manipulation. If you as an Agency, want a challenge, you can "Sort this All Out" and all of "Her Deceit" in regards to "Her Hiding Things". Obviously you will need to send someone out here with this understanding.

32. This Idiot, The Clown, has gotten my Kid seriously injured, and at least on one occasion, I thought my Kid had been killed. Never mind things done to Me. As recently as August 2010, Your Idiots did something that nearly got my Kid blind in one eye. TIP: Get someone Professional and Responsible Out Here, with a Brain and Common Sense, before something "Permanent" happens. If you do not "End This Mess", do not get the Facts and find out I have Passed Very Many Times, and My Kid is Seriously Injured again, I will absolutely come after Your Agency. I will "Lock down on your Throats" like an enraged Pitbull or Wolf—clear? How you handle Your Clown, which is "This Mess", will determine any "Problems down the Road". You did not know what is going on with Whorbat The Clown,

NOW You Know. Minus you as an Agency Totally-Screwing-Up at this point, there should be No Problems. Read my Letter to You, to get this.

33. "WHO I AM". Any Agency wondering who the hell I am, to be "Giving Tips and Pointers", never mind my "Attitude", can Kiss My Ass. I'll tell you "Who I Am": I AM the person who has "Caught this Woman", I AM the person who has accurately Profiled Her, and I AM THE ONLY ONE who knows what is going on around here. And I AM the person paying a price for it, along with my Family, including our pets. I KNOW THIS CLOWN BETTER THAN EITHER AGENCY, THAT IS "WHO I AM". If you don't like this, then next time don't Fuck-Up as Agencies. As it is, I AM DOING YOUR DAMN JOBS FOR YOU. YOU CAN BOTH PAY ME FOR IT, OR SHUT UP IF YOU HAVE A PROBLEM WITH MY "APPROACH". You Idiots have not "Suffered as a result of Your Mistakes", I and MY FAMILY have: I have two dead pets at this point, both of which were Loved and who were Family.

You plan on paying me and my Family behind the scenes for YOUR MISTAKES? No? MILLIONS of dollars have been spent at this point for Protecting a Pedophile Cop and Racist Redneck, so why not? You have nothing but "Money to Waste", and in my case this is a matter of "Damage Done" and "Balance". Maybe I should contact a Politician with all of this, No? Then you can keep Your Damn Mouths Closed and DO YOUR GOD DAMN JOBS as they are actually meant to be done, and "Legally and Justly" at that. I AM the person who was Forced Into writing a Book because of Your Corrupt Agents who are Out Of Control, I AM the person who had no choice because no Agency is "Monitoring this Mobbing and Employees", and I AM NOT THE ONE who Screwed-Up for several different Agencies, Your Employees did. THAT is "Who I Am". Clear? You had better be.

CHAPTER 33

"Don't Tread on Me"

A Letter of Chastisement to Three Different Government Agencies

Well, this Letter goes out to Three Different Agencies, all of which to some varying degree, are obviously Clueless, never mind INCOMPETENT. I don't plan on using "Pleasant Wording Here", as quite frankly I've earned it, and all Three Agencies look like a bunch of Non-Professional-Idiots, because they ARE. If you already do not like my wording, then I suggest you take a "Look at my Life" which quite obviously NONE OF YOU HAVE DONE, because you DO NOT MONITOR OR SUPERVISE YOUR EMPLOYEES, or anyone else's for that matter. This last statement is for the Agency of this Country, who has allowed The Clown "Free Reign" within their Land. If you still have problems with my "Tone" then I suggest you put someone "In Charge" capable of "Empathy". For that matter I recommend that you actually start acting as Agencies, and start performing Competently, never mind PROFESSIONALLY.

I have "Tried being Diplomatic" both in America, which got me Framed, and here in Europe, in which I was Disbelieved because of Corrupt-Agents doing a Personal-Cover-Up, of their Own Mistakes and Crimes: to avoid being Exposed and Caught. So this is what you Agencies get instead. This

Letter is based on Observation and Experience, and Spells-Out the Reality of Agents, Agencies, the "Justice System", My Life and My Families, and "This Mess". If you do not like this then BE Responsible AND Professional and Change This. I am "Still being Diplomatic" here, but my "Attitude" in regards to "All of This", has most certainly changed. Quite frankly at this point, if you "Want Respect" you will "Have to earn It". Quite frankly I do not give a shit who you are, or that you are "Agencies"—this "Title of Agency" is a Farce. Quite frankly, all Three of You, have been, and are, Sloppy and Incompetent to say the least.

If you don't "Like the Tone", then you can come on over and "Wear my Life" for a few days. A "Few Days" is the most you could handle. You had better be clear on your end, and you had better get your god damn facts straight, never mind "Get a grip on Your Employees". This is not, and has not been about Mobbing, for many years now, never mind that I was Framed-Into-It.

If you want a "Feel" for what My Life has been like, see the movie "In the Name of The Father".

As I have noted, from observing "Whorbat", The Clown, there is obviously very little "Supervising and Monitoring" of Employees. I see that there is "Little Contact" between "Free-Range-Agents" and Agency—as a result, Agents are allowed to "Do as they Please" and "Write it up any way they Want". WITHOUT THE AGENCY ACTUALLY KNOWING WHAT IS GOING ON, OR WHAT THE TRUTH IS IN REGARDS TO MANY A THING. Brilliant. In the meantime you Idiots are "Handing Out" Exotic-Technology to many a Corrupt-Lowlife. Where are your god damn minds? You are all like "Mission Impossible" meets "Forrest Gump". I have news for the Superiors of this Clown: if She knew WHO will be looking into her, as a result of this Book, She would use this Technology to "Spy On You" at the least, to try to Blackmail You, and no doubt would "Tweak Your Thinking" with this Techno, for the sake of Her Own Ass. We'll get to this in a bit, but why wouldn't She? At this point, I know your Employees better than you do. Need I say more? If this does NOT bother you, it should.

As it is, this Letter has taken me weeks to write as opposed to days at the most because of this Technology. Things such as Emotion-Pushes, which are "Left On Me" and "Thought Blocking" have resulted in "Several Versions" of this Letter which are "Not Me". The only copy I had even resembling me on many a point, "Disappeared" from our computer the day I wrote it. Obviously to "Frustrate Me" and to get me to release a more "Harsh Letter" and one which no doubt Influences Superiors within Agencies in a way that Benefits Whorbat. In the meantime I and My Family Suffer, and at this point it is all gratuitous. Thus I release this as it is. You determine as Agencies whether you are going to let these Idiots Manipulate You. At this point if this nonsense is going to end, this Information has to get out. Be Responsible and Adults, and there should be "No Problems". Once you have all of the Facts, I don't see how there could be. At this point for Agencies, if you want "Respect", you are going to have to "Earn It": so far none of you look very Intelligent here, or Responsible.

Having said that it is because None of these Agencies really know what is going on, that I am "Diplomatic". Do not make me regret this. I get "The Bigger Picture" here, and so I do Not really blame you: that said you can "Put up with my Tone" and take anything said "Into Perspective". Quite frankly all involved have "Earned It". Take a "Look at my Life" to see what I mean. With that in mind. Lets begin.

For the Agency in America

That said I will start with the Agency back in America first. I have no doubt this Book will reach you, and yes I am aware that Agencies "Keep Tabs" on New Books in the Market for various reasons, Information being one of them. Consider yourself "Informed". We will start off with the "Idea" that whatever "Records and Files" you have, like The Clown's, have been "Doctored and Edited", for the "Sake of Looking Good", and for the sake of not getting in Trouble or Caught. As a result you no doubt do not have a Clear Picture of the Truth. I feel my Book has straightened this out a bit. Look at the "Tactics Section"—your Agents also put two other people in this Mobbing for no damn reason, other than I was "Good Friends with One", and the other a Co-Worker, because your Agents did not like him—just like with me. Very Professional. As I am beginning to realize the Stereotype of "Agencies and Agents" is one great big Pile of Shit. Take a look at my Life, this Book, and this "Letter", how does it look to any of you? Professional?

Here are three of your Geniuses from this time who you can look up, for my readers I have changed the names. "Jay" from the Diner, "Tim Devilmen" from the Diner, and Margaret X. from Virginia Beach—a "Chinese Woman" who was "Bright Enough" to wear a "Name Tag" at the time of visiting me in the hospital for Heat Exhaustion. How much you want to bet she used her real name, and how much do you want to bet she didn't think I would "Notice or Remember"? These first two were involved with the Framings, one of these Framings which put me in this "Government Program of Mobbing", which a Law-Student can see the Truth of. And for, and because of, a Lowlife Piece of Shit pedophile-cop, and a racist redneck dispatcher, who delayed an ambulance for a kid who was shot, because this kid, was not part of the "Master Race". I naturally knew of the RESULTING COVER-UPS, so I guess that makes sense of everything, doesn't it? Do you as an Agency know what your Employees are Doing? Do you as an Agency, know what the fuck YOU ARE DOING? From this information alone, we can make some basic assumptions here.

First of all, we can see that you as an Agency do NOT SUPERVISE YOUR EMPLOYEES. The only "Crime" is getting "Caught" and being "Held Responsible" apparently. As an Agency, YOUR ABILITY TO MONITOR AND SUPERVISE EMPLOYEES IS A DISGRACE. Where are your god damn Priorities? You are giving Employees "The Power of Government", and a License to "Do as they Please", and many will be "Tempted by the Power" and become Corrupt, if they are not "Watched Carefully". Never mind giving these people Exotic-Technology which adds to "The Heady Mix" and "Temptation". And as you will note from this Book, I am a "Magnet for this Type of Person".

YOUR AGENTS obviously were, and are, Absolutely Filthy. Apparently Your Agents from my "Past" endorse Pedophilia, Being Racist, and Corruption. Because I knew of these Cover-Ups, I was "Targeted" and put in this "Mobbing" among other things. Is this the "Reputation" you want?

Having said that I was, and AM, INNOCENT, and YOUR EMPLOYEES belong in Prison.

We can all clearly see that being a Racist Asshole, and being a Kid-Fucker, are not "Qualifiers for this Mobbing". Which I find interesting, as "Knowing of these Things" DOES QUALIFY ONE FOR BEING FRAMED, AND MOBBING. From this we can all see that the World is Safe, and people need not worry with your Agents in Charge. As long as people ARE WHITE, AND NOT A LITTLE GIRL, NO ONE NEED WORRY. Thank god for YOUR AGENTS. We also can see, that being a Corrupt-Agent, a Criminal with a Badge, is NOT a qualifier of Mobbing as well. Apparently "Somehow", "Having a Badge", seems to "Negate being a Criminal"—which it does NOT. Who the hell do your Employees think they are? Tax-Dollars are paying YOUR CRIMINAL EMPLOYEES. Tax-Dollars are paying for YOUR AGENCY. You don't recognize the "Temptation of Corruption" that your Agency and the Job-Titles it "Hands Out", poses?

I have a "Tip for You", it is called a "Psychological Evaluation" BEFORE hiring Employees. Especially since your Agency Wields Authority and has Power. Places like Gas-Stations do it, simply because they are worried about "Employees Stealing". Never mind a Job, where the Employees start "Feeling and Acting" like they can do and get away with anything, because they are "High on the Power" that comes with the Job, never mind "Their Training" which enables them to do this. Many of these Employees are

"Taking their Training and Knowledge" and Using It for THEIR OWN SELF-GAIN and Using It for "Looking Good" and "Job-Advancement", never mind "Self-Fulfillment". Quite frankly you should be more Intelligent and more Properly Motivated—as better "Supervising" will result in "Less Mess". How hard is this to "Figure Out"? What is this, Anarchy?

Another point is that I KNOW MORE than your own Agency in regards to your Employees from this time, and WHAT THEY DID. You are what, going to "Rely on their Files"? You want to go into a Courtroom "Based on these Files"? As The Clown would tell you if She were not like your own Employees, if she were not "Her Own Agency", I have a "Near Photographic Memory" in regards to "Past Events"; and as She would tell you, even with this Exotic-Techno, She has not been able to "Dig Up" or Find Out Everything, because I understand how this Techno works, and quite frankly would be Stupid to "Pull Up all Information" for Corrupt-Phonies. Never mind Agencies which do not have a Clue as to what "Their Priorities" should be. Or what Their Employees are Doing for that matter.

How many of you are "Not White" and how many of you "Have Kids"? One thing is for sure, your Agent Margaret, who was Chinese, would have bled to death in that parking lot, along with a good number of your Field-Agents who I worked beside at various Jobs. Very "Logical".

Of course as I have "Come to Realize", thanks to The Clown, no doubt your Agency did NOT know of what these Employees were doing, but that does NOT change the Fact that I was Framed Several Times, it does Not change the Fact I was "Framed into this Mobbing", it does NOT change the Fact that I have "Lost Twenty Years of my Life", it does NOT change my being Angry about this, and it does NOT change the Fact that all of you Agencies DO Engage in this Criminal-Government-Program of "Mobbing". Having said that, I "Assume" that what your Agents did, was not "Sanctioned and Endorsed" by your Agency. And yes I do feel "Safe" in this "Assumption" as quite clearly, the Agency here in Europe has absolutely no idea of what It's Clown is doing. As a point, how do I know this?

And this is obvious, never mind that they have obviously not Figured Out that this Clown is Literally a Text-Book Psychopath. Again, this is called A PSYCHOLOGICAL EVALUATION, AND "SCREENING EMPLOYEES", and preferably BEFORE GIVING THEM ELECTROMAGNETIC-TECHNOLOGY. It is called "Supervising and Monitoring" as well. None of you as "Agencies" look very Intelligent or Responsible here. Quite frankly you all Look Non-Professional because you have been, and you all look like a Bunch of Fucking Idiots because this is how you have all "Performed".

I base my "Commentary" on Experience and Observation, which is more than Any Agency does with it's own Employees. Clear? What am I, "My Own Agency"? How many Millions of Dollars have been spent on me, how much Man-Power and Time, and for what? Have you all lost your Minds? And now my Family is Targeted as well. Because of a Pedophile-Cop and Racist Lowlife: because of Corrupt-Agents as well. Quite simply, all Three Agencies look like the Biggest Jokes on the Planet, because you God Damn Are. Take a good god damn "Look" at all of this—you want to argue that You Three Agencies look "Good or Intelligent" here?

Now I have "No Interest" in going to Court, but I have news for you, much of what you "Feel you Have" is not as you think it is, and this can be Taken care of in a Courtroom, ANY TIME YOU NEED IT. Which neither I, nor You, want. And this "Goes out to all Agencies". If I ever return to America, if I catch "One Whiff" of Corrupt-Agents sticking their noses up my ass, because they are Misusing Your Agency, and based on what THEY HAVE DONE, A PHONY-FILE, You will have "Problems". You are Responsible for Your Corrupt-Employees Manipulating Your Agency-NOT ME. Clear? As it is, your Agents tried getting ME KILLED on several occasions, and I take this personally: I'm "Funny that Way". As one example Your Agents had a Car-Mechanic NOT change my Timing-Belt for starters, which fortunately another Mechanic caught later, and before it was Too Late. I am quite sure of this. Is this even in your Files? If it isn't, it is in The Clown's at this point. At any point in the Future, if I catch Your Agency "Manipulating my Life" I will take it as the "Threat It Is". Who would not? Would You if the Roles were Reversed? Are we clear here?

What I "Recommend" is that you get a Good Look at the Truth here, and you start "Sorting Things Out" before, and if, I ever return to America. So that you do not Screw-Up, due to Corrupt-Employees who have Misused Your Agency. If you do not "Sort Your Mess Out", and do not "Sort Out" your Criminal Employees, AND I have problems in the Future, I will make myself an "Agency Name"—and it will be known for "What comes out in a Courtroom". So if I ever return to the States, keep the fuck away from me, as quite frankly, at this point, I take it as a threat to my life. AS WELL I SHOULD. What would YOU DO if the roles were Reversed? You as an Agency will have no reason to be Interfering with my Life, in the Future. I will take anything as a Threat to my Life, and as a Threat of Framing Me, which no doubt you do NOT GET, because you no doubt have NO IDEA of what has occurred in My Past—and with, and because of, Your Corrupt-Employees. You support Known Pedophiles and Known Racists? How good do you look here?

To be clear I AM NOT going to "Start Anything", but I most certainly will make sure that I am not "Framed Yet Again" on a return to the States, because Corrupt Agents Misuse Your Agency, and "Investigate me for Jay-Walking". Clear? That would be YOUR Responsibility.

I do not "Blame You" as an Agency, but it is time you knew the Truth so that there are not "Further Complications" down the road here. YOU are the ones in a position to straighten Your Employees out, NOT ME. You did not know before, NOW YOU DO. So straighten it out.

Take Responsibility for Your Mess, PUT BLAME WHERE IT BELONGS, AND GET TO THE BOTTOM OF YOUR CORRUPT EMPLOYEES, AND THIS "MESS" THEY HAVE CREATED BY LYING, MANIPULATING, AND FRAMING, BEFORE THERE IS MORE SCREWING-UP. This means "Details as Well", this means "Evidence as Well", this most CERTAINLY MEANS CONTEXT, this means knowing your Own Employees, and this means KNOWING WHAT THEY HAVE DONE.

Where are the "Brains of these Agencies"? You Train Your Employees, and the Corrupt-Ones turn around and Use Their Training for Manipulating Your Agencies. As it is, a Total-Idiot can read this Book, and see there is something "Not Right" with the way your Agents have "Handled Things". A Law-Student can "Sort out the Truth of This". I have been in this Mobbing for TWENTY YEARS almost, and for WHO AND WHAT? And at what cost? How many MILLIONS OF DOLLARS have been spent at this point? You call this Justice? This is Bullshit, and it "All Started" with Your Employees. This is the biggest god damn waste of Tax-Dollars on the Planet—but thanks "For the Training". At least it is being put to "Good Use", it most certainly is better used by me, than Your Own Employees. At least Tax-Payers are getting something for their Wasted-Money.

Having said all of this, I will end this part of the letter to your Agency on more "Level Ground". I will say this: I get that your Agency IS NOT, and WAS NOT, aware of what happened because of these Employees, and what happened to my Life as a result. I DO NOT BLAME YOU AS AN AGENCY, BUT NOW YOU KNOW. As a point, your Agents "Acquired this Exotic-Techno" because I "Set Them Up", and they became "Paranoid" I had "Figured everything Out"—as in they were WORRIED I had figured out THEIR MULTIPLE FRAMINGS. As it is, I had "Figured part of it Out". They merely Used Your Agency, for the Sake of their Own Asses, which makes you as an Agency, look Brilliant, never mind "Used". Quite frankly you look Stupid as an Agency. I wonder how THIS is written up in your Files? No doubt MANY THINGS in your Agency Files are written very differently, than the "Truth of the Matter". Read my Book, how "Unclear" does any of this look to you? Now you can get a few things straightened out. And make sure you do it before I return to the

States. I EXPECT PEACE AND QUIET, AND YOU CAN AS WELL. Take care of this on your end, there is no problem, as your Corrupt-Employees and their Twenty Year Long Farce will be "Ended". Minus Attempts on my Life and Framings, I could care less about you.

DO NOT even think of Dumping Your Responsibility, and Your Mess, on my Shoulders. The "Solution is Simple" on Your End, and I will not be "Your Patsie" in all of this to "Make Your Life Easier". I guarantee you I am Finished with That. Quite frankly I have had enough of that Bullshit. You straighten out Your Corrupt-Employees, and you stay the fuck away from me. I do not think of Your Agency as "Legal, Professional, 'Safe', or Responsible" at this point, and I'll "Believe It" when I experience it. Based on my Experience I think of Your Agency as "Irresponsible and Dangerous". So "Change This", as well you can.

As an Agency I expect Responsibility, an Intelligent Approach, and I expect you to do the "Right Thing". What I actually expect, is for you to act PROFESSIONALLY, as you are god damn supposed to. Is this too much for anyone to ask? I sure as hell hope not. In fact, if you take "Investigating this Mess" seriously, you can feel free to contact me, and no doubt I can straighten some things out for your Agency in regards to it's Files. Having said that "Mutual Agreements" will have to be made on paper first. My "Approach", regardless of "My Past", and "Tone", and "Minus some Precautions", is "Diplomatic". Do not make me regret this.

THE AGENCY OF THIS COUNTRY

The Agency of this Country and the "Most Clueless of the Three" in one sense. Having said that you obviously do not have "All the Details", as these "Slight Points" were not "Provided to You". And as quite obviously, you did not bother getting The Facts. You can read this Book, and I recommend you read the Electromagnetic-Technology Chapter.

I am disgusted with you as well, as I have already tried approaching two people who are "Involved in this Mobbing", and who I am certain are with "Law Enforcement": who I believe are with Your Agency—yet they remain "Unconvinced and Clueless" as to what is going on around here under their noses. In fact they obviously have not "Communicated anything to Your Agency". As a result of Everyone Acting and Performing as Idiots, I am this way in this Letter. This "Tone" is not about Mobbing, it is about the Stupidity, and Enormousness of it, going on and around, in my life. This is about me Trying to Expose Corrupt-Agents at this point, and Dangerous Agents, yet everyone is absolutely Clueless to this: and in the meantime my Family Suffers. Clear? You better be, not one Agency has the right for getting an Attitude with Me at this point. Anyone at any Agency, with an "Attitude" about me, needs a good case of Mobbing. What each Agency is going to do, is get it's Facts Straight, and this Letter and Book will do it for you. Lets begin.

Well. What kind of Agency are you? I have had no "Experiences of Corrupt-Agents" from your Agency, merely "Experiences with Clueless Agents", but that would be because YOUR AGENCY IS CLUELESS as well.

We live on a Continent famous for "Hitler's Last Stand", and look at what your Country and Agency have ALLOWED, and take a good look at WHY AND FOR WHAT. Do any of you Agencies even bother trying to VERIFY INFORMATION?

Who do you think you are? How do you think The People of this Country would react if they even knew that your Country Engages in this Type of Government Program and Behavior? Look at "How my Mobbing Started" in the first place. It is called Corrupt-Agents and Framing Someone, and after Hitler, you would think that European Agencies would be "More Vigilant" of "Mobbing", which is the "Creation of a small Fascist Movement against a Target" (Do Not try and Bullshit me about this Point), never mind that you would expect European Agencies to "Get all the Facts" before engaging in this Corrupt-Bullshit. You think what, that this Continent needs another "Wave of Fascism"? No? Then why are you Using It, and Experimenting with it? Brilliant.

As it is, as you may have noted, I CURRENTLY am having problems with Corrupt-Agents, who are "Framing Me with Technology", never mind by Flat out Lying and Manipulating. To "Keep Me In" this Mobbing, so that they are NOT EXPOSED for "Breaking the Laws and Rules" of YOUR COUNTRY AND THEIR AGENCY, never mind the "Rules, Protocols, and Standards" of this Mobbing. As it is you do not have CURRENT-FACTS STRAIGHT, NEVER MIND PAST FACTS FROM TWENTY YEARS AGO. Naturally you Look Brilliant, and this is based on Observation. It is called, "Read this Book".

You are not even being "Vigilant" within the Borders of your own Country, and this "Mobbing" in my Life, is Out of Control, and CONTINUES TO GO ON because of Corrupt-Government-Agents, within your Land, who do not want to be Exposed by Me. You call this "Professional" on your end? You are finding this out from a Book because two of your people, who were dumb enough to believe these Corrupt-Agents, did nothing to "Pass this Information On" to Your Own Agency. Brilliant.

This Clown is Out of Control, Misusing Her Job-Title and this "Mobbing" just like any study of "Fascism" that has been conducted by Government. This "Mobbing, Mess, and Situation" are Out of Control, because the "Person in Charge" resembles someone from any Study about Fascism where the "People In Charge" become OUT OF CONTROL—never mind This Clown being a Psychopath. Like the movie "Das Experiment", or the movie "The Wave". This woman The Clown, is Absolutely-Corrupt, and High on the Power of "What she can do, and what she can get away With". IT IS CALLED HAVING AUTHORITY AND POWER, AND CORRUPTION. It is YOUR JOB to Monitor People like this, it sure as fuck isn't mine (It only looks like it is My Job). Naturally as a result of the Power and Rank given to Her, She remains NOT CAUGHT. She is the Biggest Criminal on the planet, yet She "Hides Behind Job-Title" knowing that people have "Blind Respect for Authority". And based on what she has done to me, what do you think she has done to others as well? And within Your Land. You might want to give this serious consideration, never mind The Clown's Agency.

THIS IS WHY GOVERNMENT PROGRAMS LIKE THIS SHOULD NOT EXIST IN THE FIRST PLACE. And YOU should know better. You think your Agency and Government have it "All worked Out" so that only "Extreme Criminals" end up in this Mobbing? "Extreme-Criminals" are "In Charge of this Mobbing" and have been "Given Control of My Life", You Idiots. Guess what? For "Extreme Criminals" it is called a god damn COURTROOM AND A JUSTICE SYSTEM, SOMETHING YOU OBVIOUSLY KNOW NOTHING ABOUT.

EXTREME CRIMINALS FRAMED ME AND GOT ME IN THIS "MOBBING" IN THE FIRST PLACE: AND FOR OTHER CRIMINALS WITH CONNECTIONS TO "LAW ENFORCEMENT". But you do not "See Them being Mobbed", do you? How much sense does that make? Apparently no "Agents" bothered including these "Details" for Your Agency. You were expecting Corrupt-Agents to "Tell You They Are", you were expecting Corrupt-Agents "To Tell You of Their Crimes"? Apparently if you have a "Badge" you can do "What you Want", and it is NOT CRIMINAL. Is that it? These people are Criminals, Criminals with Badges, but Criminals nonetheless, and yet THEY REMAIN OUT OF A COURTROOM: never mind Mobbing.

This Clown is Misusing Technology that YOUR AGENCY KNOWS ABOUT, to Kill Your Citizens, yet you remain unaware because you have not enough sense to pay attention. You tell me, regardless of this "Techno" being a "Secret from the Public", is this "A Crime in your Eyes"? Is THIS Corrupt enough for you? Is this Illegal enough for you? Is this Dangerous enough for you? Is this a big enough Threat to the Public? In the meantime, I am "Kept In" this Mobbing with this Techno, and to "Convince Others", like Two of Your Agents, by "Way of Appearances". And these two Agents are "Spied On" with this Techno, by The Clown, as I will get to. Is that all "Disturbing Enough" for you?

I am literally being Framed and "Held In" at EVERY END TEST HERE, and if you need any "Proof" that this has happened in my life before, why don't you start with the PRESENT. This is happening right under your noses, I have "Passed this Information On" to two of your Employees, and you still don't know. Read my Electromagnetic-Technology chapter, for "Clarity on the Misuse of this Techno". In the meantime, two of Your Agents that I have approached and given information to, in regards to The Clown and Her Corruption, what She is Doing and has Done and is getting away with, and the FACT that She is "Holding Me In" this Mobbing, have "Disregarded this Information" I have given them. This is because they no doubt have no knowledge of this Technology, have no knowledge of My Past, and have been Programmed by The Clown who has done so for the sake of Her Own Ass.

They have been Used by Whorbat like any other Idiot-Citizen, for Whorbat's own Cover-Up, of Her Mistakes. They have been Tricked into "Being Quiet". Your Two Agents have no knowledge of the Psychology, Tactics, Programming, and Techniques, used in Mobbing—which is obvious, as they have been Programmed like a couple of Average-Citizens. Whorbat has used her Training in Psychology, Programming, Dirty Tricks, and I guarantee you this Exotic-Technology, for TRICKING YOUR TWO AGENTS, for the sake of not being Caught, Exposed, and in Trouble. So Your Agents have Information, and have NOT PASSED IT ON, as a result. Brilliant. If they understood this Mobbing better, the Techniques, Tactics, Programming, and Techno, this would have been less likely to happen.

As it is Your Agency, has NOT provided them with enough Information, for "Watching and Monitoring" this "Mobbing" in the first place. On this Continent, I expect Agencies to be much more Vigilant of this Type of Misuse, in the first place. Your Agents involved in this Mobbing know of this Techno, or don't they? You obviously have NOT provided them with enough Information about this Mobbing, Psychology, Technology, and Programming, and how this all works—or else they would not be Programmed in the first place. As a result they are not capable of Good Judgment Making. I understand more about all of this than they do, which is ridiculous. What am I, "My Own Agency"? And to be clear here, for "Variety" The Clown does not always use this Techno for "Keeping Me In", but She does ALWAYS Lie and Manipulate to do so, and THESE DAYS Techno is MISUSED EVERY TIME. In fact This Clown is Misusing this Techno on many people around here, at this point: including Your Two Agents.

As it is, The Clown has Programmed your two Agents (At the least two people involved in Law Enforcement and this Mobbing—and probably Your Agents), made fun of this fact and what she is getting away with, by

"Mocking their Intelligence". As is typical of Whorbat, The Clown is absolutely enjoying the fact that She is Absolutely Filthy, knows I know this, knows what She is getting away with, and has "Programmed and made two of Your Agents Blind", so that this all continues to go on. In the meantime, Whorbat knows that She will eventually be Caught and Exposed by this Book, BUT UNTIL SHE IS, She is "Making the Most of It". You were "Expecting her to turn herself In"? What is Your Rationale here?

My Mobbing continues because of Criminals with "Badges", who do not want to get caught for Breaking many Rules and Laws, and so "Perpetuate this Mobbing" and "Hide behind It", so as to DELAY THEIR BEING IN TROUBLE. And this is happening in Your Land, by Another Agency. ACTUALLY THEIR AGENTS, as quite obviously The Clown's Agency is just as Clueless as you are. Which makes BOTH AGENCIES look Brilliant, as in STUPID. I have news for Both Agencies, I "Passed this Mobbing" years ago, YET I REMAIN IN. Look at why.

This is what happens when You Train Corrupt-People, and in this particular case, a Psychopath, in Spy-Tactics and Techniques, Psychology, Programming, Dirty-Tricks, and give them Exotic-Techno. You end up with a "Mess", a group of Individuals who "Act as their Own Agency", and who create a "Mess" because they have not enough sense to realize that these Protocols, Standards, and Rules were put in place BY THEIR OWN AGENCY, to avoid THIS VERY TYPE OF THING. Of course the other "End of It" is that these Corrupt-Types are too busy doing what they want, too busy Misusing Job-Title for what they want, to be paying attention to Anyone's Rules. They took the Job, some at least, like The Clown, for this VERY REASON: for having THE POWER THEY WANT, TO DO ANYTHING THEY WANT, WITHOUT GETTING CAUGHT. It is a "Heady-Mix" to begin with, never mind giving this Power to a Psychopath, who is even "More Heady" from the "Power of Exotic-Technology". What were you expecting "Hannibal" to do with this Exotic-Techno?

And currently I, a Father with Family, have to deal with YOUR INCOMPETENCE, AND THE OTHER AGENCIES, AND "PAY FOR IT" DAILY because of what you DO NOT KNOW. How good a job are YOU DOING? You call this JUSTICE? YOU CALL THIS HAVING YOUR FACTS STRAIGHT? You don't know what is going on NOW IN YOUR OWN COUNTRY, NEVER MIND TWENTY YEARS AGO. And now two of Your Agents, are "LISTENING AND BELIEVING CRIMINALS WITH BADGES", BECAUSE OF BADGES. Brilliant.

You what? You Support Pedophiles, and Racist Rednecks? You "Support Corruption"? This is an International Agreement to support Pedophiles and "The Master Race"? You don't even know who I am or how this

"Mobbing Started": I on the other hand can make a Judgment about you based on Observation and Experience, never mind Your Two Clueless Agents. I told your Two-Agents this Whorbat is Corrupt, they were expecting what? That this Corrupt-Psycho would tell them the Truth? They Believed this Corrupt Liar because of "What"? Whorbat lied through her teeth about "Something", what were they expecting Her to do, based on the Information I "Passed On"?

The "Most Logical Cause" and Conclusion, based on the "Lack of Action and Results" in regards to the "Info I Passed On" to these two Agents (Their Not Listening and Looking at the Information), that I can Find or Understand, is that they remain Unaware of this Exotic-Techno which I speak of in the "Letter" I gave them. Which makes me look "Great" and "Believable" in "Their Eyes" obviously. How believable is it that this Clown is "Manipulating Emotional-States, Behavior, Actions" with this Techno, and THEN "TAPING IT" TO FRAME ME, if they do not even know it exists? How are they even going to be able to "Question Properly" the Information I have already given them, then? It is no wonder they have not "Passed on the Information" I gave them, when they do not "Know of this Techno" or "Believe it Exists". In the meantime people close to me, know of it, and because Whorbat TOLD THEM. It is the only "Logical Answer" I can find in regards to Your Agents. But Your Agency DOES KNOW of this Techno, and this I DO KNOW. So "Fix Your Mess" and "Educate and Inform" Your Agents, of Reality. I will get to "If they do know of this Techno" but believe Whorbat anyhow, in a few moments: first I have something to say about "Agencies" here.

ALL OF YOU AGENCIES LOOK LIKE A BUNCH OF FUCKING IDIOTS BECAUSE YOU FUCKING ARE. Even when you are "Co-Operating and Working Together" you can't get your shit straight. How are you going to do against "Other Agencies"? What are you going to do, "Hire Me"?

Whorbat is Deceiving Your Agency and Her Own: She is Stabbing Everyone in the back for "Her".

Everyone Whorbat has LIED TO, has been COMPARTMENTALIZED so that She is not caught Lying and Manipulating. And The Clown is LYING to Everyone, and MANIPULATING Everyone, in regards to "My Failing End-Tests", This Mobbing. Why? Obviously if this was being done by "The Rules, Protocols, and Standards", I would already be FREE. At this point, by "Doing So" She would be Caught and in Trouble, which from a Corrupt-Agent's Perspective, is Stupid, given the circumstances.

To top it off without proper knowledge of this Techno, your two Agents would be "Too Embarrassed to Pass-Information On" at the least, obviously. In the meantime THIS CHANGES NOTHING ABOUT THE REALITY I AM DEALING WITH HERE, never mind that Your Agency obviously DOES KNOW of this Technology. I am not so ignorant here. I know, and people close to me in my life know, what your own Field-Agents do NOT. Brilliant. AND IF these two Agents do know of this Techno, and believe The Clown anyhow, then I do not even want to "Know about It" at this point. If THAT is the case, when you find out the Truth, they should both be Fired. If THAT is the case, then they are two of the Dumbest Agents on the Planet. That means they believe a File filled with Lies, an Artificial-Profile as well, believe Whorbat is honest because of Her Job-Title no doubt, and have been Programmed by Current and Recent Framings made with Exotic-Techno. Like a couple of Average Citizens, like Total-Idiots. I get "Held In this Mobbing" by a Corrupt Agent doing a Cover-Up of Her Corruption, and I get to look like "The Bad Guy". Quite frankly if they are this Incompetent it is beyond my belief, and I do not even want to know.

And in the meantime, The Clown has been WATCHING these two very carefully with this Techno, ever since I "Passed Information On to Them" about Whorbat. No doubt The Clown has "Affected/Tweaked their Thinking" as well. What were you expecting someone in Trouble with Countries and Agencies, to do, when they have this Techno at their disposal? Never mind a Psychopath with this Techno.

You as an Agency do NOT have all of the Information, and you as an Agency HAVE NOT PROVIDED YOUR AGENTS with enough Information, which is YOUR FAULT. You are Both Clueless, and Incompetent, as a result, and your Agency Remains Unaware of what is going on NOW within it's Borders. Last time I checked, this is called being STUPID.

And BASED ON MY PAST IN AMERICA, why do you think I approached Two NON-CORRUPT-AGENTS in the first place? READ MY BOOK. Why the hell do you think I did this? We are not even talking about Cops here, we are talking about Government Agencies, and "Problems between the Two" based on Corrupt-Employees acting as "Their Own Agency" within a Foreign-Land. What the fuck do you think this Clown's Agency will do if they are THE ONLY ONES THAT KNOW? You think that after America, I was ONLY going to tell The Clown's Agency? Forget it.

I do not Trust The Clown's Agency and what IT WILL DO if it is the "Only One that Knows" aside from myself. Neither of you Agencies know what the hell is going on, and when The Clown's Agency is the only one to know, what will they decide? THINK I AM THE DAMN IDIOT HERE? I have news for you, you are FINDING THIS OUT THROUGH A GOD DAMN BOOK, BECAUSE I AM

NOT. I have learned the HARD WAY. I have LOST TWENTY YEARS OF MY LIFE, FOR "BEING DIPLOMATIC", never mind My Families Suffering.

What do you think I have surmised here? I know Their Clown has Misused this Techno to "Commit Murder" (It wasn't from lack of Trying) in a Foreign Land, a Host Country, yet if I only "Pass this On" to The Clown's Agency, they'll say "Thanks" and then DO NOTHING TO ME? That is NOT what happened in America, and this "Clown Situation" is far bigger than that: they'll say "What this guy knows will get us in Trouble", and then do something Extreme. TO MAKE THEIR LIVES EASIER. Which is what really happened in America.

Hard-Core-Framing or Death obviously. Having said that, thanks but I'm "Full from Both" at this Point. I'll get to "My Close Calls" and MY KIDS as well, later. In the meantime your "Vigilance and Monitoring" is "Much Appreciated". I am currently "Receiving Death Threats" and Innuendos of The Clown leaving something FATAL in my body with this Techno, before She Goes. And SHE IS the Type of Person to do this. Very "Vigilant" of Your Agency. You wouldn't know, Her Own Agency Wouldn't Know, because no one has all of the fucking Facts, and no one has INVESTIGATED what is going on with this Clown and Her Corrupt-Agents. Again from my perspective and experience, "Agencies" are not worth a Rat's Ass: let me know when this "Sinks In".

Both Agencies have been Incompetent. This Clown has "Screwed Three", one Indirectly, and Two by Deliberate Manipulation, and this Idiot is the "Sloppiest Agent of Rank" I have ever encountered. Which makes two of these Agencies "Look absolutely Stupid". I am what, James Bond? Who am I, "My own Agency"? I need to "Put myself on the Market" and "Hire Myself Out" as an Independent-Contractor and Agency? I have one for you both, it is called "It is Obvious", and it is called "Find out what is Going On", it is called "An Investigation", it is called "Corrupt-Agents with Training". It is called a Psychological-Screening BEFORE you hire Employees.

You were thinking I am "Making this all Up"? You think I and "The Jews" all suffered, and suffer, from "Delusions of Persecution by Corrupt-People"? Why don't YOU LOOK AND FIND OUT. That would be the first step to actually "Looking Competent and Professional" and to "Solving This".

As it is, your Two Field-Agents have been Programmed by This Clown, who has been Misusing Exotic-Techno "TO CREATE TAPES" which are "Convincing", but which are not the Truth in regards to me, or "How I Think, Act, or Feel". Regardless of whether they "Know of this Techno", they have obviously been Lied to and Manipulated in THIS WAY. If they have Not Been Informed of This Techno, then this is YOUR FAULT. If they know of this Techno and believe The Clown anyhow, then they are Absolute Fucking Idiots. This Clown is Misusing this Techno to "Manipulate Me, and Force Me" into Mental and Emotional-States (Never mind "Scripted-Speech"), WHICH CLEARLY ARE NOT MY OWN, for "Convincing Tapes", which this Clown has then turned around and Used to Convince Your Two Agents otherwise, never mind Others, and that I "Must be Lying". The Clown has PROGRAMMED TWO OF YOUR AGENTS, and in the Simplest and Easiest of ways. I understand the "Tactics, Techniques, Programming, Dirty Tricks, and Technology" being Used on Your Own Agents, BETTER THAN THEY DO. Brilliant. Maybe they should read my Book. As it is, I will give one a copy when it is done. I ought to bill your Agency for it.

I "Passed this Mobbing" the first time almost eight years ago, and "Exposing this has been Difficult" and Impossible based on the Power this woman has been given, never mind Technology, Training, and her Knowledge of Dirty Tricks, Psychology, and Programming People. You were thinking it is "Easy on my End"?

This Clown as an example, has Delayed My Book, for TWO YEARS (and with Techno) because she is obviously Exposed in this Book. Think this is Coincidence? Think this is about "Following the Rules" of Mobbing? Can you start to "See that something is not Right Here", and that it DOES NOT have to do with Me? Do you understand why I am approaching Your Agency as well? Do you understand why I am Using a Book to do this? I am supposed to be FREE. Instead this woman Manipulates YOUR AGENCY, AND HER OWN AGENCY, and does whatever She wants, AS IF SHE IS "HER OWN AGENCY". And in the meantime, this Corrupt-Psychopath is threatening to KILL ME, and with this Technology. Which Your Two People think is simply a Trick to Deceive Me, to "Stress Me Out"—Absolutely Brilliant on their end. Naturally I am the "Clueless One" here. I'm glad they got all of the Facts.

This is a result of Clueless Agencies, and Illegal Government Programs. And it is ALL happening in Your Country, under your noses. This is WHY Government Programs like this should not exist: They Exist for Misuse and Abuse. These Programs are a Result of Corrupt-Minds, and Corrupt-People are "Attracted to Them". I am not the Bad Guy here, I was not guilty in America, and this woman plans on "Killing one more Person" in your Country, and it is ME.

This Clown has "Programmed" two of your Field-Agents HOPING THAT if you look into this Mess, that they will be "Reliable Witnesses who Support Whorbat" because "They are Convinced" based on what THEY DO

NOT KNOW. It is called "Programming", why don't you look it up. And seeing as how I DO KNOW WHAT IS GOING ON, it is safe to say that they have been Programmed and Rendered-Stupid. They are not "Reliable Witnesses", they are Dupes and Suckers, Programmed Idiots, who actually have no idea as to the Truth. They have been "Programmed for Whorbat's Benefit", for THE SAKE OF DECIEVING AND TRICKING YOUR AGENCY, "By Way of Them... And I "Watch This" while Your Two Idiots, have no Idea. I know all of this because I DO KNOW what the Truth is. My question is, IF I DID NOT POINT THIS OUT TO YOU, would you have caught this Deceit? I bet I don't want to know what the answer is.

WHEN WILL YOU AS AN AGENCY LOOK INTO THIS WHORBAT, AND HER CORRUPT-AGENTS?

Everything this Clown has done, and is doing, is a good example of everything I have said about my "Experiences in America", the exact same thing is being done to me all over again, and to Your Two Agents. They are being Lied To, Manipulated, Programmed, just like people in America that I knew, were. Because the same Book of Tactics is being Used and Misused on them, never mind being used on the Public, Idiot-Citizens.

To top it off This Clown has been THE SLOPPIEST PERSON "IN CHARGE" of this Mobbing, and because of "Her Personality and Profile" yet She has Fooled Her Agency and Your Own. I HAVE CAUGHT HER, I HAVE ACCURATELY PROFILED HER, YOU TWO AGENCIES ARE NOT WORTH A RAT'S ASS. Clear? No doubt I am not for both of you. It is called giving a Psychopath a Job-Title of Rank, and Exotic-Technology, for starters. Is that all that is needed, is to have a "Badge", is to be a part of the "Legal-System", is to be a part of an Agency, to be "Listened To and to be Believed"?

If THAT is the case, I will join an Agency. HOW ABOUT YOURS? Now what? You going to "Look with any Degree of Logic and Intelligence" now that "I have done So"? NOW DOES WHAT I SAY "MATTER", NOW WILL YOU LOOK? Now that I am "Part of a Team", now that "I am Part of the System"? Well? Is everyone in these Agencies three years old? Now does it make sense? NOW I am "Less of a Threat", is that it? A "Badge and Agency" makes everything all right? The victims of Corrupt-Agents might disagree with you on this, I sure as fuck do.

If you are going to actually BE AN INTELLIGENCE AGENCY, then perhaps you should start acting as one. Simply "Saying It", and a "Badge and Affiliation with One" does not actually make you all "Super Spies": if you believe this then you have been watching too much "Mission Impossible". Based on my Life Experience None of You Agencies are very "Intelligent", you all look like you are being run by Forrest Gump. PRETENDING TO BE INTELLIGENT, IS NOT THE SAME AS ACTUALLY BEING IT.

This Clown, is Misusing Her Training, against Intelligence Agencies, hoping that the "Stereotype of being with One", and being "Affiliated with One", will make "Everyone Blind"—she is Playing Everyone. She is hoping to "Hide Behind Job-Title and Agency" if and when someone has the brains to Investigate what I write of here. I am sure that one of The Clown's Favorite Things is to "Play On" her "Job-Title and Rank", to "Throw Her Authority Around", and to Hide behind "Appearances". And in the meantime I happen to know, beyond a shadow of a doubt, that she is a Psychopath. It is called "Look at the Profile for a Psychopathic Personality Type" for starters.

We are talking about a Type of Person, who Manipulates and Uses All, for oneself, because "They are Bubbled within Themselves". "Teams, Agency, and Loyalty" do not exist for this Type of Person. They are "Their Own". They are Predators. They are Self-Serving, and ALL PEOPLE are "Things to be Manipulated", FOR WHAT THEY WANT—which is always about SELF-GRATIFICATION. And I have news for you, Agencies are a "THING TO BE USED", a means for "Power and Self-Fulfillment", a way of "Doing All with Immunity and Impunity" which SO FAR has worked for The Clown. She OBVIOUSLY has been getting away with this for a VERY LONG TIME and in Your Land.

Think She is "Happy about my Book"? THINK SHE IS "HAPPY" WITH THIS "ALL COMING TO AN END"? Her "Little Fantasy Land" is about "TO END". Think it has been "Pleasant around Here"? Her Officers have been "Riding on her Wave of Success", think they are "Happy"? I have been absolutely SCREWED BY THESE CORRUPT-LOWLIFES, AND IT IS TIME FOR THIS TO CHANGE. It is time for Your Agency to LOOK AND INVESTIGATE as to what is going on in Your Country, and it is time for you to realize WHO THE CRIMINALS AROUND HERE ARE. A Criminal with a Badge, is STILL A CRIMINAL. It is about time your Agency Acted as an Agency.

As far as I am concerned, their time is OVER. I am ONE MAN against a small army of Corrupt-Spooks with Exotic-Techno, and I could use some help here. Look at what I know, what I have Figured-Out, and what I and My Family go through. INVESTIGATE AND GET THE FACTS. The "Legal-System", Agencies, Agents, Job-Titles, "Law Enforcement", and the word "Professional", are all one great big PILE OF PHONY BULLSHIT. No one knows what they are doing, and Everyone is an Incompetent Jackass (Be Fair, Honest, Competent, and Professional, and it won't be that way).

This is about "Getting the Bad guys" is it? Well where is "My Agency Paycheck"? So far I have pointed out a Pedophile-Cop, a Racist Murdering 911-Dispatcher, and a small army of Corrupt-Government-Agents, all of whom "Work for the System" and all whom work for "Law-Enforcement": and this 911-Dispatcher obviously was "Connected with the Police". So go get them. As I understand it, this is what "Mobbing is For", so what are you waiting for? What is the Difference Here? A Hard-Core-Criminal is a Hard-Core-Criminal. Have I made a point? I "Need a Badge" for you to do this? As it is you have a Psychopath and small Gang of Agents, armed with Technology, that have "Taken part in a Murder" of a Citizen (Near Murder—but it is "The Thought that Counts"), and who have committed other crimes as well: we are talking about Hard-Core-Criminals. Well what is everyone waiting for? I am not the lowlife here, I came here to raise a Family. Why do I have to "Clean Up" everyone's Mess? Where is "My Agency Paycheck"? Me? I can't even be free, and I have already "Passed this Mobbing" more times than The Clown's Agency wants to know about.

Let me explain this to You. You will take a look at all of this bullshit. If you do not, I will join an agency. I will do so to get you to "Look", to get you to do so—and I am telling you this. You "Worried I will Use You"? For what "Purpose" (The Truth, this "Worries You")? The Clown is already doing this, and I am honest, so what are you "Worried About"? Look at what I "Deal With", Look at the Odds, and "What I have Figured-Out". What is the "Problem"? I'm not "Intelligent Enough"? Take a good god damn look, I already know more than Three Different Agencies, I am already "Doing Better than Three of You", so what is the "Problem"?

You can not tell me I am the Idiot here. If you are "Hoping to win an Argument" based on Logic, you can forget it. Quite obviously you can not. You "Won't Hire Me" based on what? What is the matter, "I'm not good enough"? Without my "Pointing Everything Out" to Superiors in Agencies, how well would you do, at Assessing all of this? You would be played for absolute fools by Your Own Employees—which they would view as "Opportunity". Well? Is it that "I am not Good Enough" or that "I am not Corrupt-Enough"? Millions of dollars have been spent on this Bullshit because of Corrupt-Employees, I know what you do not, and two other Agencies as well, and you are "Telling Me", you can not "Hire Me"? Based on what? Well, lets hear it. I have some news for you on this point, you can not make a "logical argument against me" in regards to this. And I have more news for you, this is not a matter of "Arrogance", this is a matter of the Truth. When I am with an Agency will you "Bother finding out the Truth"? And having said that, I guarantee you "Some Agency" would. Problem here?

Well? Or is this really about no one giving a shit about people Breaking the Laws in very serious ways, as long as they have a "Badge" and "Agency Affiliation"? That is how it looks to me, how do you think it looks to my Readers, or a Lawyer for that matter? How does it look to YOU? It looks like a Pile of Shit, BECAUSE IT IS. All Three Agencies look like a Disgrace because THEY ARE.

MY QUESTION IS, WHAT WILL YOU DO, IF I JOIN AN AGENCY? Then what? What if I join Your Agency? You going to claim "I am Playing Sides"? What do you Care About, there are Total-Assholes within Agencies already "Playing Sides" and Their Own Agencies—specifically The Clown for starters, never mind The People who Framed Me and put me in this. Then are you going to listen? The problem is you will still have "Two Different Versions of Reality", and only one is True. I have no idea of the amount of Lying and Tricks that have occurred, for that matter I have no idea as to "What extent I have been Deceived" for Whorbat's Campaign of Discrediting Me with Agencies, She has had long enough to prepare (Never mind Advantages), but I DO KNOW, she is Corrupt, I do know She has been "Deliberately Keeping Me In" this Mobbing, and I do know She is Out of Control and Breaking all the Rules, and Laws as well. So when I am an Agent of Your Land, or I join Whorbat's Agency, then what will either Agency do? Get my point? YOU WILL GET YOUR FACTS STRAIGHT, or I will do this, "Join an Agency", and YOU WILL GET YOUR FACTS STRAIGHT ANYHOW.

I assure you based on "My Skills" and "My Performance" someone will "Hire Me" based on My Intelligence alone—ask Your Clown about that one. What will you do? Well? You "Worried"? Take a "Look at my Situation" here, how much sense does "Your Logic" make? You are already a Disgrace, and You Already Agree with the "Logic", so what is your Problem? Or is this about how you at Agencies "Feel" in regards to a "Target"? Quite frankly you do not "Practice Logic" which is why I "Win at an Argument Here". ARE YOU CLEAR HERE? Having said that I would join another Agency, neither one of you look very Intelligent. You can deal with the Truth of the matter, or the Embarrassment of the matter—you decide. My "Tip to You" is to be Responsible: no matter what you "Want" you will have to deal with Reality, and will find out the Truth in the end.

In the meantime, I and My Family, are Tortured Daily with this Techno, and I am being threatened to be KILLED WITH IT "INDIRECTLY". And YOU would take this seriously if you Truly understood WHO AND WHAT I am dealing with here. Which you wouldn't because you haven't Looked.

And of course in the meantime, Whorbat "Engages in her Favorite Game" of "Mocking Others", which is what she likes to do to people she does not like, who she has Programmed. I specifically refer to Your Two

Agents here. She has gotten these Two Idiots to "Make Fun of Themselves" without these Two even knowing it: and I have news for you, this is another Sick-Game and Favorite of Whorbat's. She likes getting those She has Programmed and who She does not like, to "Mock Themselves", it is part of The Clown's Psychopathic Charm. This was a "Shared Joke" between Whorbat and Me. She enjoyed this, making Fun of Your Agents, and it is a "Common Theme and Habit" of Whorbat's, it is a part of her Profile. For Whorbat, this is Standard Operational Procedure, and it is Her's. No doubt for The Clown, this is "Performance Art", never mind "The Cheap Thrill". She likes to "Illustrate how Stupid People Are", no doubt this is all part of her "Childish need to be 'better' than Others".

Whorbat had one of these Idiots say to me "Don't be stressed Out", and weeks after I had given this Idiot a copy of a letter, exposing Whorbat, which is of course a "Sick Joke" on Whorbat's end. When you get that Whorbat Misused this Techno to get a man to attack his girlfriend with a knife, get that this was in the letter I gave to Your Idiot, and get that Whorbat has "Tried this with Me Before" and is now threatening to Kill Me with this Techno, then you get that this Idiot's Intelligence was being "Mocked". Naturally This Idiot is Unaware of "Slight Details" like this. Never mind any "Techno Manipulated Tapes" that were Used for Manipulating and Programming Your Agent. Your Brilliant Blond Agent, obviously is "Convinced" based on What She Does Not Know, based on Context Removed, that "I have Stress Due to Mobbing" or that "Whorbat is Faking It" in regards to "Threats". This is funny. Does it look like I have "Bigger Issues Here" other than Mobbing? Does it look like Whorbat was "Faking It" with this couple?

Never mind the amount of times I "Passed" which Your Idiot does not know about, and never mind that "My Stress Levels" are NON-EXISTENT. Try wrapping your heads around that one. You can not. This Book is in the process of coming out as I finish this Letter, and as you are Reading this, these Corrupt-Lowlifes are Exposed: now try to "Understand". I have "Waited and Survived" for years at this point. I have slowly but steadily been "Approaching this Point", like a Wolf, Wolverine, or Pit Bull locked in on it's "Kill", oblivious to my surroundings, and near to "Locking down on the throat of my Target". Excuse the "Cheesiness" but I hope "The Imagery" helps give you the Feeling here. THEY are what I am focused on, Not Mobbing. Mobbing is meaningless at this point, Exposing these Corrupt-Agents is Not—especially when you get that I am being "Held In" by them because they do not want to be In Trouble. How else will I have freedom when this is the case? This Blond couldn't begin to understand, she does not have enough information to make an informed judgment.

She expects Liars to tell her the Truth. She is Ignorant, and I already do a better Job than she does.

When you get "Certain Basic Realities" here, then you get that The Clown had this person "Mock Herself". Brilliant. Having said that the other irony is that much to Whorbat's Chagrin, I am NOT STRESSED OUT, because I am about to "Expose Her", and this is even in regards to Death-Threats. Obviously I do not want to Die, but that is to the side, as I am about to Expose these Corrupt Jerks: they should be so lucky. As it is, it has not been from "Lack of Close Calls". I have "Worked Hard" and "Waited" a VERY LONG TIME for this "Moment". But your Two Brainless Jackasses, could not even begin to get this, as they are Clueless. Corrupt-Agents "Forgot to give them this Relevant Detail". Especially this Blond, who has no doubt been "Played Tapes" of me being "Affected with Techno", for THE SAKE OF PROGRAMMING. Never mind a "Past" that "Exists on Paper Only", never mind "Programmed People", which they are both a good example of. And this is what happened in America.

When "I get There", they can both read this Book. I will bill Your Agency, okay? This can be my "Payment" for doing "The Work" of Three Different Agencies. Thanks tons.

With the Other Agent, Whorbat had this Individual Tell Me, "She will get in Trouble if I mail a copy of this Letter to the Embassy" which is what I originally gave her—a letter to the Embassy, and as it specifically names Your Agent, and how this Techno was Misused on her. I simply gave your Two Geniuses a "Copy of this Letter" as it contained all relevant information in regards to what has been going on, and relevant information in regards to Whorbat. This Genius "Told me she would Get In Trouble" if I mailed this, because she is mentioned in it, and people would "Ask her Questions". You will note, that this is all in my Book anyhow, and has been for some time. In fact at that point, I already had an Edited Version of this Letter, with names changed, in my Book, and Copyrighted. Thus at some point people would be "Asking her Questions" anyhow, never mind "Why she did not Report Anything" in the first place. Thus Whorbat, got her to "Mock Herself" like the Other Agent, and because she was gullible enough to Trust and Believe The Clown. The Clown is laughing because I am telling the Truth, and She got this Honest Field-Agent to take part in a Cover-Up of Whorbat's Corruption. No doubt Whorbat's Bold-Faced-Lying was so "Unbelievable for Her", that it "Had to be True", and no doubt "Whorbat" played on Her Rank in regards to this Field-Agent.

Obviously Your Second Genius, Kylie, was not Told of what is in this Book so that she could be "Mocked by Whorbat". Whorbat had this woman

"Mock Herself". Whorbat knew this woman is in my Book, Whorbat knew and knows this woman will be asked questions, and Whorbat knows that She will be "Scrutinized" by Agencies. At the time this woman did not know this (no doubt) and so "Looked Ridiculous" for saying this to me, and because she Did Not Believe Me. Which she should have. This Genius chose to believe the Corrupt-Liar instead. And as "Added Insult" I used to be friends with this woman because she is a good person. Now, god only knows what she thinks. Thus Whorbat is Totally Enjoying Herself. Yet Kylie has yet to figure this one out. She has no idea of what is in my Book, no doubt no idea of this Techno, and Whorbat did not "Share the Truth of all of This" with her. In the meantime some people around me know of this Techno, because Whorbat Told Them. Whorbat of course KNEW EVERYTHING at the time she told Your Agent to say this to me. She has been Manipulated and Programmed, by Trusting Whorbat, and is a "Jerk and Idiot" with me, but "Friendly" with the person "Stabbing her in the Back". Brilliant. How do you expect to "Monitor what is going on in your Country" when your own Agents understand less than I do? What is this, Anarchy?

So what do I expect of Your Agency?

How about an INVESTIGATION, and how about "Looking into Whorbat and her Officers"? How about Finding Out what is going on around here in Your Own Country? My KID is being Targeted by these Corrupt-Lowlifes, and I guarantee you, you would like My Kid more than you would The Clown and Her Corrupt Buffoons. How about acting as a Professional Agency, and how about going after Hard-Core-Criminals with Badges and Training in Psychology, Programming, Dirty-Tricks, Tactics, Techniques, and who have Electromagnetic-Technology? How about doing the RIGHT THING before these Idiots get me Killed? Maybe they should be "Mobbed". Do you actually think I am going to allow myself to be Continuously-Framed, and eventually Killed, for a Clown and a Lowlife Pedophile Cop, and Murderous Redneck from twenty years ago in my Life? You had better get Your Facts Straight, and Your Priorities Straight.

You will need someone on "Whorbat's Level" to do this, someone with "Her Level of Training". And you will need someone familiar with this "Non-Existent-Techno". Obviously Field-Agents aren't worth a Rat's Ass.

I am tired of seeing Criminals get away with BEING CRIMINALS because they have a "Badge and Job-Title": AND I AM TIRED OF PAYING FOR IT. THIS IS NOT GOOD ENOUGH. It is no excuse for getting away with anything. People given Power, should be Responsible with it, and HELD ACCOUNTABLE FOR

MISUSING IT. How about Investigating, and Finding-Out what is really going on? What are you waiting for?

In the meantime I continue to be Deliberately "Held In this Mobbing", I continue to suffer in ways unimaginable thanks to a Psychopath with this Exotic-Techno, and my Family continues to suffer, for the sake of a bunch of Selfish-Corrupt-Agents. This is no longer about "Mobbing" this is about a COVER-UP going on in Your Land, because of Crimes Committed in Your Land, by Criminals with Government Training in Your Land, who are "Misusing Mobbing as a Cover-Story", for "Keeping Things Quiet". And all in Your Land, and under your noses. I hope we are clear here. I assume you will read the next letter to The Clown's Agency, as well you should. There are of course more details.

Last but not Least, The Clown's Agency

You are going to have to realize that my writing is being affected by your Agents with this Techno. "Thought Blocking" of "Important Relevant Points" and "Words and Memories", "Extended Feelings", and "Spiking and Substitution of Words with their Own". Read my Exotic-Technology chapter as this contains a lot of "Clown History". With that in mind lets start.

Well as you have no doubt gathered by now, there is a problem here, and this problem is Your Out-of-Control Agent, Whorbat, also known as The Clown. I have been "Held In this Mobbing" by her, even though "I Passed", initially for Personal Reasons on Her end, which has now changed to "Other Reasons" as well. In Typical Clown Fashion, "Her Solutions" have only "Escalated Things". Say what you will, but I have come to learn "Who and What" I am dealing with. As this "Mobbing" would not end, I was forced into finding answers. For all intents and purposes I had, and have, developed a "Profile" of this Clown. I will get to this in just a bit, because if you are going to "Look at Things" you are going to have to take a few things into consideration on your end, and take things seriously. You ought to be paying me for this, I am doing Your Work.

I have been "Forced into Writing this Book", because Your Corrupt Agents are doing a Cover-Up of their Mistakes, and are Deliberately "Keeping Me In" this Mobbing, to do so. What were you expecting me to do? Sit here and suffer, while I continue to "Pass this Mobbing", so that Your Corrupt Employees can be "Comfortable with their Lives"? Forget it. Your Agents have absolutely Screwed-Up, and so have Multiple Agencies here. I WILL NOT WEAR THE BLAME. When you "Get There" I expect you to END this Mobbing, seeing as how I have already "Passed", and because

Whorbat WILL NOT "End This" because of the "Mess She Is", and WILL NOT because of "Who and What" She is. When you "Get There", why don't you try Respecting Me as opposed to Attacking Me.

I will remind you that I am an American Citizen, and I am not asking you, I am TELLING YOU, that this Agent The Clown is "Out of Control", Corrupt, and Sloppy, and I know this. And this is even by "Your Standards". I am also TELLING YOU, NOT ASKING YOU, that She is a Text-Book Psychopath, and you have given This Idiot Exotic-Techno. And my Profile of Whorbat is Correct, this is not a matter of Bigfoot, or "Amateur Profiling", especially since I am "More Properly Motivated" than you are, and whatever this has been about "My Profile of Whorbat" is not, and has not been about, a "Hobby". You can "Place your Bets" and "Put money on the Table", and you will lose on this. I know your Agents better than you, and most certainly The Clown, based on Experience and Observation. How does that make you Feel?

After 18 years (April 13th I believe) of this "Mobbing" I am "Slightly more Knowledgeable" than the average Citizen. I also know what your own Agency does NOT in regards to this woman, and she is an Employee of yours. She is not following "Rules of this Mobbing", never mind Your Own. I "Passed" years ago, by the "Rules" and I have a major god damn problem that I am still in this. And for "Personal Reasons" based on Whorbat's Profile, not Professional Ones. I am quite clear about this as well, as I have had a long list of people "In Charge" before Her, and am quite aware of what this is about, and how this "Mobbing Works". Which makes all of this "Situation" continuing, and the fact that I have to "Contact Her Superiors", her Creation and Problem not mine, and until I see what you do, not yours. I expect Intelligent and Responsible Adult Behavior, unlike Your Employees I deal with. My question is, are you capable of acting Intelligently, Fairly, Justly, Competently, Responsibly, as an Agency in regards to this "Mess" which Your Agent has Created and Perpetuates?

Do not give me some "Looking out for one Another B.S.", I have experienced enough of that. Look before you make that decision. She has screwed up royally on many levels. She is also trying to Hide behind your Agency, and trying to play "Both Sides", trying to get me "Embroiled with Personal Issues" against "Agency", and vice versa, when this is about HER. So do NOT be this woman's Dupe and Sucker, and Get Your Facts First, before making a decision about me. I am not a "Threat to You or Agency", and if this is Your Viewpoint, you had better get a good god damn look at the Facts, because for anyone not an Asshole, Your Agents are a Threat, and they are a "Threat to Me and My Family". Just recently they killed one of our pets again, YOU explain this to my Kid, and explain "How much sense it Makes". Furthermore this Clown knows I was Framed in America, for the sake of a Pedophile and Racist, so what kind of sense does any of this make?

In regards to The Clown and Buffoons we are talking about "Show-Offs" and the Incompetence that comes with it: which is how I caught them many times. Whorbat "Loves To Show Off", and has been CAUGHT AS A RESULT of Her Showing-Off, repeatedly by me. THIS IS WHY I CALL HER "THE CLOWN". As I have "Experienced Life" a "Large Ego" usually translates to "Large Mistakes", Your Clown and Her Buffoons are perfect examples of this. Out of all of The People that "Have been in Charge" she has been the sloppiest and it is a long list of "People who have been in Charge".

Having said that the LAST PERSON IN CHARGE, a man from Your Agency, was better and more Professional, as opposed to Professionally-Incompetent and Sloppy. She has taken a "Situation" that was supposed to have ended years ago, "Kept Me In", and made a Mess of everything. And this "Mess" happens to be my Life: and my FAMILIES. As a result of Whorbat "Holding Me In", I have had NO CHOICE BUT TO WRITE THIS BOOK AND LETTER. That is a "Clue" for anyone who is "Thickheaded".

Having said that I can clarify a few points for your Agency in regards to The Clown, and I most certainly hope that you take it seriously. As well YOU NEED TO, and especially in regards to dealing with The Clown. As a point, you have given her this Exotic-Techno, She has no "Recognition of Boundaries", and no doubt would use this Techno on YOU, HER SUPERIORS, if she could. If you think that She would not, then you are Absolute Imbeciles. As it is, my "Approaching People" on an Individual Basis, has been "Fruitless" in no small part because of Whorbat's "Monitoring Their Thoughts" with this Techno, to find out "What Buttons to Push" and "How". This Techno has ENABLED HER TO "PERFECT HER LYING", AND TO PEOPLE I COMMUNICATE WITH ABOUT WHORBAT. As I will get to, you have "Loaned this Techno" to a Psychopath, and you had better start "Thinking In These Terms", never mind had better start "Thinking of the Benefits and Advantages Whorbat Has" as a result.

A Brief Review and Summary of "The Clown" and her Officers "The Buffoons" A Profile of "The Clown": Points which Show She is a Psychopath

Also: A "Warning for Her Superiors"

Back in my home town, I had the "Good Fortune" of working for a business owner who was a Text-Book Psychopath. Regardless of not having a degree in psychology I can state this safely. Having said that I have had a few psychology classes and sociology classes, and know a bit more than the average person in regards to subjects like this. Including studying "Abnormal Psychology" and "Deviant Behavior Types". And in regards to The Clown this is the same case, and this is based in no small part on Observation and Experience, in addition to psychological insights from University Courses, and my former boss.

Having said that, with a bit of looking on your end you will find this is True about Whorbat. You will also find that this leads to many mistakes, a lot of "Unnecessary Risks", "Showing Off", and a lack of awareness in regards to "Consequences", and a Sloppy Employee, never mind "Leader", as well. We are talking about someone with a known Mental-Disorder working for your Agency, and all of the "Mistakes" that come with it. I have had many people in charge over the years, and no one has Screwed Up the way She and her Officers have. Why? This has to do with the quality of these people and The Clown, and obviously it is about time they were pointed out to you. So you are duly informed.

Having said that they have "Made it this Far" because they are Corrupt, and have been Hiding this no doubt, to a very large extent from your Agency. They of course Misuse their Training, to Manipulate Your Agency, and to Hide "What they are Really Doing". This is a matter of what they have gotten away with, a matter of "Looks". A matter of Misusing their Authority to "Sweep everything under the Rug", which you will note, is what they have been doing with Me. But of course with Me, they are Caught, and they are Exposed: something neither The Clown or Her Buffoons are "Happy About". This is why my Book has been Delayed, two years.

But back to my boss a Psychopath, and Predator, from my hometown, for a few moments and valid points. I have had time to observe this Type of Person, have had close contact with this Type of Person, and you can not simply "Read About One" to "Get It". I suggest you make yourself FAMILIAR with this "Personality Type" before having dealings with this woman, and make sure that you do indeed "GET IT". Having said that she has known that I would eventually approach your Agency for more than a year, and has had "Time to Prepare". I do not doubt that "She Has", and her Officers have, and I do not doubt that anything pointing to the Truth has been "Hidden". It makes sense that they are all going to try and "ride it out together": their being in Trouble that is. Having said that I do not doubt that your Agency will get to the bottom of it all, mostly. At least in regards to "My Mess" which Whorbat has "Made and Perpetuated".

If you are going to "Look into Things" and look at this Clown then you are going to have to be familiar with "Psychopaths". And excuse me, but I worked with one, and if you do not "Get It", you will be played for a Fool by this woman. Which she WILL DO, she is not about "Agency Loyalty" she is about her Power, and her means to "Violate People" and get away with it—YOUR AGENCY IS A MEANS FOR THIS. Your Agency is about "Her Warped Self-Fulfillment and Gratification". She is Corrupt and KNOWS that I am going to point this out to you, why would she not try to "Fool You"?

Her "Job" is about Self-Fulfillment, what she Personally gets from it, Power and the "Misuse and Abuse of Power", it is not about Agency, it is not about being American, it is not about a "Team", it is about getting what SHE WANTS and having it. It is about what she Enjoys, and anyone "Threatening This" is an "Obstacle" regardless of you all being "Agency", or regardless of you being "Above Her" in Terms of Rank. And I might point out, that I AM "An Obstacle" to her, on this point, never mind others—like Bosses. I have experienced NOTHING BUT a Misuse and Abuse of Authority, and I might also point out, have suffered immensely because of it. So has my Family. "Boundaries and Rules" mean nothing to Her. Like the "Rules and Standards" of this "Mobbing". At this point, Mobbing does not even exist, none of this is about Mobbing anymore, it is about Whorbat: and by HER CHOICE.

Quite frankly, "Superiors" no doubt sit where she would like to be—again More Power. She, like any Psychopath, is a Predator, and ANYONE in "Her Way" is an obstacle: you have to be absolutely stupid to think that "Agency" or "Being with the Agency" makes any degree of difference to her. Never mind "Being her Boss". As an important example "Your Looking into Her", and "Questioning Her" is only going to Piss-Her-Off in regards to YOU, Her Superiors—I guarantee it, and if "She is Not" it is because She will be HIDING THIS FROM YOU. If you do not "Get This" then put someone "In Charge" who does: this woman makes it a point of "Attacking Those" who "Question Her" like you will be doing. From Her Point of View you are an Enemy—we are talking about the "Hostility Issues" that come with being a Psychopath. My Former Boss back in the States was this way. If you are not "Familiar with this Temperament" then you will need someone who is, to "Deal with Her", otherwise you will absolutely Screw-Up.

And you should probably have "Several Superiors" deal with Her, a Small Collective, I sure as hell do not like the idea of someone who "She has Influence Over" (Blackmail), "Calling the Shots" in regards to Her. If anyone is "Lenient in Regards to Her", LOOK AT WHY. We are talking about someone with a known Mental-Disorder within Your Agency, and a Mental-Disorder Characterized by "Manipulations and Manipulative Behavior", you figure out "My Concerns" here. I have seen Her get away with Manipulating Everyone, everyone has been Her Dupe and Idiot. So far this includes Your Agency as well.

So you had better be clear on this, on your end. If you do Not Get what I am speaking of here, then I have to ask what you are doing in an Agency in the first place. If she can get away with Lying to Superiors, or Manipulating anyone else for that matter THEN SHE WILL. She likes, and is addicted to, Her Power and being able to Misuse and Abuse it. And anyone "Questioning Her" is no doubt looked at AS A THREAT to her Power, "Her Lifestyle", and Her Self-Fulfillment: including Superiors. If you do NOT get this, you had better brush up on your "Profile of Psychopaths". This is a point to be "Played Against You" if you do not "Get This". If you do not "Get This", then you will have to have someone with "The Training and Knowledge" deal with Her, or else be "Played for an Absolute Fool". So don't Screw-Up. Quite frankly if Your Agency was "Aware of all of This", I would not be "Informing You" through a Book. Which I have had to do, because Everyone I "Spoke To" was too busy being Manipulated by Her, to see the Truth.

She could care less about Agency, and the Level and Degree to which she has "Run Around" Out of Control, speaks of this, and shows this (It also shows your Agency has no idea: and why am I not wrong about this point?). What is this, Anarchy? And seeing as how She and her Officers are in "Their Mess Together", no doubt they will "Stick Together". What are they, their own Agency? The next question is, are you going to allow your "Agency to be Played" by these people? You are supposed to be a Government Agency and you are already being PLAYED FOR FOOLS by these people.

To Whorbat you are a "Thing" regardless of what you want to believe. She is "Her Own Team", everyone else and everything else is something to be "Used and Manipulated", and doing what it takes to "Get away with Something" is No Problem. "Boundaries and Rules" mean nothing to this woman which is why you are reading this. She has the Power, Freedom, and "Immunity" to do what she "Wants and Likes", and get away with it, simply because of her Agency Job-Title which she Misuses and "Hides Behind". Which is why She Chose this Job. The last Agent in charge from your Agency had far more Self-Control, and so did the various Feds in charge back in America, minus Several Exceptions, obviously. As I have "Caught Her More" she has only become more Out of Control in regards to me, my Kid, and Girlfriend. And this is not a "Clever-Ploy" for "Psyching me Out", she and her Officers are "Too Late for That", although perhaps they try to "Sell It" this way later to you. If that is the case it is a Lie.

If you have questions, ASK, as one of The Clowns favorite things to do, is to take things OUT OF CONTEXT. I see her getting away with Lying and Manipulating the Good Agents involved in this "Mobbing", and Everyone Else in this manner. I suggest you have someone who is well Trained, and who wants to "Get to the Bottom of This" Investigate what these Agents have been getting away with. I also suggest that you send the Last Agent in Charge as I will eventually get to. YOU WILL HAVE TO CONFIRM WHAT THIS CLOWN "SELLS YOUR AGENCY"—you obviously can not Rely on Files, or Her Lying Mouth. And you will have to use someone "With Her Level of Training", obviously.

As a result of Her Psychology and Personality, She has "Shown Off" many times, and made many "Blunders and Mistakes" as a result. Her "Showing Off" has been more important than "Thinking Things Through", and being Intelligent, and as a result I have "Caught Her Repeatedly". Thus She is "The Clown". A "Big Show Off", and someone who "Misuses and Abuses Power to an Extreme", who uses Extreme Force to compensate for a Lack of Intelligence. And of course this is a result of Her Profile, a result of Her being a Psychopath. I "Caught Her Repeatedly" at Lying and Manipulating, knew she was Corrupt, and eventually began "Profiling More Deeply". Eventually as I looked at all of Her Mistakes, a lot of them Obvious and Stupid, I began to "Note Patterns", and began looking for "Personality Types" from my life: different "Types of People" I knew. I began "Looking for a Comparison". Eventually I "Found this Psychopath" from my hometown, and began comparing: there were many things in common. With this in mind, I "Rewound on Whorbat" totally, and found that everything "Fit". All of Whorbat's "Past Actions and Blunders", could be explained this way. Then eventually I "Re-Updated my knowledge of Psychopaths" by way of computer, and found that Everything Fit, and even "Some Points" I had missed. This explains Everything about The Clown. Including "Her Blaming Others" for Her Mistakes, and Her "Attacking Them" for Her Mistakes.

And none of this, I might point out, was a "Deliberate Ploy" on Whorbat's End. In the beginning years, before we were "Acquainted with one Another", she made Many Mistakes based on Her Personality, which I caught. To say that "She did it on Purpose" to "Throw Me Off", is absurd. It was not until years later that I "Reached this Profile", she most certainly "Had no reason or motivation to be Faking It" so early on, and it is in "Rewinding on all Past Experiences" with Whorbat, that I "Found all of the Answers". Thus by the time I did Profile Her, it was TOO LATE for Her to be "Faking It" to "Throw Me Off". In fact Her Mistakes are all a "Part of Her".

Here are some points and examples of Whorbat which illustrate this. Her "Personality Quirks".

This woman "Likes to Show-Off" as an understatement. This is all part of her "Self-Glorifying", and Self-Aggrandizement. She is Extreme and Heavy, this has to do with Her Lack of Self-Control, Her Lack of Empathy, and Her Inability to "Perceive Consequences". An example of "Her Inability to Perceive Consequences" is her "Misuse of Techno on Me" to the "Point of it's Influence being Obvious", and then LATER REVEALING IT TO ME AND OTHERS. Thus I am not the only one to know of it. There are many examples, I do not list here.

She is also "Bubbled within Herself", incapable of "Realizing or Seeing Consequences", she lives in "Her own little Disney Land" and "Does as She Pleases"—which is WHY SHE IS "PERSONAL NOT PROFESSIONAL". I have referred to Her for years as a "Princess", this is why. An example of Her Showing-Off, of Her "Personal Attitude and Behavior", is a Lesbian Sex-Show She put on for me, The Target. This is a good example of Her "Professional Demeanor", and "Showing Off" for Her Officers; she put on a lesbian sex-show for me one night, with a complete view from our kitchen window, and quite deliberately at that. As is typical, these Spooks advertised the "Upcoming Event" heavily, up to the night it occurred—they love "Innuendos". There is nothing like PROFESSIONAL BEHAVIOR. Maybe She should TRY being a Professional Agent, along with Her Buffoons, and try being a Leader, instead of a "Professional Lesbian". Maybe She and her Buffoons should actually BE PROFESSIONAL as opposed to simply speaking about being Professional and taking a paycheck for "Saying it" as opposed to "Being it". It doesn't take much to talk about it, does it? And you will note, Whorbat KNEW I HAD THIS BOOK coming out, but at that point I believe She was "Hoping it would not get Finished". Again "No Thinking Ahead" in terms of "Consequence". And You Hired these Idiots.

Unless you want to call this "Professional Behavior", I think you get several points from this.

I found this one, well after the fact, and for a long time this did not make sense to me, in regards to Whorbat. I had already Profiled her and knew she was a Psychopath, but "This Psycho-Logic" Escaped Me for some time, until I "Re-Updated my Knowledge of Psychopaths". Whorbat's "Blame Game". Whorbat "Blames Others" for Her Mistakes, AND "THEN ATTACKS THEM". This is based on Her "Anger Problem", Her Psychopathic Temper as well, which she "Channels through Her Job": Her Job-Title here being an Excuse and Cover-Story for more PERSONAL BEHAVIOR. This woman has been getting away with this Behavior, Hiding Her Psychological-Profile under a Job-Title of Your Agency, RIGHT UNDER YOUR NOSES. I know this, and I am INFORMING YOU through a Book. How God Damn Brilliant are you "Looking" here? What am I, my own Agency?

Whorbat has an ABSOLUTE INABILITY TO TAKE RESPONSIBILITY FOR HER MISTAKES. Her idea of "Responsibility" in regards to "The Mess She has Created", as an example, has been to "Delay my Book Two Years", knowing full well she will be "Exposed Anyhow". And of course this gets down to her "Psychopathic Charm" of being Spiteful and Vindictive as well. And again there are MANY examples which point to this. How do you think She will feel about You "Looking into Her"? Try extremely Pissed-Off to begin with, and WITH YOU. In regards to Me, Her anger has already been made apparent, repeatedly. With YOU She will obviously have "To Wait".

I have seen her do this with Everyone who did not "Achieve what Whorbat Wanted Them To Achieve". And of course when Whorbat is "Calling the Shots", and "Preparing People", and "Programming People", and "Prepping Them for Staged-Events", this is of course Her Fault. In fact many times, she simply does not think of possibilities, and does not "Think things Through" in regards to the "People She Uses". Seeing as how She is "Profiling these People for Misuse", who's Fault is it? I have seen her "Attack Everyone" who has "Failed Her", and in the end, this is Her Responsibility, and Her Fault. After "Holding Me In" for the Last Eight Years, it is ridiculous for Her "To Blame Anyone" at this point. This "Slight Fact" of "Responsibility and the Bigger Picture" seem to escape her Daily.

Here are examples of "Her Blaming Others and Her Vindictiveness". One woman She put in this Mobbing. She killed the cat of another woman, because this woman "Did not let her dog BITE MY KID". Another woman, a Field Agent for this Countries Agency, she Misused Techno on, twice that I saw, and once no doubt "As an Attack on Her" for "Failing Whorbat". As a result of this woman's "Remoted Behavior" I could see the "Gears of Confusion, Wondering, and Shock" as she tried to "Come to Terms with Behavior not Her Own". Whorbat also wanted to "Show Me" what she could get away with—"Flaunting Herself" being typical of Her. In September 2010, Whorbat put my girlfriend in the hospital with this Techno because my girlfriend saw that she was Being Lied To and Used Heavily by Whorbat. In January 2011 Whorbat Misused this Techno on my Girlfriend's Parents. All of this with "Knowledge before Hand" by me, because of your Spooks and "Innuendos". In regards to my girlfriend, I got her to see that Whorbat had Lied and Manipulated Her (and nastily), and so Whorbat put her in a Hospital for seeing the Truth and then later attacked my girlfriend's parents with it, to Scare My Girlfriend and to "Beat Her Down with Fear". In spite of my Girlfriend "Helping this Clown", Whorbat was Stabbing Her in the back, and when my Girlfriend saw the Truth, Whorbat attacked Her. We are talking about the Selfish-Stupid Temperament of a Spoiled Three Year Old here, who has to "Have Everything her Way" or else "She Throws a Temper Tantrum": except in this case Your Brat and Snob is more Dangerous. And as you no doubt have already read, Whorbat also Misused this Techno to "Get a young Man" to attack his girlfriend with a knife. This was "Unrelated to Me" and "My Mobbing" except of course, Whorbat gave Innuendos for a week—and it was no doubt "Personal and Vindictive" on Whorbat's End as well in regards to this Couple. And this was done, "Giving Me Innuendos of this Knife Attack and Remote Controlling It", to Shut-Me-Up, and to Scare Me into Stopping My Book. Very Effective Obviously.

This "Woman" Misuses Force, and has "Power Issues", and Responsibility Issues. She "Inflicts Traumas and Pain" to get what she wants, and does so to an EXTREME, not based on "Job" but based on "Personality". She Uses Excessive Force, to "Get her Way" and "What she Wants". SHE IS EXACTLY LIKE A RAPIST: Rapists rape because they have "Power Issues", not "Sex Issues", with Whorbat, only "The Act" is different—it is "Hidden". She is Out of Control. What do you think she would do to You, Her Superiors, if it meant "Avoiding Consequences"? Get my point and warning? And think of this Technology you Idiots have given Her. Do you LITERALLY GET what I am saying here? She would Literally use this Exotic-Technology on YOU, HER SUPERIORS, to get what She Wants, and it would be UNPLEASANT FOR YOU. Are you Fucking Stupid? Have YOU AS AN AGENCY absolutely lost your Minds? Who am I, "Super-Spook"?

Another point, and "Common Habit and Tell" of Whorbat, is that She likes to "Mock Others", especially those she does not like. She does this Constantly, usually with people She has Fooled, Tricked, and Programmed: and once "You are In" with Her, the more "She Pushes This". Whorbat likes to see "How Far She Can Take This"—again "Extremes". It is all a part of "Her Campaign of Self-Greatness", She enjoys this, and gets "Cheap Thrills" no doubt from this. It is all a part of "Her Childish Sense of Being Better Than Others", and of "Winning". For all intents and purposes, Your Agency has Hired a Spoiled-Three-Year-Old-Brat, and given this Snob Dangerous Technology.

In regards to Her Lack of Empathy, who TARGETS A CHILD TO GET TO THE TARGET? MY KID has been through MORE EXTREME THINGS AND EVENTS, than YOU, and a lot of Adults with a "Full Life".

Most of this done with Techno, although Whorbat has also Used Programmed Idiots, to "Take Chances with MY KID". One example is when my Kid went to the playground with someone "Who had been Programmed" about "My Paranoia", and so this Idiot was "Told to Stand Away from my Kid" while she was "Trying Dangerous Monkey Bars" at an early age. I had already "Told this Person" to "Stay Close and within Catching Distance and Reach", WHICH IS COMMON SENSE NOT PARANOIA, yet as a result of "Programming and being Convinced" that "I was Paranoid about This" this Idiot Deliberately Stood Away. My Kid fell TWO METERS onto HER BACK as a result of Whorbat and this Programmed Idiot. Needless to say the Injury was Extreme. This is "Mobbing"? This is Bullshit. YOUR Employees are doing this to my Kid because of a Racist Lowlife and a Pedophile. AND HOW MUCH MONEY IS BEING SPENT ON THIS, AND HAS BEEN SPENT ON THIS?

As a result of "Targeting My Kid", and being Sloppy, my Kid has been SEVERELY INJURED REPEATEDLY, and on one occasion, because of YOUR FUCK-UPS, I THOUGHT MY KID HAD BEEN INJURED TO THE POINT OF DYING. My "Hair Whitened" the next day, and You Idiots call yourself "An Agency"? YOU CALL THIS MOBBING? This woman not only has No Empathy, but is Dangerous as a result of "Risk Taking", a LACK OF SELF-CONTROL, and a Lack of "Thinking Ahead". In August 2010, these Idiots came close to Blinding My Kid in one eye, by "Taking a Risk". Professional? Mobbing? Try Risky/Sloppy, try "Not Giving a Shit" on their end. And "This" will fall on YOUR HEADS, as an Agency, if something happens.

Here is something From a Letter I mailed to someone that involves Me and My Kid, and the "Same Theme": "They have tried 'Machining me Strokes', indirectly, by 'Giving me Heart Speed Ups' of 170 beats per minute and putting my blood-pressure so high, I could literally feel the blood-veins in my temple 'pulsing/beating/throbbing' with my heart beat. In fact my 'Eyes were Throbbing' with each heart beat: you have to be a Fucking Idiot to think this is not dangerous and taking a chance with my life. Being 'PARANOID' is based on experience, and when it is real it is not 'Paranoia'."

"Last night they gave my kid a headache, a migraine, to the point that my kid felt like throwing up—again migraines are dangerous because 'The veins in your head become constricted and Dilated' (Which is what they are doing to her BRAIN with this Exotic-Techno)—you can get a stroke from these too. This Clown got the idea from a 'Friend of my girlfriend's Parents' who had a stroke a few years ago. This is Dangerous and Reckless, and if something happens it becomes 'Permanent or Fatal'. When it becomes Fatal this is called Murder. And what will they say then? 'Whoops'? 'We made an error in judgment'? Or will they say, 'We did not do a thing, it is only a coincidence'. A Lie and Excuse of course. In the meantime they have 'Put several mysterious bumps on Me' (In addition to a Wart), after finding out someone I know has had skin cancer before, and in spite of the fact, that minus their Exotic-Techno, I and my Kid are 'Healthier than the Average Person'. And this is only 'Tip of the Iceberg'—when it is YOU having the experiences, then it is not simply 'Paranoia'—and keep in mind the things I have said that you do not believe about these people."

That is from a Letter, and as you see these "Risks" are Unnecessary, Stupid, and Dangerous, never mind that my Kids brain is being affected with this Techno. This goes "Beyond Mobbing" and has been going on for years, and when you realize that Whorbat is "Holding Me In" this Mobbing, this simply is about "Torture for Tortures Sake" not "Mobbing, Stress, or Tests"—get my point? And the examples I give here, are not all of them, in regards to my Kid. It does not begin to be. Keep this in mind. And yes I have a problem with this. This is Dangerous, Gratuitous, and has nothing to do with Me or Mobbing. It is Torture for Tortures sake when you have all the Facts.

What does my Kid have to do with any of this? Do you have a good answer? Would you like to explain the "Logic of This" to my Kid? To a child? To your own kid, or perhaps to the child of a fellow officer? Are you listening, and do you get me? My Kid was "Given Pneumonia for five Weeks" recently. And the list of "Things done to my Kid", as my Kid has been repeatedly Targeted, is longer than you even want to imagine. Never mind the amount of suffering my Child has been through. You personally would not want to go through what my Kid has been through, and in no small part due to this Technology, and MY KID HAS been through it: never mind me. Clear? Who Targets a child, to get to another? This is "Agency Protocol"? Are you all sick? This is the "Rules of Mobbing"? A "Situation" that started because of a Pedophile-Cop and a Racist-Jackass. What does your Agency choose to do? I will be Waiting and Watching to see what your Agency's Reaction will be. And I will be watching you closely.

AND I will say here and now, that if Your Agency DOES NOT TAKE CONTROL OF "THIS MESS", AND IT'S IDIOTS, YOU WILL ABSOLUTELY REGRET IT. I currently have no money, but that is about to change, and I will absolutely Crawl Up Your Ass. I will do this for MY KID, never mind Me and my Girlfriend. I WILL KNOCK YOUR FUCKING TEETH IN. I have years of this "Mobbing Documented": a long list of "Weird Crimes" that stinks of Agency for starters. I know Your Clown's real name, where She lives, Her Address, the car She drives, Her license plate, know that her phone number is not listed, and know that if I write down the names of people living in these buildings around us, we will find a "High Percentage" of Your Agents. IF ONE THING HAPPENS TO MY KID BEFORE YOU TAKE CARE OF YOUR CLOWN I WILL COME AFTER YOUR AGENCY AND WILL "LOCK DOWN ON YOU" LIKE A PIT BULL ON YOUR THROAT. And at the end of it I WILL MAKE SURE THAT I AND MY FAMILY ARE COMPENSATED FOR YOUR

GOD DAMN MISTAKES. And while You Clueless Jackasses are at it, I suggest you start looking at the Media. If you are paying any damn attention, the Public-Sector will very soon be "Caught Up" in regards to this Electromagnetic-Technology. Think about this. And I have been in this "Mobbing" eight extra years: because of what you don't know. And Tortured.

You had better get things "Straightened Out", and if you can not "Relate to my Attitude" then I suggest you start trying. Whatever this is about these days, it is NOT ABOUT MOBBING. This had better be Sinking into Your God Damn Heads. YOU TAKE CARE OF THIS PROFESSIONALY, AND RESPONSIBLY, AND YOU TAKE CARE OF YOUR GOD DAMN CLOWN, BEFORE SHE ENDS UP DOING SOMETHING PERMANENT. AND DO NOT FUCK UP. Are we clear here? You had better be.

There is no "Excuse for Any of This", the woman is OUT OF CONTROL, DANGEROUS, VINDICTIVE, SPITEFUL, SLOPPY, PERSONAL, NON-PROFESSIONAL, AND QUITE OBVIOUSLY "NUTS". She is "a Snob, a Princess, a Three Year Old, A Clown", all words I have used for years to describe her, even before my Profile of Her, and as you can see these "Fit Her Profile" as well. If you as an Agency think She is "Cleverly Deceiving Me" here, then you are Dumb Beyond Belief, and You are Being Deceived. If that is the case then maybe I should be running your Agency. And I might point out, that The Clown is angriest with ME, and I have not begun to put anything here that she has done to Me. I have a "Page Count" to think of here, in regards to this Book. OBVIOUSLY, "This Woman" is a Psychopath, obviously this Woman's Profile is Accurate. Obviously this explains Her Personal Behavior, and Lack of Professional Behavior. How Ignorant do you plan to be here?

And you have given this Dangerous Idiot Exotic-Techno. Think of "Hannibal" from the movies, or Ted Bundy, the Zodiac Killer, or Charles Manson while you are at it. Brilliant. Are you starting to understand why I am saying you had better be "Familiar with this Personality Type"? If she knew WHO SPECIFICALLY she would be dealing with, from Your Agency, and I speak of Which Specific Superiors here, She would use this Technology to Spy On You, at the least. Never mind use it to Affect Your Thinking. Or Blackmail. Clear? I sure as hell hope so. This says something about "Her Officers" as well. You as an Agency, and Her Superiors, literally have to think of this woman as being with another Agency, and Hostile, in regards to You—because that is how SHE IS. THIS IS THE TYPE OF PERSON YOU HAVE HIRED. As long as we are clear on this above, we can move on to other things.

In regards to "Other Things", Whorbat has left Permanent Damage on My Body, and My Kids Body, with this Exotic-Techno. I expect YOU as an Agency, to remove this. She is also threatening to LEAVE SOMETHING FATAL IN ME, before She goes. I expect you to look into this and to make sure this

DOES NOT happen, and I expect you to be Intelligent Enough to take this Threat Seriously. You had better not "Simply take Her Word for It". If I get anything down the road here that is Fatal, I will make sure Your Agency regrets it—what are you going to do at that point, "Kill Me"? Whorbat has also taken Irreplaceable Personal Items from me, along with stolen money (Hundreds of Euros) from my Kid, I expect this back as well. You can give Your Agents "Innuendos" that if they do not pay back the money, and return the stolen items, they will "Have Problems". I recommend "Mobbing" for them—no doubt after this they'll take working for Your Agency, more Seriously and Professionally. As it is they are Total-Fuck-Ups.

How you handle this as an Agency, how Intelligent and Responsible you are as an Agency, and putting "Blame where it Belongs" will determine, any "Problems". As I said this is a matter of "Principles" and I will no longer wear "The Blame" for other people's Actions and Decisions because of a "Badge or Agency Title". Anybody "Hiding behind a Badge and Agency" at this point, can absolutely Stick-It. From my Experiences and Observations, these are Absolutely Worthless. WHO YOU ARE AND HOW YOU ACT IS WHAT COUNTS HERE. So far I am "The Only Adult" around here. Just send someone out here who is Professional, Non-Corrupt, and Decent, and no doubt this will settle everything.

I do not have "Issues with Your Agency", I have Issues with Your Employee, this "Mess" She has created, and "Problems that may Result" because of Whorbat and "Who and What" I know She is. So DO NOT Screw-Up. If I end up with something "Fatal" down the road time wise, or Experience one more Deliberately Sloppy Attempt on Me or my Kid, I will take you to court for "Safeties Sake", never mind if you do not Investigate Your Clown, and she "Continues to Deliberately Hold Me In this Mobbing", I will do so as well. Clear? You do not Hire Amateurs, then turn around and Blame Me. I AM DOING YOUR WORK, and no doubt have done "Her Superiors" a "Favor"—I am not so Ignorant here. You do not Hire a Psychopath with the Maturity and Mentality of a Spoiled Three Year Old, and then Blame-Me. I am finished with that, and my Kid and Girlfriend are not "Responsible Either". SO YOU CAN SHUT THE FUCK UP, TAKE A RESPONSIBLE APPROACH, OR PAY ME FOR DOING YOUR JOBS. Anyone "Getting an Attitude" with me, at this point, is a Disgrace to America, Democracy, and Justice. I AM the person who has had to deal with Everyone Else's Mistakes, and My Family as well, NOT YOU.

All three Agencies have Absolutely Screwed-Up. Do the right thing, End this Mess, and Investigate and Review Your Employees.

Sincerely,
Dante Kali Das

Made in the USA
Monee, IL
08 November 2026

67380765R00275